Altered Sensations

Archimedes

NEW STUDIES IN THE HISTORY OF
SCIENCE AND TECHNOLOGY

VOLUME 24

Archimedes has three fundamental goals; to further the integration of the histories of science and technology with one another: to investigate the technical, social and practical histories of specific developments in science and technology; and finally, where possible and desirable, to bring the histories of science and technology into closer contact with the philosophy of science. To these ends, each volume will have its own theme and title and will be planned by one or more members of the Advisory Board in consultation with the editor. Although the volumes have specific themes, the series itself will not be limited to one or even to a few particular areas. Its subjects include any of the sciences, ranging from biology through physics, all aspects of technology, broadly construed, as well as historically-engaged philosophy of science or technology. Taken as a whole, *Archimedes* will be of interest to historians, philosophers, and scientists, as well as to those in business and industry who seek to understand how science and industry have come to be so strongly linked.

For other titles published in this series, go to
www.springer.com/series/5644

David Pantalony

Altered Sensations

Rudolph Koenig's Acoustical Workshop in Nineteenth-Century Paris

Springer

David Pantalony PhD
Curator, Physical Science and Medicine
Canada Science and Technology Museum
Adjunct Professor, Department of History
University of Ottawa
Ottawa, Canada
dpantalony@technomuses.ca

ISBN 978-94-007-3061-8 e-ISBN 978-90-481-2816-7
DOI 10.1007/978-90-481-2816-7
Springer Dordrecht Heidelberg London New York

Softcover reprint of the hardcover 1st edition 2009

Cover image source: Guillemin 1881, p. 65

Printed on acid-free paper

Springer is part of Springer Science+Business Media (www.springer.com)

For Trevor Levere, who introduced me to the history of scientific instruments.

Acknowledgments

This research began as a small instrument cataloguing project initiated by Doug Creelman at the Psychology Department at the University of Toronto. It led to research in the Koenig collection at the Physics Department (one of the largest in the world), followed by the discovery of the Koenig-Loudon correspondence in the U of T archives and then to research in the Koenig collection at the Smithsonian Institution. I have since tracked down Koenig's materials and instruments in collections across Europe and North America. It has been an adventure and privilege to trace the instruments and history of one of Paris's more prolific instrument makers. I am deeply thankful to Trevor Levere for first seeing the value of doing this project and to Randall Brooks (Canada Science and Technology Museum) for supporting its completion in this form. I would also like to thank Jed Buchwald for supporting the publication of this book in the Archimedes series.

I would like to acknowledge the following people and institutions for their key support in this undertaking:

University of Toronto: Trevor Levere, Sungook Hong, and Ian Hacking for their supervision of the first part of this project, the doctoral dissertation (2002).

National Museum of American History, Smithsonian Institution: Steve Turner, Debbie Warner, Roger Sherman, and Karen Lee.

Dartmouth College: Rich Kremer.

Dibner Institute, MIT: George Smith, Myles Jackson, David Cahan, Erwin and Elfrieda Hiebert.

I am indebted to Julian Holland (Australia) and Myles Jackson (Polytechnic University, Brooklyn) for carefully reviewing the entire manuscript.

I would also like to acknowledge the generous research assistance and manuscript suggestions from a number of people at museums and universities throughout North America and Europe: Paolo Brenni (Fondazione Scienza e Tecnica, Italy); Doug Creelman, John Slater, Harold Averill, Louisa Yick and Rob Smidrovskis (University of Toronto). Tom Greenslade (Kenyon College); David Cahan (University of Nebraska); Marta Lourenço, Gil Pereira, Catarina Pires,

Marisa Monteiro, Ermelinda Antunes (Portugal); Michael Kelley, Sara Schechner, Jean-François Gauvin, Marty Richardson and Samantha Van Gerbig (Harvard University); Roland Wittje (University of Regensburg); Ralph Gibson, Tom Kenyon, Kellen Haak and Debbie Haynes (Dartmouth College, NH); Elizabeth Ihrig and David Rhees (Bakken Museum); Sylvie Toupin (Musée de la Civilisation du Québec, Québec, Canada); Elizabeth Cavicchi, Debbie Douglas, Markus Hankin, Yinlin Xie and Sam Allen (MIT); Neil Brown (Science Museum, U.K.); Bill Fickinger (Case University); Michael Wright (London); Thierry Lalonde (CNAM); Jean Barrette (McGill University); Anna Giatti (Fondazione Scienza e Tecnica, Italy); Fulvio Medici (University of Rome); Kathy Olesko (Georgetown University); Barnaby Frumess and Ennis Pilcher (Union College); Dennis Alexander (Aylmer, Quebec); David Murray (Queen's University, Canada); Mike Allibon (Toronto).

I am grateful to Eberhard and Reinhild Neumann-Redlin von Meding of Bückeburg, Germany for opening their home and family archives.

This project was funded and supported by the following agencies and institutions: Institute for the History and Philosophy of Science and Technology, U of T; Massey College, U of T; School of Graduate Studies, U of T; Smithsonian Institution Pre-Doctoral Fellowship; Ontario Graduate Scholarship; Social Sciences and Humanities Research Council Grant, Government of Canada; Munk Centre for International Studies; Dartmouth College, NH, Post-Doctoral Fellowship; Dibner Institute, MIT, Post-Doctoral Fellowship.

With particular thanks to Mom and Dad, my family (the Pantalony Foundation), and Rebecca and Dominic for their continued and generous support.

Contents

Notes on Acoustical Terms

Today we use the terms Hertz (Hz) or cycles per second (cps) to refer to the frequency of a vibrating body. One Hz represents a complete sinusoidal vibration. In the current text, frequency is described in Hz except when quoting from an original text or instrument.

"V.S."

In the nineteenth century the French had a tradition of referring to frequency numbers in terms of half a cycle, or "vibration simple" (v.s.). Alexander Ellis, the translator of Hermann von Helmholtz's *Sensations of Tone*, added the following explanation of the French system: "French physicists have adopted the inconvenient habit of counting the forward motion of a swinging body as one vibration, and the backward as another, so that the whole vibration is counted as two. This method of counting has been taken from the seconds pendulum, which ticks once in going forward and once again on returning." Ellis (1954, p. 16).

"V.D."

The French also adopted the term "vibration double," (v.d.), which was equivalent to a complete vibration (1 Hz) and corresponded to American, English and German traditions.

The French notation, used by Koenig on his instruments – UT, RÉ, MI, FA, SOL, LA, and SI – derived from a Latin hymn in honour of Saint John the Baptist written by Paulus Diaconus: "*Ut* queant laxis *re*sonare fibris *mi*ra gestorum *fa*muli tuorum, *sol*ve polluti *la*bii reatum *S*ancte *I*oannes" (Loosen the guilt of the unchaste lip, O Saint John, so that with relaxed throats your servants might seek to resound the wonders of your deeds).

American, German and English systems used variations on the letters: *c, d, e, f, g, a* and *b*. Each of them are referred to in their in their original context. In addition, where appropriate, I have added the modern notation with capitals, for example, ut3 is C4; the rest of the French scale in this octave, ré3, mi3, fa3, sol3, la3, and si3 would be D4, E4, F4, G4, A4 and B4 respectively.

Each octave (the interval between a tone and another tone having twice the number of complete vibrations) had a corresponding number referring to its height on the scale. ut3 referred, for example, to 256 v.d. (256 Hz), or what was middle "c" on the piano (c' in German notation). For this particular note, Koenig's tuning forks were marked, "UT3, 512 v.s." ut4 was the next octave up the scale at 1024 v.s. ut5 referred to the next octave, at 2,048 v.s., etc. ut3 = 512 v.s. = 256 v.d. = C4 = 256 Hz.

Archives Consulted (Abbreviations)

ASQ	– Archives de Séminaire du Québec, Québec City, Canada
AUC	– Archives of the University of Coimbra, Portugal
DCSC	– Dartmouth College Special Collections, Dartmouth College, USA
IAMIT	– Institute Archives, Massachusetts Institute of Technology, USA
LC	– Library of Congress, Washington, DC, USA
MCQ	– Musée de la Civilisation du Québec, Québec City, Canada
MELSC	– Daniel Coit Gilman Papers – Milton Eisenhower Library and Special Collections – Johns Hopkins University, USA
NFA	– Neumann Family Archives in Bückeburg, Germany
SIA	– Smithsonian Institution Archives, Washington, DC, USA
SIA-JHP	– Smithsonian Institution Archives – Joseph Henry Papers, Washington, DC, USA
UARCUP	– University archives of the University of Pennsylvania, Philadelphia, USA
UTA-JLP	– University of Toronto Archives, Toronto, Canada – James Loudon Papers, B72-0031/004

Other Abbreviations Found in Text and Notes

CR no. 27 refers to number "27" in the Catalogue Raisonné of Koenig's instruments at the back of this book.

List of Figures

List of Tables

Introduction

Between 1859 and 1901 a Prussian immigrant named Rudolph Koenig ran one of the more popular scientific ateliers in Paris. It was a place singularly devoted to sound. Visitors bought instruments, performed experiments, learned about acoustics, discussed the instrument trade in Paris, witnessed demonstrations, and stayed for an evening of food, drink, music and literature. Many of the apparatus which adorned his atelier became the foundation of modern acoustics. There were graphical instruments for recording sound, manometric flame instruments for making sound waves visible, sirens, tuning forks for precision experiments, and a variety of demonstration instruments. Henry Crew of North Western University visited in 1900 and later recalled the "atelier up by Notre Dame.…A visit of an hour or two there…only a few months before the old gentleman's death, is one of the high spots in my recollections of the last 50 years."[1] In 1898 the Canadian graduate student, J.C. McLennan, spent a week of afternoons at the atelier writing home that "the Doctor" had "impressed on me that I had heard things with him that nobody else had heard."[2] McLennan, like many science students at the turn of the century, learned classical physics using instruments made in Paris (Fig. 1).

This book is a portrait of Koenig's atelier and what it tells us about the nature of instrument making in Paris, one of the centres of nineteenth-century scientific culture. When the American physicist D.C. Miller visited Koenig's atelier on Quai d'Anjou in 1896 he described it as part workshop, showroom, laboratory and living quarters.[3] I use these four spaces as themes for exploring Koenig's role in the history of acoustics, and also for examining broader issues that characterized science during this period. *Workshops* entailed the spaces where instruments were actually made; *showrooms* (or boutiques or studios) involved the business activities of instrument making; *laboratories* supported experimental activity; and *living quarters* related to the daily aspects of life as a scientific artisan.

These four themes played a major role in shaping one of the central events of late nineteenth-century acoustics, the 1863 publication of Hermann von Helmholtz's *Die Lehre von den Tonempfindungen als physiologische Grundlage für die Theorie der Musik*, otherwise known to English readers as *On the Sensations of Tone as a Physiological Basis for a Theory of Music*. The title of this work, *Altered Sensations,* follows from the familiar English title of Helmholtz's book to emphasize the role Koenig's atelier played in transforming the study of acoustics, including its teaching,

Fig. 1 Rudolph Koenig about 1880. Source: Miller (1935, p. 84)

research, and applications; it also played a part in unsettling fundamental questions about the nature of acoustical sensations themselves, which opened new conceptual spaces in psychophysics at the turn of the century.

In the introductory sections below, I discuss how these themes have been used in the history of science and their potential, through engagement with historic instruments and collections, to offer new directions in this field. In the chapters that follow I shall elaborate on the impact of these themes on the development of acoustics.

Workshops in the History of Science

For over 400 years, instrument makers' workshops have been creative engine rooms for producing and honing effects, and yet, in covering some of the most fundamental events, historians have paid surprisingly little attention to the details of these spaces. The seventeenth-century Florentine instrument makers Divini and Campani, for example, waged famous duals to test and promote their lenses, but aside from some specialized investigations of these maker's workshops and techniques, we know little about the details and full context of their lens-making spaces and, more importantly, the impact of their workshops – in material and social terms – on the development of optics during that period.[4] In contrast, art historians have long seen value in exploring, right down to purchase records and names of workers, the inner details of the Renaissance artistic workshops.[5] Martin Wackernagel's classic work of 1938 on the *Lebensraum* (environment or habitat) of Florentine artists looked at the social and business context of the artists. He also paid particularly close attention to the artist's workshops or "immediate environment," where he described extensive drawing studies that led to a "methodical sharpening of the powers of perceptions" or the introduction of chalk or red ochre (produced by ground cinnabar) that led to bolder, simpler styles of representation.[6]

Workshop literature is not as robust in the history of science, but there have been several novel studies that show the value of digging deeper into these spaces. Anita McConnell, for example, has looked closely at the eighteenth-century London makers documenting the shift from specialty craft to factory-like manufacturing, with specific reference to the workshop of Jesse Ramsden.[7] Alison Morrison-Low has studied workshops in eighteenth and nineteenth-century provincial Britain, documenting the changing relations to London makers, organization of labour, skills, materials, products, mechanization and use of female labour, and instrument making as it related to the industrial revolution.[8] Myles Jackson, in his book on optics, explored the processes of making lenses at the workshop of Joseph von Fraunhofer in the early nineteenth century. He situated these activities in the local artisan and social conditions surrounding a Benedictine monastery near Munich, which had a large impact on the theory and practice of optics and the scientific community as a whole.[9] From the same period, Klaus Hentschel has described the work of instrument maker Moritz Meyerstein on precision verification in the Kingdom of Hanover. Meyerstein used his instruments and methods to build trust with scientists and local government officials in order to promote metrological reforms.[10] Stuart Feffer has detailed the manner in which the workshop activities of Ernst Abbe generated knowledge about optics that came to shape practice and theory and influence the microscope market.[11] Looking at the same workshop, David Cahan's history of Zeiss ultramicroscope shows the merging of theory, practice, academic institutions and industry in the *Zeiss Werke*.[12] Jed Buchwald has studied the details of Heinrich Hertz's lab notes and writings to reconstitute the tacit knowledge that played a part in Hertz's early electromagnetic inventions and experiments. His findings take us into the details of instrument creation to elucidate the relations between theory and experiment.[13] Other historians have paid attention to scientific workshops, providing details about their products and inventories, tools used, methods of production, division of labour, and larger role in their fields and society.[14]

When it counts, however, the details and significance of workshops are largely taken for granted or passed over. In Steven Shapin and Simon Schaffer's classic account of the air pump in the seventeenth century, we see how the integrity of the instrument became a central issue in a battle surrounding the legitimacy of experiment.[15] And yet these issues did not impel the authors further into the spaces where the instruments were actually made. Surely if debates centred on the integrity of the instruments and their function, questions would have been raised about the merit of various makers, construction skills and materials; Surely local makers who collaborated with Royal Society members would have been involved in these debates, or even actively influenced the debates, exposing the potentially deep social function of workshops in scientific controversy. In the absence of written records of these shops, historians have to examine historic instruments from that period to appreciate the contemporary workshop culture in all its forms and connections.

Philosophers also have much to learn about science from the workshop. Davis Baird has argued that instrument making is an independent form of knowledge production, one that should be taken seriously on its own terms.[16] For Baird, scientific instruments represent material knowledge which is on par with theoretical knowledge. The achievement of a "reliable signal" is similar, he argues, to the predictive power of a theory. "Where truth serves as one regulative ideal for theory construction," he writes, "the regularity and dependability of a phenomenon serve for instrument construction."[17] Makers hone these phenomena until they produce a dependable effect, thereby constructing specialized knowledge about the natural world. This knowledge is passed on through instruments and construction skills. Baird not only shows how instruments are made, but how this knowledge behaves in the wider world – shaping practice, changing conceptions about phenomena, and affecting industry and economies. His analysis shows how important it is to do serious workshops studies in order to obtain a critical understanding of knowledge production at this level.

The relationship between sound and workshops, the specific subject matter of this book, is particularly fruitful for historical study. The purposeful manipulation of sound in precise ways (mostly for musical instruments) dates back to early human civilization – an artisanal history that is arguably older than all other areas of physical knowledge such as optics or electricity. We find remarkably sophisticated acoustic artifacts scattered throughout time, geography and cultures that literally speak to us through the centuries. To take a recent discovery, a team of archaeologists have uncovered the oldest known multi-note flutes in Henan Province, China.[18] The flutes, dating from 7000 to 9000 BC, were made of bone from the red-crowned crane and have a carefully crafted eight-note scale. Researchers played one of the instruments and it produced surprisingly pure notes, resurrecting sounds of the "Central Yellow River Valley" in Neolithic times.[19] How much of our musical traditions can we trace to the possibilities and limits of early materials, listening and construction skills?

Some of these ancient acoustical artifacts have a complicated cultural story to tell. In *The Sounds and Colors of Power* MIT material scientist and archeologist Dorothy Hosler examined ancient Mexican bells bringing to life intersections of music, metallurgy, politics, and economics in Mesoamerica. For Hosler, the microstructure of the artifacts carries a "human fabric" that reveals "technical imagination, errors, experiment, inspirations, and business as usual."[20] She is particularly interested in the choices made by artisans, and what these choices tell us about relations between technology and culture.

Closer to the time period of this book, Myles Jackson has studied the workshop culture of musical instrument makers in the early nineteenth-century German territories.[21] As with his earlier study of optics, he found that the production of instruments took place in a variety of conditions – material, social and cultural – that had a considerable impact on music, scientific acoustics and society. The seemingly simple organ pipe, for example, was the focus of intense artisanal and scientific interest. By looking at its immediate workshop context, Jackson's study brings to life a story of artisanal and theoretical interaction.[22] In a more modern version of these interactions, Trevor Pinch has studied the development of the Moog synthesiser in the 1970s. Similar to Jackson's work, Pinch details the context of early synthesiser experimentation and studio work to draw lessons about the interactions of science, technology, music and wider culture.[23]

How do these sorts of studies relate to the present book? If, in thousands of years, archaeologists discovered a steel tuning fork signed "RUDOLPH KOENIG À PARIS/UT2 256 v.s./ RK," what would they learn from it? Would they look at it as musical artifact, scientific artifact, interesting steel sample, product of labour and artisans, something distinctly Parisian? In sections of this book I discuss the commercial, educational and experimental context of Rudolph Koenig's instruments. But what do we know about how the instruments were made, where they were made, and the people who made them? How does information about these activities broaden our understanding of acoustics, science and history?

Hundreds of Koenig instruments in collections around Europe and North America document, in material form, his workspace and, more generally, the vibrant workshop culture in nineteenth-century Paris as a whole. The artifacts contain several kinds of wood (oak, mahogany, pine, walnut), clockwork mechanisms, precisely graduated dials, brass workmanship, high quality steel, rough-looking cast iron molds and stands, musical strings, delicately insulated electrical windings, shellacs, oils, leather bindings, cardboard parts, optical pieces, influences from cabinet making, ivory piano keys, turned wooden handles, manufacturing marks, and skilled glass blowing. They represent a fertile and dense ecology of materials, industries, skills and ideas circulating on the Left Bank (the academic, artistic and artisan district) of Paris at its height of production and popularity; they tell us how material knowledge about sound literally came into being, and how it came to be expressed and used. They also document a variety of connections to events in the German territories, France, England and North America.

How do historians access this fascinating world? In the history of science and technology, several scholars have begun using collections and actual objects in order to develop new themes and venture into alternative historical spaces.[24] In the present book, I have tapped material culture methods to stimulate questions, fill gaps and provide a fresh interpretation on workshop culture. As a general guide, I refer to a classic account of material culture studies given by E. McClung Fleming where one interrogates an artifact based on object history, materials, construction, function, design, and ornament, along with a broader analysis which includes identification, aesthetic evaluation and cultural analysis.[25] Each category relies on a systematic set of questions for guiding examinations. In the cultural analysis, for example, one compares the artifact with other objects in the same field or period to place it within a broader material culture context and not just an intellectual context. Another strategy is to separate the purely functional v.s. unnecessary aspects to reveal choices made by a maker; where there are choices there is culture and history. Artifacts also display features that convey values, status, meaning and ideas. This questioning method, otherwise known as the Winterthur model, has recently been used by Rich Kremer (Dartmouth College) and myself for teaching the history of science.[26] We have found that historic instruments and collections are one of the best ways (sometimes the only way) to access forgotten and significant historical spaces such as workshops.[27]

Showroom: The Business of Instrument Making

Instrument history is also about economics and business. Promotion, client cultivation, adapting to market trends, demonstrations for clients, operating costs, payment to workers, shipping costs, duties, and surviving economic fluctuations had a profound impact on the development and overall scope of science in the nineteenth century. In an essay about shopping for instruments in eighteenth-century London, Jim Bennett described scientist's informative and entertaining visits to scientific boutiques and workshops within a vibrant street-level commercial context for instrument making.[28] Roger Sherman has written about one of the many colourful itinerant scientists that spread electrical demonstrations in post-revolution America.[29] In the nineteenth century, Alison Morrison-Low has looked at instrument businesses in the context of supply and demand economics in industrial Britain. The rise of merchant shipping, for example, gave impetus to adaptations in the local trade in marine instruments.[30] Richard Dunn, to take another example, has written about the relations between design, consumption and profit embodied in historic instruments. He studied artifacts from the Renaissance, the eighteenth century, the early twentieth century to show how instruments are connected to economic and cultural forces outside science. Instruments, he contends, are designed to attract specific customers.[31] Similar issues emerge in histories of more recent instrument making. Cyrus Mody's description of the instrumental origins of nanotechnology, between garage workshops and big corporate research laboratories, reveal convergences of tinkering and commerce remarkably similar to those of London in the late-eighteenth century and Paris in the mid-nineteenth century.[32] Other studies of these issues in recent history reveal comparable patterns.[33]

Closer to the immediate context of this book, Paris in the nineteenth century was at its core a commercial city, famously characterized by Walter Benjamin as a vast display of consumption.[34] He quoted an illustrated guide to Paris which singled out the indoor arcades as examples of modern commercial activity:

> These arcades, a recent invention of industrial luxury, are glass-roofed, marble-walled passages cut through whole blocks of houses, whose owners have combined in this speculation. On either side of the passages, which draw their light from above, run the most elegant shops, so that an arcade of this kind is a city, indeed, a world in miniature.[35]

Selling scientific instruments was part of this evolving commercial culture. The optical maker, J.B.F. Soleil, was one of the first vendors in the Galerie Vivienne in 1824.[36] (Fig. 2). The decades between 1830 and 1880, which Paolo Brenni has called the golden age of French instruments, were one of the more commercially active periods for instrument makers in France.[37] Previously, London had been the centre of this trade; from the 1880s Germany would take over. But for the span of 50 years between 1830 and 1880, following the blossoming of science under Napoleon, and the industrialization of France, Paris became the central destination to buy

Fig. 2 Soleil's storefront mosaic at Galerie Vivienne, Paris. c.1825. Photo by author, 2001

scientific goods. Throughout this period, the Latin Quarter or Left Bank and other areas were crowded with makers of instruments for optics, electricity, heat, mechanics, horology, astronomy, surveying and medicine. They had a strong retail presence on the streets throughout the school district. Scientists came from around the world to visit shops and purchase instruments. They often worked through agents, visited local institutions and laboratories, and spent evenings in workshops witnessing demonstrations.[38]

Many visitors to Paris fell unexpectedly into what one could only describe as a science monopoly. In his study of Adolphe Ganot and the story behind the production of his long-running editions of physics textbooks, Josep Simon portrays the thriving connections between science publishers, writers, book dealers, instrument makers, natural-history specimen dealers and medical instrument and model makers.[39] Instrument makers and book dealers, for example, jointly attended local science lectures. Makers even participated in lecture-demonstrations promoting their goods and helping the lecturers illustrate their concepts. Catalogues appeared with attractive engravings of the makers goods.[40] Texts included engravings of instruments with references to specific makers (often using the engravings from the maker's catalogues). Paris, in short, was a self-contained, self-perpetuating scientific culture that promoted its own research, and more importantly for the market, defined the content and boundaries of science education with the active promotion of its products.

Entertainment was a vital element of this commercial environment. Gabriel Finkelstein has called nineteenth-century Paris "the Broadway of scientific performance."[41] Spectacle had long been a feature of science teaching, but Parisians made it central to their scientific culture. Amphitheatres and showroom/workshops became stages for explosions, light shows, electrical experiments and mechanical demonstrations. Teachers made their reputations with big classes; research, good and bad, lived and died on the stage at the Académie des Sciences. Studios of instrument makers were famous for lively, attractive demonstrations.[42]

The greatest showcase and public venue for French artisans, however, were the international fairs. Historian Bruno Giberti has argued fairs were a massive classification project for modern consumption and capitalism.[43] One visitor to the 1876

Exhibition in Philadelphia remarked that the "wealth of the world is before us."[44] From 1851 until 1889 the French dominated these occasions with impressive, attractive booths, especially in the sciences. For scientists and the public, the isles were lined with beautiful products and displays offering a concentrated version of Left Bank boutiques. People gathered around the booths, talked to makers and fellow scientists, placed orders, and debated recent developments. Juries judged the goods and handed out coveted awards that would make and enhance reputations.

Throughout the second half of the nineteenth century, French makers personified this growing mixture of commerce, materialism and science. Driven by growing demand from educational institutions, buyers became lost in a competitive frenzy to fill their new laboratories with the same goods as their colleagues in other institutions. In an attempt to separate himself from the image of growing consumerism in the instrument trade, Koenig adamantly portrayed himself as a scientist and maker for pure motives, and not just for commercial gain. In the 1880s, he even provided several examples to Loudon of instrument makers who worked for love and not profit (Chapter 7). One finds a similar tension with other instrument makers, particularly in Britain.[45] The growing prevalence of commerce in the sciences strengthened ideal notions of "vrai science."[46] By the time the American scientist Henry Augustus Rowland gave his famous address on pure science in 1883, science and commerce had become inseparable.[47]

Laboratory: Instrument Making and Experimentation

Julia Loudon: You surely did not experiment on Sunday?
Rudolph Koenig: Why not? Le bon Dieu – he loves a good experiment.[48]

Now, how can an experiment be wrong? Richard Feynman.[49]

The context of experimentation changed dramatically in the nineteenth century. Precision instruments proliferated, there was a move from private laboratories to larger institutional laboratories, and teaching laboratories emerged modeled on the reformed German education system. Instruments became intertwined with notions of trust, class and morals.[50] Since the 1980s there has been an extensive literature on the subject of experiment,[51] but few sources have dealt directly with the role instrument makers have played in defining experimental culture. There is much to learn from the creative collaborations between instrument makers and scientists,[52] and, just as important, how they differ on fundamental issues. Koenig, who was involved in several controversies on the nature of sound and hearing, would have answered Feynman's question (above) very differently than his main rival, Helmholtz; the former believed that experiments were only as good as the instruments used to perform them; the latter viewed experiment and instruments as limited in their ability to provide final answers to stubborn challenges (Chapter 7).

These issues were even more pronounced in acoustics where long-held mathematical and artisan traditions, running parallel for centuries, finally merged into an uneasy alliance in the mid-nineteenth century. In his essay, "The Essential Tension," Thomas Kuhn described how mathematical and empirical traditions came together in the eighteenth and nineteenth-century science.[53] Previously, the classical sciences – astronomy, optics, geometry, mechanics, and harmonics – had been almost entirely mathematical; while parts of chemistry, natural history, optics and electricity, following the scientific revolution, centered on an empirical fact-gathering model. During the nineteenth century the two traditions combined in fields like optics and electricity, where powerful mathematical descriptions merged with rigorous laboratory research.

In acoustics, where experiment and new instruments infiltrated the ancient study of "harmonics," the tension between theory and practice remained prominent. Consequently, the nineteenth-century acoustical laboratory became a battle ground for experimental and instrumental legitimacy. Victor Regnault, whom we encounter in Chapter 4, brought his obsession with expunging error into the study of sound.[54] Koenig, influenced by his exposure to Regnault, spent months, even years in his private laboratory focused on a single problem or series of instruments. The blending of his workshop and laboratory reinforced these views. Artisan training (tools, knowledge of materials, traditions, skills, as well as artisan values, ideals and standards), social status and education played into these tensions and transformed controversies into personal issues related to livelihood (Chapters 1, 5 and 7). Koenig's focus on experimental and instrumental integrity reflected a broader reaction against those who he portrayed as relying too much on theory and limited idealizations of complicated real-world conditions. Amidst both theoretical and experimental developments in acoustics, he continued to be an ardent proponent of the empirical tradition and the idea that knowledge about sound was best obtained from close study of instruments and experiment. In addition to his discomfort with theory, Koenig was just as worried with the transformation of his trade into a factory-like production model. He was an artisan in the old-fashioned sense with deep suspicion of instruments made outside the master artisan model. Hand-made instruments could be trusted more than instruments made on a large scale.[55]

The boundary between objective and subjective observations was another battleground in nineteenth-century experiment. Certain phenomena tested the limits of instruments, methods and the understanding of human observations. Did observed or measured phenomena reflect a reality in the physical world, or a distortion caused by the nature of the observer? One problem that developed in astronomy, for example, was the "personal equation." The discrepancies of measured transit times, even with precise chronographic instruments, came to revolve around the action of the person recording the event. Some recorders were slower than others at marking the beginning and end of the same event. This "reaction time" in turn became a psychological/physiological problem in its own right.[56] As more came to be discovered about the brain, sense organs and psychological processes, there was an increased desire to clarify the definition of subjective and objective observations. Helmholtz, for example, grounded sensations in objective physical and physiological processes, which in turn were governed by psychological processes.[57] Ewald Hering and Ernst Mach (and Koenig, as we will see later), redefined the nature of observations by treating sensations as realities in themselves, thus opening the door to a new status for psychological phenomena that had a wide ranging impact on psychology, physics and the nature of knowledge obtained from experiment.[58]

Finally, there was a broader shift towards visual observation in the nineteenth century.[59] It became the favoured mode of making and recording observations. This shift is a major part of acoustics today, as scientists rarely use their hearing or "expert ear." This trend appeared in several observational fields – physiology, meteorology, medicine, acoustics, and astronomy. A classic example occurred in medicine with the replacement of the stethoscope by the x-ray for probing inside the body. Medicine has since become dominated by visual technologies.[60] Koenig's graphical and optical instruments became central to shaping his practice and concepts;[61] they also became important for teaching, business and, as we see later, winning over sceptics to this controversial views.

Life as an Instrument Maker

What can we learn from the everyday lives of instrument makers? There are only a handful of book-length biographies of scientific instrument makers. They rely on letters, purchasing records, trade literature, articles and, of course, instruments, thus revealing important facets about artisanal life in the scientific realm.[62] These biographies highlight interactions between makers and patrons, suppliers and other makers; they describe the educational background of makers, movement between trades and details about family history; they also show the immediate culture and preoccupations of being an instrument maker and how these activities had an impact on their work. Anita McConnell, for example, has written about the importance of Jesse Ramsden's personal charm in his relations with Jean-Dominique Cassini (Cassini IV).[63] To some historians this may seem trivial or irrelevant, but it often turns out to one of the main reasons why an institution purchased large numbers of instruments from a particular maker. In short, even a gifted maker such as Ramsden relied on his skills as a salesman and promoter. Other biographies of famous eighteenth-century makers such as Jan van Musschenbroek and George Adams have provided revealing glimpses of the private lives and networks of instrument makers.[64]

What was daily life like for an instrument maker in nineteenth-century Paris? Paolo Brenni has described the careers, major instruments and accomplishments of the key Parisian instrument makers in a series of articles, but there are few detailed biographies of these artisans.[65] This has not been the case for their glamorous contemporaries in the art world.[66] In fact, even though it is rarely noted by historians of art, Parisian instrument makers and artists existed in close proximity. From 1864 and 1877, for example, Koenig lived at 30 Hautefeuille right next to the studio of the realist painter, Gustave Courbet (32 Hautefeuille). On the other side of his atelier was Andler's Brasserie (28 Hautefeuille), the famous meeting place for Courbet and his followers (Fig. 3).

Fig. 3 Andler's Brasserie as sketched by Gustave Courbet. In the mid 1860s, Koenig lived between Courbet and Andler's place on Rue Hautefeuille. Source: Delvau (1862)

This fortunate coincidence has provided the only information we have on Koenig's immediate neighbourhood and potential contacts during the early part of his career on Hautfeuille. We know, for example, that during this time Courbet and his friends met regularly at Andler's "Brasserie des Réalistes." (Fig. 3).[67] Among many notable names – Corot, Champfleury, Daumier, Baudelaire – were the musician and painter Alexandre Schanne and the scientist/demonstrator Ignace Silbermann, who, in the 1830s had assisted Félix Savart with his acoustical research.[68] In 1860 the art critic Jules Catagnary described Andler's as the "baptismal font" of Courbet's realism where he held court from 6 to 11 in the evening.[69] The brasserie was the "*véritable atelier*" of Courbet, claimed Jules Champfleury in 1872.[70] It operated in the rustic German style with a dark interior and no windows, wooden tables and benches, a billiard table, "hams hanging from the ceiling, garlands of sausages, rounds of cheese as big as millwheels, [and] barrels of appetizing sauerkraut."[71] It must have been a welcome place for a young Prussian artisan. Only a few years into Koenig's stay at this address, however, Courbet and his group had started to frequent another brasserie around the corner.[72] The two men remained neighbours until the demolition of their buildings in 1877 to make way for a medical building (Chapter 6).

Even though he has not been the subject of countless books and exhibitions like Courbet, Koenig's life was equally rich with accomplishments, personalities, interesting connections, struggles, financial pressures and scientific triumphs.

He emigrated from Prussia following the 1848 revolution. He apprenticed in Vuillaume's well-known violin workshop. He collaborated with scientists such as Victor Regnault, E.J. Marey, William Spottiswoode, and Hermann von Helmholtz. He was one of the top makers in the famed Parisian precision instrument trade winning awards throughout Europe and North America. He was involved in scientific disputes that influenced the field. He survived the turmoil of the Franco-Prussian war, Paris Commune, anti-German prejudice and the fluctuations of the French economy and scientific market. He spent many years in isolation researching controversial questions that challenged instruments, theory and observation. Even in his last few years, he continued to create instruments and explore the limits of mechanical acoustics just prior to the emergence of electrical acoustics. The everyday activities of his life, therefore, with accounts of construction, purchases, struggles with money, social interactions, entertainment, financial fluctuations, status, eating, and travel shall provide important context for understanding the background of his work.

Sound in History

The main diet of Koenig's life, however, was sound; and it seemed particularly good at taking him across many national, social and disciplinary boundaries. In the 1850s and 1860s, as the sciences were becoming increasingly specialized, Hermann von Helmholtz combined developments in physiology, physics, mathematics, music and philosophy to create a new conceptual and experimental framework for studying and manipulating sound. In the midst of rapid scientific, technological and industrial development of the late nineteenth century, John Tyndall, Lord Rayleigh, Lord Kelvin, Alexander Graham Bell, and Thomas Edison added ideas, inventions and directions to the study of sound. By the early twentieth century, electroacoustics had transformed what Emily Thompson has called our modern "soundscape."[73] Its impact was felt everywhere within reach of electricity from laboratories to musical performances. Roland Wittje has written about the instruments of early radio and electroacoustics, their social, political, cultural, engineering and scientific context, and how they spread practices and techniques into surprising areas such as early atomic physics or mass political rallies.[74]

Myles Jackson's recent book *Harmonious Triads* offers particularly rich lessons about the rich relationship between sound and society. From 1800 to 1850, a period of which very little has been written about acoustics, Jackson found active and fertile interactions between scientists, musicians and instrument makers.[75] He gives equal weight and depth to these three groups showing the details of their developments and how they interacted and influenced each other. The events covered take place in the post-Napoleonic German territories, which contributed to and were changed by such seemingly specialized knowledge as the theory of adiabatic phenomena and the details of manufacturing organ pipes. Jackson's work combines industrialization, artisans, scientific societies, overlooked scientists, musical performers throughout the German territories and Europe, musical pedagogy, the politics of standardization, and philosophical context. His work demonstrates that the history of sound is naturally interdisciplinary.

Chapter Summary

Like many instrument makers, Rudolph Koenig brought a diverse and unusual background to his career as a maker of precision acoustical instruments. In Chapter 1, I describe his education and upbringing in Königsberg and his move to Paris in 1851. I survey his work in Vuillaume's violin workshop from 1851 to1858 which included immersion in the skills and culture of violin making. I then describe his transition to scientific instruments within the famous precision instrument trade of Paris.

Much of Koenig's career was a response to Hermann von Helmholtz *Sensations of Tone*,which derived from a vastly different scientific and cultural context. In Chapter 2 I look at Helmholtz's background and training to show the different path he took towards "reforming" the study of acoustics. I situate his study of acoustics in the wider context of German academic science, and in particular in the experimental, physical, physiological and psychological context of the time. Finally, I describe how he combined these elements into an overall theory of harmony and music.

In the late 1850s and early 1860s Koenig's atelier also became a centre of acoustical innovation. Barely 30 years old in 1862, Koenig was at the forefront of two movements that remain major components of acoustics today: the introduction of graphical acoustics and transformation of Helmholtz's ideas and instruments into an entire line of acoustical products for teaching and research. Chapter 3 describes these developments from the perspective of the workshop, where Koenig designed, constructed, tested and sold his instruments. I look at some of the main products and describe how he invented and modified them in his workshop.

The market for scientific instruments had a profound influence on acoustics. Chapter 4 looks at the early years of Koenig's business activities and how even at that time he was a leader in the Parisian precision instrument trade. In 1862 he attended his first major exhibition at London. Five years later he participated in the international fair in Paris. During this period he began actively promoting and selling graphical and Helmholtz's instruments. I high-light changes in the market as they related to Koenig's business; I also look at specific customers in the United States and Europe to illustrate the needs of some of his major clients. The story of two customers – one from MIT, and the other from Portugal – informs us about the commercial context of the famous nineteenth-century instrument trade in Paris, and point to influences on the rapidly growing acoustical market.

Chapter 5 covers key experiments with which Koenig became involved during the period 1866 to 1876. Making trustworthy instruments was at the heart of these efforts. In the mid 1860s he tried to tackle two of the more elusive targets in acoustics, the velocity of sound and the nature of vowel sounds. In the 1870s, he began a lengthy series of precision experiments on combination tones and timbre. Both series of experiments eventually brought him into conflict with Helmholtz. In the midst of his combination-tone experiments, he constructed his large tonometer. I describe the process by which he made hundreds of forks and how this came to influence his views on disputed phenomena.

By the mid 1870s, the North American market, particularly that for science teaching, became a driving force behind the Parisian instrument trade. Chapter 6 describes Koenig's business challenges following the turmoil of war in 1870–1871. The market fluctuated but he continued to experiment, invent instruments and seek customers. I focus on his relations with two clients, Joseph Henry of the Smithsonian Institution and James Loudon of the University of Toronto. The Centennial Exhibition held in Philadelphia in 1876 was one of the highpoints of his career due to his award-winning display; it turned into personal turmoil, however, as the expensive research equipment he brought for display became the centre of a controversy with the University of Pennsylvania. His struggle to sell and then reclaim this collection provides a glimpse of the intense commercial pressures and daily stresses of being an instrument maker in the nineteenth century.

The final years of Koenig's life were spent navigating the fluctuations of the instrument trade and controversies with Helmholtz. In Chapter 7, I portray life and business at 27 Quai d'Anjou between 1882 and 1901. I then look at how his workshop became a theatre for promoting his findings and winning over colleagues and clients. English scientists sided with Koenig, seeing him as a "Faraday of sound." The German response, on the other hand, was not as warm. In his final years, amidst waning interest in these debates and, for that matter, basic acoustics, Koenig embarked on his last experimental quest – to record on paper and measure frequencies far beyond the human threshold of hearing.

Notes

1. Henry Crew to Miller (1935), Miller Papers, Physics Department, Case University.
2. J.C. McLennan to James Loudon, Sept. 4, 1898. UTA-JLP.
3. Miller (1935, p. 91).
4. Bedini (1961, 1966). In addition there are studies that place Divini and Campani in the context of seventeenth-century astronomy, see Bonelli (1981) and Helden (1994).
5. Ames-Lewis (1983), Cadogan (2000), and Ladis (1995).
6. Wackernagel (1981, pp. 308, 316, 320).
7. McConnell (1994, 2007).
8. Morrison-Low (2007, pp. 175–201).
9. Jackson (2000).
10. Hentschel (2007).
11. Feffer (1996).
12. Cahan (1996).
13. Buchwald (1994).
14. Cattermole (1987), De Clercq (1997), Fox and Guagnini (1998–1999), Klein (1996), Mertens (1998), and Turner, G. L' E. (1996).
15. Shapin and Schaffer (1985).
16. Baird (2004).

17. Ibid., p. 127.
18. Zhang et al. (1999).
19. The flute can be heard at http://www.bnl.gov/bnlweb/pubaf/pr/1999/bnlpr092299.html
20. Hosler (1994, p. 3).
21. Jackson (2006).
22. See for example, Jackson's chapter on pipes and adiabatic phenomena, pp. 111–150.
23. Pinch (2002).
24. For an overview of this innovative approach, see the essays in Taub (2006) or Lubar (1993).
25. Fleming (1982).
26. Pantalony (2008).
27. See Pantalony (2005b) for this method applied to the King Collection of Historic Scientific Instruments at Dartmouth College.
28. Bennett (2002).
29. Sherman (1991).
30. Morrison-Low (2007, pp. 263–267).
31. Dunn (2006).
32. Mody (2004).
33. Baird (2004, pp. 211–237) and Joerges and Shinn (2001).
34. Benjamin (1978).
35. Quoted from Benjamin's essay "Paris, Capital of the Nineteenth Century" in Ibid., pp. 146–147.
36. Brenni (1996). Date of Soleil's shop at Galerie Vivienne, personal communication with Paolo Brenni.
37. Brenni (1993–1996).
38. Pantalony (2004a).
39. Simon (2004).
40. Brenni (2002).
41. Finkelstein (2003, p. 261).
42. Pantalony (2004a).
43. Giberti (2002).
44. Ibid., p. 106.
45. See, for example, Spaight (2004) for an account of Herschel's instrument making enterprise.
46. Rudolph Koenig to James Loudon, Jun. 22, 1883. UTA-JLP.
47. Rowland (1883). For more context on Rowland's life and work, see Sweetnam (2000).
48. Loudon (1901b, p. 11).
49. Feynman (1995, p. 2).
50. Gooday (2004), Olesko (1991), and Warner (1992).
51. See Hacking (1983), Galison (1987, 1988), Gooding (1989), and Buchwald (1994). Recently, there has been a growing literature on the replication of experiments. Blondel and Dörries (1994) and Sibum (2000).
52. Levere (1994) and Sherman (1988). The famous Dutch physician, Boerhaave, and the instrument maker, Fahrenheit, collaborated on the making of thermometers, see Golinski's chapter in Holmes (2000).
53. Kuhn (1977). Also see Buchwald (1994, 2005), and Heering in Blondel and Dörries (1994).
54. Dörries (2001) describes this approach in meteorology.
55. Benjamin presents aspects of this tension in "The Work of Art in the Age of Mechanical Reproduction." Benjamin (1968). For an American context of these changes, see Hounshell (1984).
56. Benshop (2000), Canales (2001), Boring (1957, pp. 134–153), Schaffer (1988), and Schmidgen (2005).
57. Hatfield (1993) looks at these issues in the work of Helmholtz.

58. Kremer (1992). See Chapter 4 in Ash (1995).
59. Schmidgen (2007), Hoff (1959), Brain (1998b), and Braun (1992).
60. Kevles (1997).
61. Hankins and Silverman (1995) and Silverman (1992).
62. De Clercq (1998), McConnell (2007), Millburn (2000), and Warner (1995).
63. McConnell (2007, pp. 142–144).
64. De Clercq (1997) and Millburn (2000).
65. Brenni (1993–1996).
66. See Milner (1988) for an overview of the Parisian art studios.
67. Courbet lived at 32 Hautefeuille from 1848 to 1877. See, for example, his letters from this period in Courbet 1992. Galeries Nationales 2007. Mack (1970, pp. 25. 57) and Nochlin (2007).
68. Lindsay, J. (1973, pp. 40–43). Silbermann was most likely Ignace Joseph Silbermann, demonstrator at the College de France. Lindsay p. 42 notes how he advised on the weather at Andler's. Silbermann also produced a series of stunning and colourful didactic paintings on optical studies, Brenni (2007).
69. Quoted in Lindsay (1973, p. 40).
70. Champfluery (1872, p. 189).
71. Ibid., p. 186.
72. Mack (1970, pp. 57–63).
73. Thompson (2002). For general account of 20th century acoustics, see Beyer (1998).
74. Wittje (2003).
75. Jackson (2006).

Chapter 1
Training

In the cluttered storeroom of the Physical Sciences collection of the Smithsonian Institution in Washington DC, amidst thousands of historic scientific instruments, it is not difficult to spot the woodwork of Rudolph Koenig. One instrument in particular, the Barbareau *grand sonomètre*, represents the high art of Koenig's mastery of wood and sound. It displays a finely-grained spruce top covered in light varnish, thin mahogany veneers on the sides, walnut ends (for maintaining tension with the pegs), and oak bridges (for strength). The upper surface has eight steel strings stretched over inlaid boxwood metric scales divided into millimetres and numbered by centimetres along with the standard French musical notations for the physicist and equal termperament scales (CR no. 134).[1] The sides have stylized, lyre-shaped sound holes. When examined closely, this instrument, like many of Koenig's instruments, carry many clues about the material and social development of acoustics in the second half of the nineteenth century. In particular, they document the coming together of violin making, precision instrument making, and scientific developments in Koenig's early career (Fig. 1.1).

Journey to Paris

Karl Rudolph Koenig was born on the 26th of November 1832 in Königsberg, East Prussia (now Kaliningrad, Russia). His father, Johann Friedrich Koenig (1798–1865), was professor of mathematics and physics at the Kneiphöfischen Gymnasium. He had been a pupil of Friedrich Wilhelm Bessel (1784–1846), the well known astronomer, and enjoyed wide connections in Prussian scientific circles. Rudolph's mother, Mathilde Koenig (c. 1806–1893), born Preuss, descended from a prominent artisanal Königsberger family. Her father, Martin Preuss (b. 1774) was a clockmaker and two of her ancestors were organ builders, Jakob and Johann Preuss, one of whom graduated from the "Albertina," the Albertus-University of Königsberg, as an instructor of organ music. In the latter part of the eighteenth century Jakob and Johann restored the organ at the Königsberger Dome Cathedral, which at the time was the largest organ in Prussia.[2] The Preuss family also made concert pianos in the early nineteenth century.[3]

D. Pantalony, *Altered Sensations*, Archimedes 24, DOI 10.1007/978-90-481-2816-7_1,

Fig. 1.1 Barbareu sonometer. Photo courtesy of the National Museum of American History, Smithsonian Institution, Washington DC, cat. no. 314, 589, neg. 2009.001. Photo by Steve Turner

Königsberg was a thriving port town on the Baltic Sea adorned with the architecture of the Hanseatic League, a castle where Prussian kings had been crowned, and the Kneiphöf island in the centre of town where the philosopher Immanuel Kant, Königsberg's most famous citizen, was buried in the Dome Church. Rudolph, who had three sisters, grew up in a stimulating, cultured environment. In later years he fondly recalled musical and literary evenings with friends such as the Dulks, a prominent Königsberger family. Friedrich Philipp Dulk (1788–1852) founded the chemistry institute at Königsberg. His son Albert (1819–1884) was an actor who became a radical journalist during the turmoil of the 1848 revolution. In the first half of the nineteenth century, Königsberg was a renowned location for studying physics, and through his father, Rudolph met several figures in German physics and mathematics. One family friend was Franz Neumann (1798–1895), the head of one of Prussia's leading seminars in mathematics and physics. Koenig's sister, Anna, married Neumann's son, Ernst, who became a prominent laboratory physician best known for his work in haematology. Franz Neumann's seminar, one of the first of its kind to focus almost entirely on precision measurement, emerged in the wake of the Prussian education reforms after Napoleon's defeat.[4] Rudolph Radau (1835–1911), one of Neumann's students, would become a key promoter of Koenig's efforts in Paris. The most famous physicist to work in Königsberg, a friend of the elder Koenig, was Hermann von Helmholtz (1821–1894).[5] Helmholtz was the professor of physiology at Königsberg from 1849 to 1855 during which time he did some of his work on optics and measurements of the nerve impulse. He soon moved away, however, one reason being that Königsberg, a relatively small town, lacked the skilled mechanics he needed so badly for his novel experiments.[6]

Rudolph showed an early aptitude for art, literature and music. In fact, he developed a lifelong passion for music inspired by friends and family members. He also demonstrated abundant mechanical skills and was encouraged by his grandfather Preuss, who had been impressed by the precision instrument trade in England and France during his travels. Unfortunately, Rudolph could not channel these abilities toward a formal education. He had great difficulty with the classical languages, a main requirement for graduating from the humanist orientated Kneiphöpf Gymnasium. The end of the year report for 1849, when he was approaching his senior year, showed that there was a large emphasis on Greek, Latin and Hebrew. Aside from the formal language courses, there were three courses devoted solely to Ovid, Homer and Virgil. In English, the students read, among several items, the *Christmas Carol* by Dickens and the *Prisoner of Chillon* by Byron. In French they read *l'Avare* by Molière and a history of Napoleon by Dumas.[7] There were also courses in history, geography, singing, penmanship, natural history and German. Professor Koenig taught French, mathematics and physics – three courses that influenced Rudolph's later career in Paris. The report did not specify the elder Koenig's course outline for physics, but we do know that he had a modest physical cabinet at his disposal for demonstrations. In 1849 he added to this cabinet a magnetic needle with a stand, a large brass concave mirror on a tripod, an achromatic optical demonstration device, and a magnifying glass. He was overqualified for his position; he not only had a doctorate, but had obtained his habilitation (post-doctoral lecturing qualification) in 1839. He was the only full professor at the Gymnasium and he frequently published articles on mathematics in leading journals such as *Crelle's Journal für die reine und angewandte Mathematik*and *Grunert's Archiv der Mathematik*.

Rudolph subsequently failed the abitur (the state regulated examinations for leaving secondary school), which was a great disappointment to his father. Prof. Koenig had put much pressure on his only son to succeed and because the family did not have enormous resources, Rudolph decided to take up a trade. Inspired by his grandfather's stories of England and France, and family traditions in musical instrument making, he moved to Paris in 1851 at the age of nineteen and became an apprentice to the celebrated violinmaker Jean Baptiste Vuillaume (1798–1875). In this endeavour he combined his "unusually good hearing" (*ungewöhnliche gute Hörfähigkeit*) with his skilled hands.[8]

Following the uprising of 1848, bohemian Paris was the ideal setting for the romantic young Königsberger. Since the thirties, his literary hero, the poet Heinrich Heine, had sent regular dispatches from Paris to German readers, and now Rudolph wanted to be part of this scene. He was also part of a larger migration of Germans out of the territories during this period (1.3 million between 1845 and 1858),[9] especially skilled workers that would end up in the trades.[10] The 1850 s and 1860 s were a prosperous and relatively stable period in Paris, especially for the sciences.[11] In late 1851, Napoleon III declared himself Emperor, thus beginning the Second Empire that would last until the Franco-Prussian War in 1870. The years following Napoleon's coup witnessed substantial economic and industrial expansion, a reversion to more conservative politics, and ambitious public projects. Baron Haussman's sweeping renovation of Paris began in 1852 shortly after Koenig's arrival, and, as we will see later, provided an opportunity for one of his more memorable series of experiments in the sewers of Paris.

Koenig began his apprenticeship at a remarkable time. His employer, Vuillaume, had just won the gold medal for stringed instruments at the Great Exhibition of 1851 in London. Also in that year, Vuillaume had invented the famous Octobass, an eleven-foot giant double bass, and demonstrated it at St. Eustache Church in Paris. In the world of French musical instruments, it was the era of Vuillaume's violins, Adolphe Sax's innovations with brass instruments and Erard's stronger and more powerful pianoforte. Koenig was suddenly at the center of a thriving musical instrument market.

Vuillaume's Violin Workshop – 1851–1858

> *It was here [Vuillaume's workshop] that he first manifested an interest in acoustical problems, an interest so keen that, on Vuillaume's advice, he abandoned violin-making, at which he had become an expert, for the work of an acoustician. Koenig, however, never lost interest in the violin. His recollections of the great violin-maker, who subsequently became a millionaire, were so interesting and entertaining that I more than once urged him to write a memoir.*[12] *James Loudon, 1901.*

Vuillaume made a lasting impression on the young Prussian immigrant. A glimpse of his workshop reveals a bustling place of varied activity – interactions with musicians, scientists, and artisans; precision artisanship, experimentation, invention, teaching, promotion and business – that strikingly resemble features of Koenig's later business. Vuillaume was a legendary character in the world of violins.[13] He descended from a violin-making family of Mirecourt, a small town in Vosges, which had been a flourishing centre of French lutherie. During his early career in Paris he made his mark through imitation, being the first to successfully copy Cremonese masters with respect to tone quality and appearance. He was also credited with innovations, such as new instruments and bows, and had a reputation for making instruments with impeccable and beautiful finish. He was an extremely successful businessman, an adept self-promoter and by 1850 was conducting business throughout Europe. Koenig's seven years of apprenticeship in the shop on rue Croix-des-Petits-Champs became an important part of his reputation and identity as an acoustician (Fig. 1.2).[14]

Vuillaume's shop operated for over 50 years, becoming one of the more prolific musical instrument businesses in the city. He displayed and promoted his works at the major world fairs; he cultivated relationships with the composer Hector Berlioz, well known players such as Paganini, Ole Bull, and local French talent (and

Fig. 1.2 Wooden resonators. Koenig's background as a violin maker is readily apparent in his instruments made of wood. His resonators are made of finely grained spruce with a light varnish and mahogany veneer on the side. CR 38a. Museu de Física, University of Coimbra, Portugal. Photo by author, 2005

future son-in-law), Delphin Alard;[15] he had a wide network of admirers, customers, pupils, agents and unusual friends (such as the reclusive Italian violin collector, Luigi Tarisio); he also held local clinics, some of which were quasi-fraudulent for eager young men who crowded the Parisian violin-making trade. During some of his demonstrations, for example, he concealed his true secrets from the paying audience.[16] He owned a vast collection of famous old violins, which served as a resource for copying and innovations.[17] One story circulated, for example, that Vuillaume had repaired a Guarnerius for Paganini and at the same time made a copy so faithful that Paganini could not tell the difference.[18]

There were many colourful stories that spread Vuillaume's legend. Koenig often recalled an incident attached to one of the most famous violins owned by Vuillaume, *le Messie*. Stradivarius made it in 1716 and it was reputed to be his favourite violin and an example of his late style and design. In 1827 it fell into the hands of the eccentric Italian violin collector, Luigi Tarisio, who guarded the violin in his apartment in Milan and often boasted of his possession on his selling trips to Paris. Upon Tarisio's death in 1855, Vuillaume went to Milan and purchased the instrument, along with several others, from unsuspecting relatives. He returned to his workshop in triumph to show the prize to his workers.[19]

In order to make faithful copies and do repairs, Vuillaume had dozens of skilled apprentices who were trained in precision woodwork. Even the smallest change in thickness or size of material had a significant impact on sound quality. Life at the workshop included a cult-like admiration for the master's keen eye, ear and feel for wood, glues, varnish, bow hair and strings. One of Vuillaume's former pupils, Delanoy, recalled his master's authority and strict supervision:

> As far as his art was concerned he had an eagle's eye. Quite often as he was watching a workman's almost completed task, he would grab the instrument from his hand, and seizing a file, a rasp or a pocket knife, would start filing here and there to the great disgust of the workman who did not dare say anything, and when he was satisfied he would say to the man: "Fix those rough edges and polish." When this was finished the workman had to recognize that his work looked better and had more chic. Vuillaume would laugh and leave, pleased with himself.[20]

Violin making was (and still is) a conservative craft.[21] Vuillaume and his contemporaries used the techniques and tools of the Cremonese masters. They had an engrained respect for history and tradition. Labels, complete with reproductions of medals and awards, projected the unquestioned authority of renowned makers. The basics of the workshop did not change over time and included many items found in a cabinet-makers shop and those found in the instrument trade (Chapter 3). There were plates with drawings of violin moulds of actual size, graduation maps (thickness contours) of model violins (Stradivarius), chisels or gouges for rough shaping, templates and forms for continuous fitting and checking of shape, oval planes for precision smoothing and shaping, rasps for shaping, gauging callipers for precision measurement of thickness, marking and measuring compasses, purfling chisels, violin maker's knives, clamps, glues and varnishes. Making a violin involved measurement, countless corrections and comparisons with models and standards. It consisted of a methodical and repetitive process of gouging, marking, scraping, planning, finishing, verification, "constantly consulting the gauging callipers,"[22] and, of course, listening.

The final product, a hollow box from 13 to 14 inches in length, 8½ inches wide, 2½ inches deep and about a pound in weight, represented a delicate harmony of over 70 pieces. In his classic 1885 treatise on violin-making, Edward Heron-Allen wrote: "The wondrous capabilities and wonderful equilibrium of all the parts may be summed up in one short sentence – it supports a tension on the strings of 68 lbs., and a vertical pressure on the bridge of 26 lbs."[23]

One of the most important skills that Koenig would have taken into his second career as a maker of scientific acoustical instruments was the experienced coordination of hand and ear. The thickness and contours of the back plate, for example, or the position of the bridge or sound post, depended on a precise, repeated series measurements and trials. As the maker shaped the maple plate to the contours of the graduation map of the prescribed model violin, he had to verify through a combination of physical measurement and a repeated tapping for the right tone. Especially with a violin, there was a constant awareness of the "life" of the wood, meaning the presence and strength of harmonics. The well-used classic violins made by the Cremonese masters were said to be alive, responsive and rich with multiple tones and harmonics. Makers such as Vuillaume or Koenig were acutely aware, at the material level, of controlling and manipulating every aspect of timbre. This acute sensitivity to harmonics would become the centre of Koenig's workshop and research activities related to timbre and combination tones.

The selection of wood was also a highly developed skill; the back, neck, sides and bridge of the violin were to be made from a specific kind of aged maple; the belly, bar, blocks and sound post were of pine; the tail-piece was made of ebony. There were specific instructions for choosing well aged wood with just the right structural and acoustic qualities.[24] Vuillaume reputedly selected the best wood from the unlikeliest places. A few stories circulated that he had used pieces of old furniture and a bridge for some of his violins.[25] He even invented a special oven to treat his wood, although he abandoned this after realizing that artificial methods did not produce results comparable to nature.[26]

> His final conclusion was that the best wood from the point of view of resonance is the wood that has been seasoned in planks for some 30 or 40 years, of about 3 cm thickness, and, most important, that the seasoning should be carried out in the fresh air and under cover.[27]

In fact, violin makers went to great lengths to achieve the right sonority by their selection of wood. The violinmaker, John Broadhouse, claimed that for the best sound qualities and brilliance of tone, the tree should be cut in December or January, when no sap flows. He suggested cutting it from the south side of the tree, as the Italians did, and to make sure it was seasoned for 7 years. Judgement and experience were crucial. "Vuillaume, of Paris, travelled in Italy and Switzerland for the express purpose of procuring pine wood, and bought chairs, tables and other articles of furniture whenever he found the kind of wood he wanted."[28]

The final products were judged by workmanship, finish and tone. The players would judge the violin by its power ("volume and sweetness are imparted to the inaudible vibrations of the strings"), delicacy ("the slightest touch of the bow draws forth a tone sweet and true and pure"), and penetration ("for the tones of the instrument, even when played pianissimo, carry further than ten times the volume of mere noise").[29] The workmanship would be judged at a glance by the symmetry of the f holes, purflings (the dark lines of plane-tree wood that outlined the plates and prevented cracking at the edges), proportions, the unique design and construction of the scroll, finish, grain structure, and weight.[30]

The master makers, however, intended their abilities to be obvious to a select few, without giving away too much. Violin makers, like all highly skilled artisans, learned the art of covering their tracks as a way of simultaneously concealing their methods while showing off their mastery – the final product was seamless and without blemish, and difficult even for experts to deconstruct the challenges, secrets and tedious steps of construction. The purfling in the corners, for example, was the mark of a master maker; it was a critical operation that had to be done as carefully and cleanly as possible without slips of the knife that could "spoil symmetry."[31] It was in this purfling, wrote Heron-Allen, "that the true delicacy of handling and workmanship really shows itself in the construction of the fiddle."[32] "It is the mark of a good workman to make these joins at the corners and ends as imperceptible as possible."[33] To take another example, the final touches of finishing and smoothing the contours of the back and belly, entailed a process of planning and scraping, followed by wetting the wood to show slight defects for correction. This was not to be done "carelessly" or "lazily," "for on it depends the entire character and beauty of your instrument."[34] This approach would be especially apparent in Koenig's tuning forks (CR no. 36).

Tools, as the above language indicated, carried artisanal values as well as specific technical traditions. Words such as true, pure, perfect, defective, sweet and false were central to the maker's vocabulary and culture, and revealed a powerful set of guiding ideals. The tests for the "truth of a string" or their "falseness" were as rigorous as cross-examination in a court of law. And the judgement was not just on the maker of the strings, but on the chooser as well. Good buyers knew to examine strings carefully for perfect homogeneity and go to the "best dealers." "They [the strings] must be true."[35] A just as rigorous set of ideals impelled violin makers to strive for the highest set of standards, well beyond appearances. Even inside the violin, not seen be players or dealers, had to be refined and smoothed with the same standards as the outside. "Remember," wrote Heron-Alllen, "it will be not be pleasant to think that in some centuries to come the repairer will find that the work you have been so careful over outside is slovenly inside."[36]

These values and traditions carried with them strong emotions and individual pride. The immersion in tough, tedious work created a deep, personal connection to the products. Violins were not just products for sale, but works of art, part of the person's "soul."[37] This personal dimension would be a central trait that Koenig brought from violins to tuning forks. Heron-Allen put it thus:

> Until he has pursued the art no one can imagine the fascination of violin-making, – the thousand pains the player never dreams of, the thousand touches the uninitiated eye never appreciates, the exquisite work of the interior which no eyes save those of the maker and repairer, ever will be priviledged to see. These are the things which make the Luthier love the work of his hands, as if it were his own child.
> Years ago (it is said) there lived in Bremen a watchmaker, whose fame was universal, for his watches were the most perfect in the world. No once could discover the secret to his preeminance. At last he sickened and died, and the secret was revealed, for all his watches stopped one by one: – he had wrought a little of his own Soul into each timepiece, and when he died – they died also.[38]

In the nineteenth century, science became integrated into these values and traditions. Vuillaume, in particular, was a leader in bringing scientific culture into violin-making. In William Alexander Silverman's fanciful, yet historically based novel *The Violin Hunter*, Vuillaume boasts: "Perhaps the world will not remember Vuillaume as a violinmaker, but as a scientist."[39] It was well-known that he collaborated with the French scientist, Félix Savart, on a series of experiments on the violin. In an act of scientific sacrifice that would horrify present musicians, they cut Stradivarius violins into segments to perform Chladni (vibration) tests.[40] He also performed several studies on strings, woods, "f" holes and bows,[41] and frequently redesigned parts and carried out experiments with other makers and players.[42] For Vuillaume, science offered a potential means to uncover the secrets of Cremonese violins. The Belgian composer, François Fétis, claimed that Vuillaume was constructing copies based on proven laws and guidelines.[43] There was also the issue of rapidly changing demands of players. In the nineteenth century, there was a growing need for adapting the violin to larger concert halls and evolving playing styles. Makers looked to acoustics for ways to promote their views on these subjects.[44] Aside from genuine concerns with improving violins, the scientific study of sound was above all a means for Vuillaume to create a progressive, cutting-edge image, thus providing an advantage in a highly competitive market place. Just as he had used the mystique that surrounded the secrets of his craft, Vuillaume used the popular image of modern acoustics and technology to build his identity.

Science and violins, therefore, were deeply connected during this period, and both Vuillaume and Koenig used these connections to their advantage. Whereas Vuillaume brought science into violin making; Koenig brought the culture of violin making into science. It would become central to his reputation. The Parisian science critic, François Moigno, continuously reminded readers that Koenig had been given "solid instruction"[45] under Vuillaume and had "acquired the skill of being able to perfectly finish wood that is so necessary in the construction of acoustical instruments."[46] Each of his instruments carry aspects of Vuillaume's workshop within.

From Violins to Tuning Forks

In 1858, after 7 years of working with Vuillaume, Koenig launched his own business as an acoustical instrument maker for scientists. While working in the violin shop he had devoted leisure time to the study of mechanics and physics, and attended public lectures, such as those of Victor Regnault at the Collège de France, the celebrated experimentalist (Chapter 5). He was integrating into the scientific world of Paris. During his apprenticeship he had already begun constructing instruments for scientists. The acoustical maker Albert Marloye had retired in the early 1850s and one of the largest makers, Pixii, and their successors, Fabre et Kunemann, continued making these instruments for a short time. Their 1855 catalogue states: "Un des meilleurs élèves de M. Vuillaume est chargé sous notre direction de l'ajustement des tuyaux, diapasons, etc."[47] That "student of Vuillaume" making "pipes and tuning forks" must have been Koenig. Three years later, he opened his shop at Place Lycée Louis le Grand, which was on the grounds of the famous Lycée in the heart of the school district.[48]

The violin market was vast and competitive; the scientific instrument trade, by contrast, was smaller but growing. After doing contract work for Fabre and Kunemann, the 25-year old saw an opportunity to develop his own business. The recent departure of Albert Marloye, who had been the first specialist in acoustical instrument making, created an opening in the market. Acoustics was a developing field and there was increased demand for a specialist maker to supply research and teaching equipment. It is not hard to imagine, therefore, how a trained violin maker, with a family background in precision instrument making, musical instrument making, some experience making scientific instruments, and an interest in physics, could move into this business full time.

Acoustics was a growing field of interest. New instruments, experiments and attractive demonstrations had started to enter the study of sound in the first half of the nineteenth century.[49] In 1802, Ernst Chaldni published, *Die Akustik*, a specialized treatise on acoustics.[50] His studies of vibrating bodies in particular gave birth to a whole line of instruments which in turn led to original studies by Savart, Michael Faraday and Charles Wheatstone, and as we saw above, the experiments by Koenig's master, Vuillaume.[51] Another significant development came in 1819 when the Frenchman, Cagniard La Tour, invented the siren, an instrument in which pressured air blew against a pierced rotating disk to produce powerful sounds. Savart created related instruments (e.g. rotating toothed wheel) and a series of studies on the quantitative conception of pitch and its upper and lower limits.[52] In the 1830 s and early 1840 s, two German physicists, August Seebeck and Georg Ohm expanded these studies with a series of modified sirens to explore the physics of sound and its relation to human perception.[53]

Even with these developments, acoustical instrument making had been almost nonexistent before 1830. It was still mostly a mathematical field with little experimental tradition and no speciality makers such as those found in optics or electricity. Most of its scientific instruments, in fact, came from musical instrument makers. This situation changed with the collaboration between the scientist Félix Savart, the demonstrator I.J. Silbermann, and the instrument maker Albert Marloye in Paris. Previous to his relations with Savart, Marloye had been a maker of geometric models and wooden mathematical instruments. In the 1830 s, he started building apparatus for Savart's series of lectures on sound at the Collège de France.[54] Savart was now putting his efforts into a comprehensive course on the subject after years of working on acoustical problems and inventing several novel instruments. François Moigno, who attended these lectures, was so taken by the instruments that he asked Marloye to make a series of them so he could repeat the experiments at the École normale ecclésiastique. The project succeeded and Marloye quickly turned this series of demonstrations into a commercial enterprise, constituting the first specialty line of scientific acoustical instruments. He released a catalogue in 1840 which had organ pipes, reeds, interference demonstrations, vibrating rods, bars, membranes and plates, and stringed instruments. There were two measuring instruments – the Savart wheel and the La Tour siren. As Moigno later recalled, acoustics suddenly became fashionable.[55] Marloye's instruments spread throughout the scientific world and established a foundation for teaching acoustics on a large scale (Fig. 1.3).

The Scientific Instrument Trade in Paris

"À PARIS," a recognizable part of the signature of Parisian instrument makers, became a symbol of the French instrument brand at its height. Scientific instrument workshops, in all their splendor and grime, were a big part of what the American physician, Oliver Wendell Holmes called the "concentrated scientific atmosphere" of Paris.[56] The extensive network of workshops in this "concentrated" scientific center formed a vast, dispersed factory of practical scientific know-how. In fact, the precision instrument trade in Paris enjoyed its golden age from 1830 to 1880.[57] It was at its peak following the 1851 exhibition. English makers still had a large presence on the market (maybe still larger in numbers than other countries),[58] and

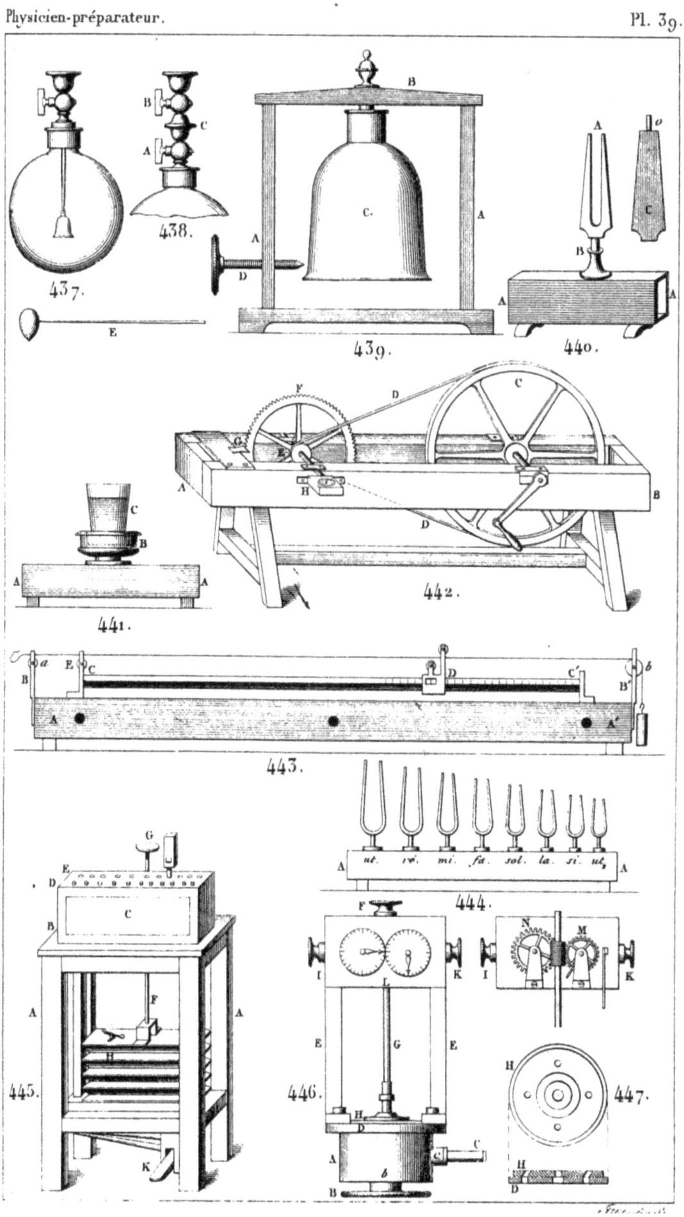

Fig. 1.3 Marloye instruments. Fau and Chevalier (1853, plate 39)

there were some well known German makers in the market, but French makers such as Soleil-Duboscq (optics), Breguet (horology, electricity), Ruhmkorff (electricity), Nachet (optics and microscopes), Gambier (dividing engines), Froment (electricity), Brunner frères (geodesy, surveying), Richard (meteorology), and Chevelier (optics)

Fig. 1.4 Koenig's signature on a pine resonator. Photo by author, 2005. Physics Department, University of Toronto, Canada

dominated the market. Buyers from every country came to Paris to make large purchases. There was also a thriving book trade. Agents swarmed the workshops of the Left Bank which were meeting places for science, business, and manufacturing (Fig. 1.4).[59]

It was an intensely entrepreneurial and creative trade. Similar to the world of musical instrument making, the precision trade was full of collaboration between clients and makers. It was not uncommon for an instrument maker to publish the results of experiments, reshape his field with inventions, and collaborate closely with researchers. One notable figure in Parisian scientific circles was a young German instrument maker, who had moved to Paris in the late 1830 s. Heinrich Daniel Ruhmkorff (1803–1877), from Hanover, developed an extremely powerful and efficient electrical coil in the mid 1850 s, the "Ruhmkorff coil," which thrust electrical studies into a new era.[60] Ruhmkorff's successor, Jules Carpentier (1851–1921), who became renowned for his excellent precision workmanship, worked closely with Marcel Deprez (1843–1918) and Arsène d'Arsonval (1851–1940) in their pioneering electrical work.[61] As mentioned earlier, Savart had worked with Vuillaume and Marloye. The scientist Jules Lissajous collaborated with instrument makers Secretan and Lerebours to design a standard tuning fork in the late 1850 s. Organ builder Aristide Cavaillé-Coll worked with instrument maker Gustave Froment, scientist Léon Foucault and astronomer Le Verrier to produce an apparatus to measure the speed of sound. The instrument trade in Paris benefited from these collaborations and became a source of much innovation in the sciences.[62]

Makers, however, had to contend with strong competition between each other. Marloye and Koenig did not have serious competitors in acoustics – Marloye was the first specialist and he dominated the field until his retirement in 1854, and Koenig quickly came to dominate the acoustics niche from 1859 onwards – but both makers did have to compete with other disciplines for space in the physical cabinets of growing departments. They had to make the case for the importance of acoustics amid a chorus of rapidly developing areas like optics and electricity. Instrument makers therefore refined strategies for attracting customers, or went out of business.

One well documented purchasing trip shows this promotional activity in action. When Charles and Ira Young of Dartmouth College went to Paris in 1853, they were clearly impressed by Duboscq (optics) and Marloye's evening demonstrations. Their subsequent purchases, heavy in optics and acoustics, reflected their favourite stops. Nineteen-year-old Charles Young described in detail visits to Marloye who "very kindly showed us many very interesting experiments upon sound" and made note of his red ribbon from the "Legion of Honour." He was also impressed with his evening visits to Duboscq's atelier where "I have been seeing the finest optical experiments I ever saw...His experiments were many of them magnificent, especially in polarization."[63]

The Youngs bought a Duboscq-Soleil spherometer (still at Dartmouth College) that also tells us much about interconnections within the Left Bank instrument trade. It rests under a glass dome on a wooden platform made by a glass blower named Rigault, who worked at 16 rue Guénéaud. There is also a hand-engraved signature on the brass stopper that fits on the central adjustable screw. It reads "roger Mathieu, 7 rue s. Severin, 1853." Mr. Mathieu most likely made precision screws for Duboscq and other makers. His signature signifies the "dispersed factory" run by instrument makers at this time. It also tells us that his work was valued enough to make his own personal mark on the instrument. Both Rigault and Mathieu were located close to the major instrument makers.[64]

Parisian makers were especially known for their attention to aesthetics. Instruments with elegant design and decorative touches attracted customers and found their way to prominent in physical cabinets. There is no doubting the sleek beauty of a Duboscq polarizer with its turned brass stand and asymmetric series of tapered brass cylinders holding optical components,[65] or Pixii's air pumps with brass pillars, finials and stylized engraved signature, or Marloye's vibrating plate apparatus with beautiful carved wood stands.[66] By the later part of the century French makers no longer decorated instruments with eighteenth-century frills like those found on Nollet's instruments,[67] but they used high-quality materials and continued to spend time on decorative features. The instruments were as much a statement about their inherent workmanship, as they were about function. There was a pronounced self-consciousness to the French products that reflected on artisanal and national pride and commercial pressures. At the end of the nineteenth century, beauty still sold instruments, but function and precision were rapidly becoming more important, representing a new direction in commercial forces on which German makers capitalized.[68]

In its "golden" period, the French instrument trade had a profound influence on the overall material structure of science; French instruments were found in laboratories around the world; students learned basic physics using French instruments. But this was a two way street: the growing scientific market just as readily influenced the French instrument trade. It is clear from hundreds of scattered collections in North America that this market kept the French businesses going following the industrial and scientific decline which began in the 1870 s. The constellation of competing new colleges, each wanting to establish physical cabinets and teaching laboratories, kept scientific Paris in business. The microscope maker Nachet, for example, told a colleague that his best customers were American and that he sold most of his instruments to American medical students.[69] In fact, students formed the basis of the instrument market. There were far fewer serious researchers in the nineteenth century, and therefore, most of the market was for education, or more precisely, classrooms of boys in their late teens. This explains the abundance of elementary and entertaining educational instruments sold in the Parisian market. It also explains the growing popularity of visual instruments.

The large education market supported the French instrument trade, sustaining the smaller market for high-quality precision instruments. In the fields of electricity, optics and acoustics, there were a number of researchers on the continent and in North America who demanded cutting-edge experimental equipment. Ruhmkorff and Carpentier in electricity, Duboscq in optics and Koenig in acoustics were makers with a reputation for producing apparatus for serious researchers. In Koenig's case, the fact that he became a serious experimenter himself fuelled and supported this dimension of this business. As his career took shape he was forced to continually balance between these two aspects of the business – research and education; the former helped build his reputation as a scholar and innovator, the latter spread basic acoustics and paid his bills.

In conclusion, in the 1850 s and 60 s Koenig became immersed in the artisan culture of Paris. He trained in the world of musical instruments and then integrated into the precision instrument trade. Both worlds shared similar values, tools, business approaches, labour traditions, and social dynamics. In Vuillaume's workshop he learned the tools and skills of violin making, the relations between hand, ear, materials and sound. He absorbed the ideals, values, standards and traditions of the violin workshop culture, the art of promotion, the need to cultivate relations with clients and the awareness of the growing trend of utilizing science for tackling acoustical problems. In the precision instrument trade, Koenig became part of commerce, collaboration, new materials, science education, experimentation, precision skills, artisan life, international connections and invention.

Even after his move to acoustics, Koenig maintained direct connections with musical instrument makers (see Chapter 3) showing that the relations between the two trades continued to be beneficial. During his later career, he was once asked by a musical instrument dealer to identify the "old master" who made a recently purchased violin. The response was a surprise to the dealer – that he (Koenig) was the maker. "A glance at the tail-piece sufficed."[70] There were other forms of collaboration. The University of Coimbra in Portugal has a Koenig apparatus for demonstrating the Doppler effect that carries a reed mechanism made by Julian Jaulin, the award-winning Parisian reed and organ maker (CR no. 78).[71] Jaulin would have been subcontracted for these specialized parts. In addition, a fairly simple looking ellipsoid bell signed by Koenig and recently found at Union College in New York, produces surprising complex sounds for stimulating patterns on membranes. Such an instrument would have been made by a bell specialist or at least with guidance from that trade (CR no. 124). The Smithsonian also has three bows from Koenig's workshop (CR nos. 3–5) linking his work to a bow maker or contractor. As well, hundreds upon hundreds of Koenig organ pipes exist around the world, making it quite likely that he had skilled help from specialty musical workshops. His demonstration pipes, for example, were made of carefully selected pine with mahogany reinforcements at the lips and feet; their dimensions (volume and thickness of wood) are remarkably consistent between each instrument in a series, with only slight increases or decreases in size from note to note (CR no. 89–116).[72] It is likely that an experienced pipe maker from a shop like Jaulin's or Cavaillé-Coll's worked for Koenig on these instruments, which were a major part of his business.

Musical instrument making continued to be part of scientific acoustics. In the next chapter, we see how the study and practice of sound went through a dramatic change from a different context. Between 1856 and 1862 Hermann von Helmholtz built a theoretical edifice that linked the workings of the inner ear, an understanding of the psychological aspects of sound perception, and the physical behaviour of sound waves. He also created several instruments for demonstrating and testing these ideas.

Notes

1. CR no. 134. NMAH, Smithsonian cat. no. 314, 589. Koenig (1865, p. 23); Koenig (1873, p. 8.)
2. Information about Koenig's early life in Königsberg comes from the personal archives of the Neumann family in Bückeburg, Germany. One of the better sources on Koenig's personal life are tributes by his niece, Helene Neumann, see Neumann (1932a,b,c). Dr. Eberhard Neumann-Redlin von Meding, the keeper of the family archives, has published an article on Koenig's background and life, see Neumann-Redlin von Meding (2001). Neumann-Redlin von Meding's article derives from a talk he gave on Koenig May 15, 2001 at the German-Russian House in Kaliningrad, Russia. Other biographical information comes from the Loudon Papers at the University of Toronto Archives (UTA-JLP) and several obituaries and profiles: Boyer (1901), Le Conte Stevens (1890, 1901), Loudon (1901b), Moniteur (1901), and Thompson (1891). For secondary sources on Koenig see Brenni (1993), Miller (1935), and Shankland (1970).
3. Gause (1965, p. 2180).
4. Olesko (1991) and Turner (1971). For more on Königsberg culture during this period see, Olesko (1994).
5. Neumann-Redlin von Meding (2001).
6. Olesko (1994, p. 230) and Brenni (2004).
7. Skrzeczka (1849, pp. 29–33).
8. Neumann (1932c).
9. Blackbourn (1997, p. 192).
10. For a colourful example of such a skilled migrant, see O'Connell (2008).
11. Paul (1985).
12. Loudon (1901b, p. 3).
13. In 1998 there was an exhibition of Vuillaume's life and work at the Cité de la Musique in Paris, see Campos (1998). For a profile of Vuillaume, see: Beare (1980). The Belgian composer, François-Joseph Fétis, wrote a book on the history and theory of Cremonese violins in collaboration with Vuillaume, see Fétis (1864).
14. Moigno (1868). Idem., 1865, p. 534.

15. Alard was a teacher at the Paris Conservatoire and one of the first musicians to introduce Parisian society to chamber music in the 1840 s, see Campos (1998).
16. At mid-century, Paris was crowded with eager young men who wanted to be violin makers. There are a number of anecdotes about Vuillaume capitalizing on this market by offering trade clinics. During his Thursday afternoon varnish clinics, to take one of the more colourful tales, he would sit at his workbench applying varnish with a special brushing technique for an attentive audience. He sold the coveted varnish to the students when the session ended. After the pupils left, to the amazement of his workers, Vuillaume wiped the varnish off and applied his own, true mixture. Millant (1972, p. 90) and Silverman (1957, pp. 158–159).
17. Millant (1972, pp. 88, 119, 134).
18. Sandys (1864, p. 247).
19. Millant (1972, p. 119) and Loudon (1901b, pp. 3–4). Today le Messie rests in a display case in the Hill Collection of Musical Instruments at the Ashmolean Museum, Oxford.
20. Campos (1998, p. 33).
21. I wish to thank Dennis Alexander, a Luthier of Aylmer, Quebec for his guidance on the art and traditions of violin making.
22. Heron-Allen (1885, p. 247).
23. Ibid., p. 126.
24. Ibid., pp. 125–139.
25. Millant (1972, p. 87). He once used the wood of an old bridge of Turin. Campos (1998, p. 25).
26. Heron-Allen (1885, p. 129).
27. Millant (1972, p. 87).
28. Broadhouse (1890, p. 10).
29. Heron-Allen (1885, p. 127).
30. Personal communication with Luthier, Dennis Alexander.
31. Heron-Allen (1885, p. 263).
32. Ibid., p. 263.
33. Ibid., p. 163.
34. Ibid., p. 244.
35. Ibid., pp. 206–208.
36. Ibid., p. 251.
37. Ibid.
38. Ibid., p. 320.
39. Silverman (1957, p. 176).
40. Fétis (1864, pp. 77–92), Maniguet (1998), and Savart (1840, p. 70).
41. Fétis (1864, pp. 77–79, 121–24).
42. Haine (1998). The Norwegian player, Ole Bull, visited Vuillaume's shop in the 1850 s to do experiments on the soundboard. Bergsagel (1980).
43. Fétis (1864, pp. 77–92, 121–124).
44. Hutchins (1997b, p. 7).
45. Moigno (1868).
46. Idem., 1865, p. 534.
47. I thank Paolo Brenni for informing me about this statement found in *Fabre et Kunemann* (1855, p. 2). Pixii was one of the first makers to offer a complete line of acoustical instruments after Marloye. Brenni (2006, pp. 15–16).
48. Le Conte Stevens (1890, p. 546).
49. Dostrovsky et al. (1980) and Hunt (1992).
50. Jackson (2006, pp. 13–44).
51. Beyer (1998, pp. 27–54). Pantalony (2005b, pp. 143–144).
52. In fact, he and Marloye had a falling out over the lower the limit of hearing. Brenni (1995a).
53. Turner (1977).
54. F.M.N. Moigno in Tyndall (1869, pp. viii–xi). Also see Brenni (1995a).

55. Tyndall (1869, pp. viii–xi).
56. Quoted in Morse (1896, p. 108). Also see Warner (1998, p. 3).
57. Brenni (1993–1996).
58. Hackmann (1985, pp. 61–65).
59. Pantalony (2004a) and Simon (2004).
60. Brenni (1994b).
61. Brenni (1994c).
62. Blondel (1997).
63. Charles A. Young *Diary 1853*, p. 45, Young Papers, DCSC. Also see Pantalony (2004a).
64. King Collection of Historic Scientific Instruments, acc. no. 2002.1.34464. Also see Pantalony (2004a).
65. Brenni (1995c, p. 119).
66. King Collection of Historic Scientific Instruments, acc. no. 2002.1.35290 and 2002.1.34026.
67. David M. Stewart Museum (2002).
68. Brenni (1991).
69. Warner (1998, p. 294).
70. Loudon (1901b, p. 5).
71. See Mach's apparatus with two reed mechanisms (University of Coimbra) signed "J. Jaulin Bte. S.G.D.G." CR no. 78.
72. The University of Toronto has an almost complete set of Koenig's organ and reed pipes. CR no. 89–116.

Chapter 2
Hermann von Helmholtz and the *Sensations of Tone*

> *Music has hitherto withdrawn itself from scientific treatment more than any other art. . . .It always struck me as a wonderful and peculiarly interesting mystery, that in the theory of musical sounds, in the physical and technical foundations of music, which above all other arts seems in its action on the mind the most immaterial, evanescent, and tender creator of incalculable and indescribable states of consciousness, that here in especial the science of purest and strictest thought—mathematics—should prove pre-eminently fertile.*[1]
>
> Hermann von Helmholtz, Bonn, 1857.

In the 1840s it seemed improbable, even offensive to some, that musical sounds could be analysed in the same way that a chemical compound could be reduced to elements, or the way light could be separated into a spectrum. Today we take for granted the notion that musical sounds are in fact a compound of simple, pure frequencies. Electronic equipment does this analysis automatically. We play a trumpet into a microphone and a spectrogram appears on a monitor. The basis for this, Fourier analysis (a mathematical description of periodic behaviour), first appeared in the mid nineteenth century. What historical circumstances made this mathematical theory so "pre-eminently fertile"? How did musical sounds, "the most immaterial, evanescent, and tender creator of incalculable and indescribable states of consciousness,"[2] enter into the laboratory to be analyzed, manipulated and measured? How was the German context of this development different from Koenig's unique Parisian training with sound?

Through a seminal book and new instruments, Hermann von Helmholtz laid the foundation for an analytic conception and practice of sound. Rudolph Koenig was the first instrument maker to capitalize on this major development. But, as we will see in later chapters, Koenig eventually rejected fundamental aspects of Helmholtz's work, in particular the mechanistic elements. Even though both men shared a love of science and music, and both had even shared the same social circles in Königsberg, they came to differ markedly on their approach to acoustics. In this chapter I look at the origins and background of Helmholtz's studies of sound, part of which carried a mechanistic (physical and physiological) view of sensations, a psychological theory of perception built on these assumptions, experiments

D. Pantalony, *Altered Sensations*, Archimedes 24, DOI 10.1007/978-90-481-2816-7_2,

and instruments that reinforced these views, and sophisticated mathematics that explained some of the more elusive phenomena in this framework. It represented a compilation of Helmholtz's famous phrase from a lecture in 1862 that science strove to achieve "the intellectual mastery of nature."[3] Different from Koenig who failed to enter the German academic world yet achieved fame in the Parisian artisan classes, Helmholtz ascended to the top of the social elite of German academic scientists.

Hermann von Helmholtz

Musical culture was central to German science in the nineteenth century; it inspired inquiry, formed social cohesion and stimulated collaboration between scientists, musicians and musical instrument makers.[4] Hermann von Helmholtz, scientist and amateur musician, was an exemplar of these traditions. Music had always been essential to his life. In 1838 when he left his birthplace in Potsdam to attend medical school at the Friedrich Wilhelm Institute in Berlin, he wrote immediately to his father about the arrival of his piano at his new quarters. His Silesian room-mate, he reported, played the piano well but only cares for "florid pieces" (*colorirten Sachen*) and "modern Italian music."[5] The elder Helmholtz responded by warning his son to beware of "Italian extravagances" (*Ueberspanntkeit*) and not to forget the inspiration of German and classical music – the former was a distraction while the latter was an education.[6] As we will see, the piano itself, and not just the music, would provide inspiration for Helmholtz's studies in acoustics.

"Florid pieces" and "Italian extravagances" were just a taste of the novelties that Helmholtz experienced in his student years and early career. There was growing social and political uneasiness that culminated in the failed revolution of 1848. There was the triumph of steam power, the beginnings of train travel, the introduction of the telegraph and prosperity brought on by industrial development. Even Helmholtz's favourite pastime, music, went through dramatic changes in this period.[7] There was the emergence of the modern, more powerful pianoforte that would change concert music. There was growing acceptance of the well-tempered scale that would alter traditional notions of harmony. There were also problems, recognised throughout Europe, related to the standardisation of pitch.[8]

Helmholtz epitomised Prussia's *Bildungsbürgertum* (educated upper-middle class), with its emphasis on cultivating a whole individual, and strong social and intellectual connections between artists and natural scientists. Intimate Sunday afternoon salon events for family and friends included concerts and lectures on literature, art or popular science.[9] The Prussian education system, which came to represent these ideals, went through upheaval during the post-Napoleonic period, becoming the first system to replace classical teaching methods with an emphasis on research and laboratory-based teaching. In turn, the decentralised German states established a dynamic network of competing research institutes.[10] Justus Liebig's chemical institute, founded in the 1830s attracted students from across Europe and North America and became a veritable factory for research in organic chemistry.

Several schools in the German territories and later around the world would imitate Liebig's model for success.[11] We see the impact of this new teaching and research emphasis in Helmholtz's medical training. In his thesis year, 1841, Helmholtz joined the laboratory of the physiologist, Johannes Müller (1801–1858), and became acquainted with Ernst Brücke, Emil du Bois-Reymond and Carl Ludwig. This circle of young researchers formed the "1847" group becoming leading advocates for a school of physiology based on physical and chemical principles. Helmholtz's first paper showed his dedication to mechanistic notions with detailed studies on animal heat and muscle contraction. To complement his schooling in physiology, Helmholtz read the masters of eighteenth century mechanics and mathematics – Euler, Bernouilli, D'Alembert and Lagrange. In 1847 he combined this background with his knowledge of physiology to develop the mathematical principles for the conservation of energy.

From 1849 to 1855 Helmholtz taught physiology at Königsberg where he started to focus on sensory physiology. In particular, he began studies on optics and colour research. As he would do in acoustics, he explored the relations between the basic elements of light (the frequencies of the spectrum) and their counterparts in physiology, the receptors and nervous tissues. He relied on Müller's doctrine of specific nerve energies whereby specific nerves performed specific sensory jobs.[12] Since Descartes there had been a notion of the mechanics of sensation (i.e. based on a reflex system), but no one had ever suggested that the nervous system had a built-in, differentiated structure that divided processing jobs automatically. Müller's doctrine enunciated a radically new architectural blueprint for the sensory system.

From the primacy of physics and physiology, therefore, Helmholtz built a mechanistic conception of sensations. Psychological processes brought order to these sensations constituting our perception of the world. In this way, there was a progression from physics to sensation to perception. Patrick Macdonald has argued that the act of experimenting itself strongly shaped Helmholtz's view of perception. Experimenting, or the active "varying of conditions" in a laboratory became for Helmholtz a mirror of how the mind ordered incoming sensations. He saw perception as an act of will that could reorder or deliberately alter the conditions of experience.[13]

Helmholtz was as attentive in the laboratory as he was on the theoretical front. In one of his first series of studies at Königsberg, he measured the velocity of a nerve impulse with delicate electrical apparatus of his own invention. Before that time, nerve transmission was thought to be too fast, even instantaneous, for study in the laboratory. He used a precise electrical timing apparatus that was connected to a 50–60 mm long nerve of a frog's leg to produce a fairly consistent figure of 25 m/s. He checked his results using graphical apparatus (invented by his friend Ludwig) to map the sequence of the nervous impulse over time.[14] In addition to these researches, in 1851 Helmholtz became a celebrity in medical circles for his invention of the ophthalmoscope that allowed physicians a novel means of studying the inner structure of the eye for the first time. The ophthalmoscope became an indispensable instrument for his studies in optics.[15]

Contacts with industry and skilled artisans enriched Helmholtz's experimental endeavours. The self-regulating interrupter used in his vowel synthesiser was initially developed by Werner Siemens.[16] Friedrich Fessel of Cologne (see below) built the actual synthesiser for Helmholtz. The Berlin instrument maker, E. Sauerwald, who had collaborated with Gustave Magnus on early electrical apparatus, made the original double siren. He also made a myograph (for studying the action of electricity on bodies) based on Helmholtz's earlier myograph made by Egbert Rekoss of Königsberg. In 1852, Rekoss also invented the rotating disk for Helmholtz's ophthalmoscope.[17]

Taking advantage of his growing fame, and his ability to harness more research time and facilities, Helmholtz took up two positions in Bonn in 1855 and then Heidelberg in 1858. In 1855, as he neared the completion of his first volume on physiological optics, he began seriously investigating acoustics. For the next 8 years, this research would overlap with work on fluid dynamics and optics, culminating in his grand treatise *Die Lehre von den Tonempfindungen als physiologische Grundlage für die Theorie der Musik* (*On the Sensations of Tone as a Physiological Basis for a Theory of Music*). However, he published little work in acoustics after 1863. Following the publication of *Tonempfindungen* he moved away from physiological problems to thermodynamics, electrodynamics and hydrodynamics. He published three more editions of *Sensations*, but each with relatively minor changes. In 1871, after productive years in Heidelberg, he moved to Berlin to start an institute devoted to physical research.[18] This move signalled his break with physiology. The scope of physiology, he believed, had become too great for one individual to master.[19] He left acoustics just as his theories were becoming part of mainstream of teaching and research.

Physical Acoustics – Theory and Instruments (Tuning Forks, Tonometer, Double Siren)

Helmholtz's acoustics was a forceful and original synthesis of instruments, physics, mathematics, physiology and psychology.[20] The building blocks of this framework, however, came from physics and mathematics, in the form of an elemental conception of sound which derived from the work of Georg Ohm and Joseph Fourier. In the spring of 1856 Helmholtz wrote to a colleague at Königsberg that he had already formulated the foundation for the reform of acoustics.[21] In the early 1840s within the context of his work on the physical nature of sound,[22] Ohm had used Fourier analysis[23] to describe musical sounds as being made of a mathematically-related series of simple sinusoidal waves, which came to be known as simple partial tones. In addition, he made the radical assertion that the ear functioned as a Fourier analyser enabling humans to sense these tones, either as part of a compound or on their own. Critics, such as fellow German, August Seebeck, claimed that one could not always detect the simple tones predicted by a Fourier series; they appeared, he argued, to be mathematical abstractions with no basis in reality.[24] Helmholtz set out to make them a laboratory reality.

Beats, a musical phenomenon long known to musicians and tuners, became the key to this enterprise. When a 200 Hz and a 203 Hz tuning fork were struck and placed next to each other, distinct pulsations of three beats per second were produced permitting one to count the number of cycles between tones; they therefore provided a means of testing the presence of specific frequencies. Another phenomenon called the third tone, difference tone, or combination tone, also long known to musicians, provided a basis for interacting with the hypothetical Fourier simple tones.[25] Similar to beats, they appeared when two powerful tones were played together. For example, the combination of 100 and 250 Hz played at a strong intensity created a "combination tone" of 150 Hz. The cause of these tones was not understood, and they had not been consistently observed, yet they provided another useful tool for studying interactions with other tones.

The proper use of these effects – beats and combination tones – relied on instruments. In Helmholtz's work, emphasis was put on purity and precision – i.e. no unwanted harmonics in the sound source, and the ability of the source to consistently produce a specific frequency. This emphasis was novel for the 1850s. As Carlton Maley has noted, "the newly revealed importance of overtones [simple tones predicted by Fourier analysis] cast doubt on all acoustical experiments done with sources of unknown overtone structure."[26] Helmholtz, therefore, performed his studies with tuning forks in place of more traditional instruments used by physicists such as toothed wheels, monochords, reed pipes or organ pipes. Musicians had used tuning forks since their invention in 1711, but they were not considered worthy of attention by scientists until the work of Ernst Chladni, who had studied their vibrations.[27] In fact, they were mostly used by orchestras and were still fairly crude instruments up to the 1830s. Even so, tuning forks offered qualities that would become valuable for Helmholtz's experiments with beats, combination tones and simple tones: the u-shape was good for counting beats as it enabled strong vibrations to continue for long periods of time without losing energy; the u-shape was also purer than other sound sources (such as reed pipes) with much fewer unwanted harmonics; finally, the steel or iron could hold the pitch consistently (compared to wood used in reed pipes) with minimal changes over long periods of time, or due to changing room temperatures. To ensure the purity of his forks and to isolate and amplify the single sound, Helmholtz combined them with pasteboard resonating cylinders (Fig. 2.1).

Helmholtz also adopted a technique that greatly expanded the range of his studies. He used a tuning-fork apparatus invented two decades earlier – Scheibler's tuning-fork tonometer – that allowed him to work in several frequency ranges with equal precision. The tonometer was a series of over 50 tuning forks, covering an octave on the musical scale, each separated by a set number of vibrations that served as a base of comparison for the sound source under scrutiny. Using the tonometer, Helmholtz was able to study interactions between beats, combination tones and simple tones at the same time, with consistent results (Fig. 2.2).

Another apparatus, the double siren, permitted Helmholtz to test the nature of combination tones under high-pressured conditions. The siren had been developed in the 1820s but the double siren, two siren disks that faced each other, created

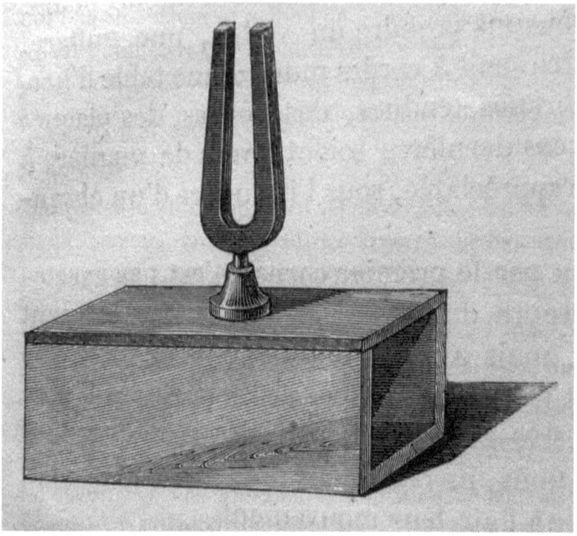

Fig. 2.1 Tuning fork and wooden resonator. CR 38
Source: Helmholtz et al. (1868, p. 54)

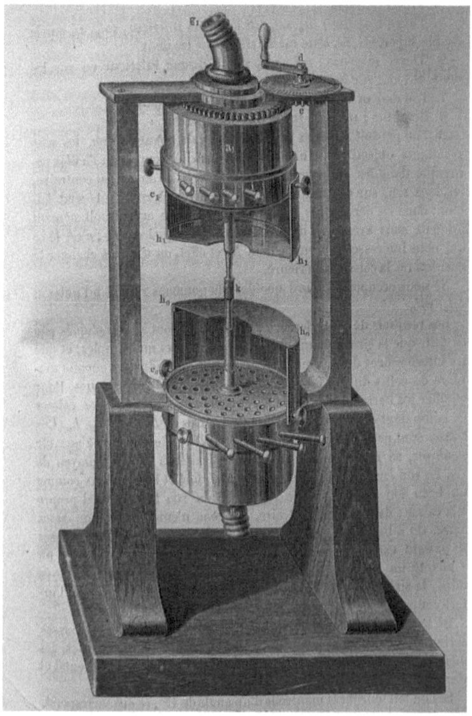

Fig. 2.2 Helmholtz's double siren. CR 27
Source: Helmholtz et al. (1868, p. 203)

a highly pressurized combination of two tones. Counting dials were placed in the middle of the two sirens for recording the number of turns per second and, with the aid of a clock for timing the revolutions, determined the frequency of a particular row of holes. A handle at the top allowed one to rotate the upper siren by degrees in order to create a shift in the phase of the upper sound source compared to the lower sources (for studying interference effects). Helmholtz also created special brass covers to ensure that the sounds were pure and without harmonics.[28] The polyphonic double siren, therefore, produced a means for investigating combinations of musical tones in a controlled fashion under intense air pressure. F. Sauerwald of Berlin constructed this invention for Helmholtz.[29]

Instruments as Agents of Change

The above instruments contributed original data for Helmholtz's studies, but they also introduced concepts, approaches and values that would reinforce the analytic conception for generations. As Stephen Vogel has pointed out, the siren, first introduced by Charles Cagniard de la Tour in 1819, introduced a radically different conception of sound.[30] Previously, sound had been viewed as a wave; the siren, with its pierced disk, created a conception based on discreet pulses. This conception made it easier to digest the analytic framework proposed by Ohm and Helmholtz. Sound could be decomposed with numbers alone, without resorting to waveforms.

Similar to the role played by the siren, tuning forks became conceptual bearers of pure, simple tones and not just sound producers. A series of them, in the form of the tonometer, became a classic embodiment of the Fourier system. A piano represented a series of notes too, but they were tones with multiple harmonics. Tuning forks, with very few unwanted harmonics, also contributed an added time dimension to experiments. They produced strong, consistent sounds for up to one minute. This made it easier to count beats with precision (beats were counted with a chronometer – e.g. 120 beats in 60 seconds resulted in 2 beats per second). Tuning forks thus began to reshape expectations and acoustical practice.

In addition to these added dimensions, the tonometer carried social values deriving from its unique industrial origins. As Myles Jackson has shown, in the 1830s Scheibler, a silk manufacturer, created the tonometer as a labour-saving tuning device and, more importantly, as a means for tuning automatically without resorting to an expert ear. Tuning was done by counting beats. The tonometer emerged from a context of automation and "deskilling." This change in the tuner's art, adopted by Helmholtz, created a practical context for objective, precision acoustics and later for visual acoustics. In the same way that the introduction of direct reading instruments introduced moral questions into nineteenth century laboratories, the tonometer and similar instruments raised issues of tuning made too easy.[31]

Finally, the siren introduced extremely powerful sound combinations in the laboratory. It was an experimental chamber in the classic sense that it created new effects, a novelty for a science with little experimental traditions. It could combine and measure sounds, which enabled the study of combination tones under controlled conditions. The double siren, therefore, enabled Helmholtz to make the connection between combination tones and their high-pressure, non-linear origins (see below).

Experimental Results

With the tuning forks, tonometer and siren, Helmholtz investigated the nature and appearance of combination tones in an unprecedented range of conditions.[32] He focussed first on these phenomena because they had been notoriously difficult to measure with certainty, and they did not fit into the analytic theory of sound deriving from Fourier and Ohm. Clarifying their behaviour would prove fundamental to his other investigations of Fourier simple tones.[33] In his initial experiments, therefore, he observed and mapped what he called first-order combination tones (the mathematical difference between two tones); second-order combination tones (the difference between the first combination tone interacting with one of the primary tones); and third-order combination tones. Each order of tones became successively weaker. He detected the more difficult tones by listening with resonators (see below) or by observing their activation of tuned membranes. In these experiments, he confirmed observations made by earlier investigators, and added several high-pitched (inaudible) combination tones that he had detected using beats. Using his apparatus, he thus claimed to map the appearance of a whole series of combination tones with greater precision and certainty than anyone previously. Furthermore, he reported a class of combination tones called *summation tones* (the sum of the two generating tones). These tones were much weaker than the difference tones and could not be heard by the naked ear. He claimed, however, that they could still be detected using "objective" methods (membranes or resonators), even though their weak presence made them controversial.[34]

Helmholtz had to explain the unknown mechanism of the combination tones and how it fit the analytic framework. He therefore proposed that combination tones were indeed independent phenomena (separate from simple tones) created under unique conditions (the production of very strong tones) from an actual physical transformation within the combined sound waves. In other words, the resultant sound waves were compounds with entirely new tones generated from the transformation. But what was the mechanism? Observation with the double siren showed that the intensity of combination tones increased at a greater rate than those of the primary generating tones, leading Helmholtz to suspect a non-linear effect. He was then able to demonstrate this mathematically. In physical terms, he viewed these tones as "accessory" phenomena that were not part of the Fourier structure of complex sound. Moreover, even though some of the combination tones were the mathematical difference between two tones, Helmholtz claimed that they were not beats that blended into a tone. They were their own objective phenomena.[35]

In addition, he used his findings to locate previously unobservable higher Fourier harmonics. He did this by making higher-order combination tones beat with higher, unobservable harmonics of a fundamental tone that had been predicted by Ohm's theory. The combination tones had themselves been generated from other combination tones that were predicted from his theory. The entire experiment was remarkable because the higher harmonics could not be heard, and, in essence, Helmholtz used beats to verify the existence of both the predicted harmonics *and* combination tones. With one experimental stroke he provided evidence for his explanation of combination tones and confirmed Ohm's controversial theory. He had also moved precision acoustics beyond the range of the ear.

The first part of Helmholtz's reform of acoustics, therefore, entailed clarifying the physical nature of combination tones as a way of verifying Ohm's theory of sound. A crucial part of this reform derived from instruments specifically designed to produce precise, pure frequencies. These developments would frame the pursuit of laboratory acoustics for the next 40 years.

Physiological Acoustics – The Piano as a Model for the Inner Ear

Earlier theories of music had stated that harmony derived from the human mind's abstract appreciation for simple mathematical ratios. According to the Enlightenment mathematician, Leonhard Euler, the mind sought simplicity and order, and therefore chords with simple relations (e.g. the fifth with 2:3) would create appealing harmonies.[36] Helmholtz, on the other hand, claimed that harmony and dissonance had a physiological basis in the inner ear. He viewed the phenomena of beats as the key mechanism of harmony. The more rapid beats became (i.e. in the range of thirty pulses a second) the more they tended to produce an irritating or grating effect on the inner ear. Such a grating effect would be perceived as discord.

But before he developed an overall theory of harmony and the inner ear, Helmholtz sought to clarify the physiological substrate for the sensation of simple tones, and how certain sounds combined to form a distinct quality, or timbre. He had verified the physical existence of the higher Fourier harmonics and now needed to investigate the physiological aspect of Ohm's theory. In November 1857, in the midst of intensive investigations on the nature of vowel sounds, he wrote to the Dutch physiologist, Franz Donders (1818–1889) that he would next attack the origin of timbre (*Grund der Klangfarbe*) in order to address what he viewed as the fundamental problem of physiological acoustics (*Grundfrage der physiologische Akustik*) debated by Ohm and Seebeck. Helmholtz agreed with Ohm that the ear must analyze compound sounds in accordance with Fourier's theorem.[37] He had done some preliminary experiments with his piano and discovered that specific vowel sounds were related to the number and strength of upper partials (simple tones).[38] He found, for example, that the piano strings tuned to specific notes responded in sympathy to the partials of a sung vowel thus providing physical evidence of the existence and strength of a partial in a vowel sound. In effect, the piano was the first sound analyser, and served as a powerful model for his emerging physiological conception of sound.[39]

These experiments and Helmholtz's belief in Ohm's theories were supported by recent discoveries of the anatomy of the ear. In 1851 Marchese Corti (1822–1876) published an intricate anatomical study of the inner ear, or the cochlea. He devised a special staining technique to improve microscope examination of this snail-like structure. The cochlea started at the oval window (where the hammer, anvil and stirrup ended) and divided into three sections – the cochlear duct, scala vestibuli and scala tympani – within which lay the organ of Corti (the seat of hearing), with the rods of Corti, the hairs, the tectorial membrane and the basilar membrane. Helmholtz now had a tantalising hypothesis for substantiating Ohm's theory of sound. He proposed that the differing strengths of Corti's rods may contribute to the sensation of different tones.[40] He pictured a whole battery of vibrating bodies (like the individual piano strings) lining the organ of Corti, each responding to a specific frequency.

The inner ear was thus pictured as a series of vibrating bodies that responded to frequencies in the same way that a series of piano strings responded to simple tones in a compound sound. To explain this idea Helmholtz proposed the hypothetical situation where every string of a piano connected to a nerve fibre. The piano strings, acting as a Fourier analyser, would vibrate sympathetically to the individual components of the sounds in the air; these vibrations, in turn, would be transmitted to the nerves and sensed independently.[41] The sensation of these dissected sounds in turn depended on the specificity of nerve cells connected to the inner ear. This was a direct application of the doctrine of specific nerve energies (*Lehre von den specifischen Sinnesenergien*) that Helmholtz adopted from his teacher Joannes Müller. Hearing, as with sight, was dependent on the specific nervous arrangements (*den verschiedenen Nervenapparaten*) of the sensing organ. These arrangements, no matter what the source of stimulation – mechanical pressure, light, sound, and electricity – produced the same, specific sensation if the nerves were activated.[42] Using Müller's doctrine, Helmholtz was able to create a strict, mechanical conception of the physiology and anatomy of the inner ear. This conception was based on a one-to-one correspondence between the elements of the inner ear and those of the physical world, the simple tones. He made minor modifications in light of physiological findings in the 1860s,[43] but the concept remained the most comprehensive explanation of simple-tone sensations until the 1930s and the work of Georg von Békésy (1899–1872) on the function of the basilar membrane.[44]

Psychological Acoustics – Resonators as Aids for Hearing Simple Tones

Helmholtz's one receptor/one tone hypothesis did not match with everyday experience. In the presence of a strong fundamental tone, upper partials were often difficult to hear, which opened the door to a psychological explanation. Shortly after Ohm proposed his Fourier analysis of sound, Seebeck criticized it on the grounds that some of the supposed simple tones could not be heard, and perhaps did not exist at all. For Helmholtz, however, this discrepancy with theory was not due to a flaw in the Ohmian conception of complex sound, nor with the physiological complement, but with the psychological aspects of sound perception. Seebeck, he claimed, even with all his experience as an experimental observer, had failed to direct his attention (*Aufmerksamkeit*) at the predicted tones.[45]

A psychological factor, termed by Helmholtz as an "unconscious inference," acted to distort basic sensations.[46] For example, after many years of hearing human voices, the ear becomes accustomed to the combined (compound) sounds and perceives them as a fused whole, making it difficult to hear the individual components. One must concentrate to pick out the elements that habit has seemingly blended into one phenomenon. In the 1850s Helmholtz had applied the same principle to his work in optics. For instance, when looking at a point in space, he asked why we see one image instead of two (with two eyes, in slightly different positions, we should see two images). Some believed that the two optic nerves physically joined making a united image in the mind. Helmholtz, on the other hand, argued that the nerves were indeed separate, yet an unconscious blending made one point from two. A similar situation presented itself in the study of sound, where a well trained ear, with proper use of attention, could pick out the elements that had blended into a tone. He developed this perspective partly from his training under Müller who had emphasised the necessary separation between the sensory and attentional processes, and partly from the confidence he enjoyed as an amateur musician. Musicians had long been trained in the art of picking out sounds that non-trained listeners could not detect. Helmholtz's friends, to take an anecdotal example, were amazed at his admirable observational gifts. They claimed that he could even pick out melodies and chords amidst the splashing and noise of the fountain at Sanssouci, sounds that they could not hear even after he pointed them out.[47] But there were still sounds that Helmholtz needed help observing.

If tuning forks were the first precision simple-tone *generators*, resonators became the first precision simple-tone *detectors*. These spherical glass or brass globes, tuned to respond to specific frequencies, were held to the ear, thus allowing an observer to detect simple tones from complex tones in the surroundings. They were a mechanical means for uncovering the underlying basic sensations that had been obscured by mental processes. According to Helmholtz, when the skilful use of attention failed to uncover the partials, the resonators could materially help the ear make this separation.[48] The observer directs his attention by using these material aids. Once having learned what to listen for, the observer can do away with external support (Fig. 2.3).[49]

For Helmholtz the resonators offered clear, indisputable proof of the existence of simple tones. In *Tonempfindungen* he argued that these partial tones, predicted by theory and perceived by the ear, objectively existed external to the ear and that they were not merely a "*mathematische fiction*."[50] He had already accomplished this with his experiments on the piano and his use of tuned membranes to detect partials and combination tones. In the winter of 1857 he introduced the resonators at a public lecture in Bonn, "the native town of Beethoven, the mightiest among the heroes of harmony."[51] The resonators were glass retorts or receptacles with two openings, one received the sound from the surroundings, and the other a glass tube that was inserted into the ear. In effect these receivers performed the same analytic task that

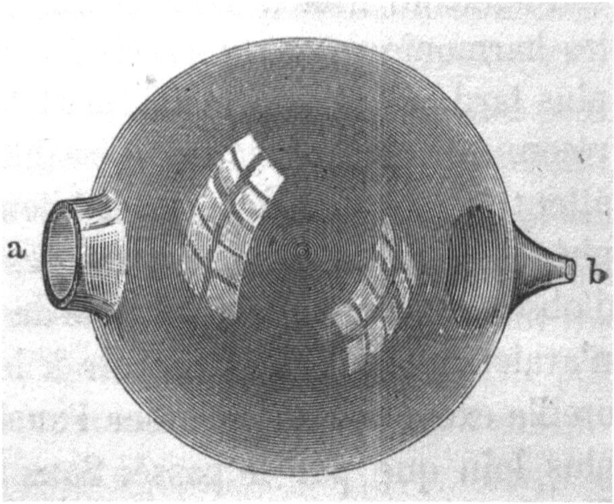

Fig. 2.3 Spherical resonators. CR 54
Source: Helmholtz et al. (1868, p. 59)

the piano had performed in the vowel experiments, except they only responded to one simple tone.

According to Helmholtz, the piano strings, tuned membranes, rods of Corti and resonators all worked on the principle of sympathetic vibration. In his lecture he reminded the audience that they had observed sympathetic phenomena in stringed instruments. After the damper is lifted from the string of a pianoforte, for example, an exciting tone causes the string to vibrate. The tone continues even after the exciting tone stops.[52] Piano strings could vibrate in many modes making them difficult for experiments intended to detect one simple tone, and membranes were found not to be sensitive for fainter simple tones. On the other hand, a globe of air could be set into its natural vibration mode through sympathy with much more precision and strength, and the ear, connected to this apparatus, could hear the proper tone with much greater intensity.[53] It was therefore much easier for someone to determine if a simple tone existed in the mass of tones making up a complex tone.

In the eighteenth and early nineteenth centuries researchers had performed several studies of aerial resonating cavities. These studies were mostly intended for refining the resonating aspects of musical instruments, understanding how they produced a specific pitch, and for developing the laws that govern organ pipes and other aerial resonating tubes. Helmholtz, however, reconceptualised resonators as *tools* for *selecting* specific tones from a complex tone. This was a dramatic reinterpretation of resonators from tone *producers* to tone *detectors*. Ohm's theory and the physiological perspective provided the framework to reinterpret the use of resonators. In a vivid illustration of this re-conceptualisation, Helmholtz attached a membrane to the open end of a bottle that responded to a specific frequency. When that frequency was present, a pith ball jiggled upon the vibrating membrane. This simple device served as a model for the glass resonators with two openings where

Fig. 2.4 1881 Portrait of Hermann von Helmholtz by Ludwig Knauss
Source: Pietsch (1901)

"the observer's tympanic membrane" replaced the artificial membrane.[54] The process by which Helmholtz invented the resonator was most likely not as neat as described in the above example from his Bonn lecture, but his description highlighted the physiological context of his reinterpretation of the resonators as analytic tools. He had provided a way to demonstrate and detect the simple tones of Ohm's theory. In *Tonempfindungen* he emphasized that the simple partial tones (*einfachen Partialtönen*) contained in the compound musical sound produced objective effects independent of the ear.[55] In the same way that the prism came to define Newton's optics, resonators became the emblem of the analytic conception of musical sound. In the 1881 portrait of Helmholtz painted by Ludwig Knauss, a spherical resonator rests prominently on the table next to a tuning fork and prism (Fig. 2.4).

Synthesising Vowels Sounds

Vowel sounds served as one of the more challenging illustrations of timbre. If *A* and *U* were sung at the same pitch, they sounded different in quality, but the same in pitch. This effect had traditionally been ascribed to a different shape of waveforms. Helmholtz, however, demonstrated that timbre could be reduced to a distinct number of simple elements at a certain intensity. As mentioned above, it was the piano, an

Fig. 2.5 One of eight electromagnetic resonators of the sound synthesiser. CR 56
Source: Helmholtz et al. (1868, p. 154)

instrument that he knew well, which served as a model for his emerging conception of the inner ear as a Fourier analyser. In a letter to Donders in November 1857, he described singing a note into the undampened strings of a piano (*wenn man in das Clavier hineinsinge)* and observed the different strings that responded to particular harmonics of the vowel sound.[56] He thus roughly analyzed the components of his voice.

In 1858 Helmholtz devised a way to test his analytic conception through synthesis, the production of sounds of different qualities by combining different simple tones. He went to the instrument maker Friedrich Fessel of Cologne with the design for a vowel synthesiser. By April 1858 he wrote to Emil du Bois-Reymond that thanks to financial help from the King of Bavaria he was able to build "an apparatus with electromagnetically driven tuning forks, reinforced by resonators, which could produce combined sounds that mimicked timbre [*Klangfarbe*]"(Fig. 2.5).[57]

The sound synthesiser was a clear illustration of Helmholtz's theory of timbre. It consisted of eight tuning forks that corresponded to B (B2) "in the deepest octave of a bass voice" and its upper partials as far as b2 (B5) "the highest octave of a soprano" comprising the notes B (B2), b (B3), f1 (F4), b1 (B4), d2 (D5), f2 (F5), a2 (A5), and b2 (B5).[58] Each fork was framed by a horseshoe electromagnet and connected in series to an interrupter tuned to 120 Hz (the frequency of B) oscillations per second. The interrupter kept all the forks vibrating at their natural frequencies. Helmholtz reinforced the tuning forks with tuned cylindrical resonator tubes made of pasteboard. The resonators could slide toward or away from the tuning fork to adjust the intensity of that tone. The mouth of each tube had a moveable cover attached by thread to a piano key. When the circuit operated, there was a slight hum to the electrical forks, but as soon as the cover was lifted from the resonator, the tone was generated powerfully and clearly.

By combining various partials, Helmholtz claimed to reproduce the basic vowel sounds. He adjusted the intensity by moving the fork away or toward the resonator.

He discovered that the imitated vowels resembled those of the singing voice more than the spoken voice. For example, *O* comprised primary note B (B2) and its powerful octave, b (B3). *E* was especially characterized by the third note f1 (F4), with a moderately sounded second note b (B3) and two very weak higher notes. (At the time he wrote his 1859 article, Helmholtz had not completed his studies for all the vowels because he did not yet have high enough forks. These were added by the time he published his book in 1863.)[59] Beyond the initial experiments with the synthesiser, he was able to confirm his results by detecting the partial tones with his spherical, glass resonators. He concluded, in line with his developing theory of timbre that the distinctive quality of vowels depended on a certain number of partials each at a specific intensity. The same harmonics could be present, but each one could display different intensities, making the overall sound distinctive.

A Comprehensive Theory of Harmony and Music

Tonempfindungen tied together all of the above findings into a comprehensive analytic theory of sound. Harmonics or upper partials, which had been observed by musicians for centuries, became the "elements of sound" with a strict mathematical, physical and physiological definition. Most importantly, as we saw earlier, Helmholtz created instruments that reflected his analytic thinking. *Tonempfindungen* was an introduction to resonators (for precision detection and analysis of simple tones), tuning forks with cylindrical resonators (for production of precision simple tones), the sound synthesiser (for producing complex vowel sounds from simple tones), and the Lissajous vibration microscope (for analysing the elements of vibrating bodies).

The second part of *Tonempfindungen* applied these findings to the structure of music as a whole. The interaction of upper partials, combination tones and beats explained, for example, the difference between flutes and violins. Minor chords, Helmholtz conjectured, obtained their distinctive character from slightly inharmonious but weak combination tones. He described the differences between various scales. The dissonances of the equal-tempered scale, according to Helmholtz, derived from "bad combination tones." He also invented what he called the justly intoned harmonium, a special reed instrument, for experimenting with the scale of just intonation and pure intervals. In his hands this instrument was an elaborate version of the polyphonic siren, designed for investigating the relations and effects of all the major scales at once.[60] It served as a powerful contrast to his piano: "when I go from my justly intoned harmonium to a grand pianoforte, every note of the latter sounds false and disturbing."[61]

Helmholtz argued in the third section of his book, for example, that musical preferences were ultimately determined by cultural taste. One culture, for instance, may not tolerate certain dissonances that another culture would view favourably. This was a radical position that recognised dissonance to be a matter of degree and not of kind, and it may have foreshadowed the freedom with which later composers made music, released from traditional notions of harmony.[62] With respect to this claim, he maintained that his contribution had been confined to establishing the "elements" and basic principles of musical sounds. Various cultures and traditions ultimately decided how these laws could be applied, he wrote, "but just as people with differently directed tastes can erect extremely different kinds of buildings with the same stones, so also the history of music shows us that the same properties of the human ear could serve as the foundation of very different musical systems."[63]

In conclusion, during the mid nineteenth century a mechanical and analytic conception of sound emerged from a grand synthesis of physics, physiology and psychology of sound. Through the work of Helmholtz there was a convergence of culture, science and instruments and that contributed to the "reform of acoustics." Helmholtz applied Fourier's mathematical theorem to sound, refined the laws of resonance, and clarified the physical and mathematical nature of combination tones. In physiology, he studied the workings of the inner ear and linked these to findings on the physics of sound. He then added a psychological dimension to explain perception. These conceptions all came together in a theory of harmony and music that continues to influence acoustical practice and theory. In the process of these studies, Helmholtz created and utilized several instruments – the double siren, tuning-fork synthesiser, spherical resonators and tuning-fork tonometer – for demonstrations and experiments.

In the next series of chapters we see Koenig's complex reaction to Helmholtz's acoustics. During the 1860s he enthusiastically transformed these ideas and instruments into a successful commercial line of acoustical apparatus. During the 1870s however, he started to question Helmholtz's basic findings, which led to completely new pathways for acoustics.

Notes

1. Translation from Cahan (1995, p. 46) and Helmholtz (1865b).
2. Ibid.
3. Quoted from Ash (1995, p. 21).
4. Jackson (2006).
5. Koenigsberger (1902, vol. I, pp. 22–24). "Colorirten" comes from the Italian word, *coloratura*, referring to ornamental flourishes in vocal music.
6. Ibid. Translation from, idem., 1965, pp. 13–14.
7. Hiebert and Hiebert (1994).
8. Jackson (2006, Chapter 7). In the nineteenth century cities throughout Europe had different standards of pitch. Also see, Ellis (1968).
9. Jackson (2006, p. 3) and Lenoir (1997).
10. Turner (1971).
11. Brock (1997) and Rocke (2001).
12. For an account of how this applied in optics, see Kremer (1993).
13. Macdonald (2003, p. 190).
14. Holmes and Olesko (1995) and Brenni (2004).
15. Arlene Tuchman in Cahan (1993, pp. 17–49) and den Tonkelaar (1996).
16. Lenoir (1994, p. 199).
17. Brenni (2004).

18. Cahan (1989).
19. Turner (1973).
20. See Kremer (1993) for a comparison of his work in optics.
21. Koenigsberger (1902, vol. I, 267–268).
22. For a review of these debates, see Turner (1977, pp. 1–11) and Vogel (1993, pp. 261–266).
23. In 1822 Baron Jean Baptiste Joseph Fourier (1768–1830) published the mathematical treatise, *Théorie Analytique de la Chaleur (Analytic Theory of Heat)* where he demonstrated that any finite and continuous periodic motion can be decomposed into a series of simple, pure sinusoidal motions.
24. Turner (1977).
25. In 1748 and 1754 respectively, the organist, Georg Sorge (1703–1778), and the violinist, Giuseppe Tartini (1692–1770) both observed that when two tones were played, a third tone resulted. For a history of combination tones up to Helmholtz, see Maley Jr. 1990.
26. Ibid., p. 121.
27. Chladni (1802, pp. 111–114). For more on the context of Chladni's work, see Jackson (2006).
28. Helmholtz (1863, p. 243).
29. Ibid., p. 241.
30. The introduction of the siren was part of the inspiration for Ohm's definition of tone. As Stephen Vogel has observed, "The essential feature of this new definition was the reduction of tone to mere periodicity and the elimination of the former assumptions about the form of the vibration." Vogel (1993, p. 263).
31. Gooday (2004).
32. Helmholtz (1863, pp. 227–236). Idem., 1856.
33. Helmholltz (1863, pp. 249–250).
34. Ibid., pp. 227–236, especially p. 234.
35. Ibid., p. 250. Idem., 1856, pp. 531–535.
36. Dostrovsky et al. (1970, pp. 666–669).
37. Koenigsberger (1902, vol. I, p. 283).
38. Helmholtz (1857). See also Helmholtz (1882, vol. 1, pp. 395–396).
39. Kursell (2006).
40. Helmholtz (1863, p. 218).
41. Ibid., p. 198.
42. Ibid., pp. 220–221.
43. With the findings of Victor Hensen (1835–1924) in the Helmholtz singled out the basilar membrane as the main substrate of sympathetic resonance. Hensen, like other physiologists of the time, had immediately set out to test the new ideas of Helmholtz upon reading Helmholtz's work. In his paper of late 1863, he singled out several potential candidates for our internal resonating systems and drew detailed diagrams of their structure, especially the basilar membrane. He performed ingenious studies observing the responses of the organ of Corti to a bugle, see Hensen (1863a,b).
44. Beyer (1998, pp. 264–267).
45. Helmholtz (1863, p. 100).
46. Hatfield (1993).
47. Koenigsberger (1902, vol. I, p. 56).
48. Helmholtz (1863, p. 14).
49. Ibid., pp. 84–85.
50. Ibid., pp. 58.
51. Translation from Cahan (1995, p. 46) and Helmholtz (1865b, p. 57).
52. Helmholtz (1865b, p. 72).
53. Ibid., p. 84.
54. Helmholtz (1863, p. 73).
55. Ibid., p. 60.
56. Koenigsberger (1902, vol. 1, pp. 282–283).

57. Ibid. vol. 1, p. 298. "Auf Kosten des Königs von Bayern hab ich mir jetzt einen complicirten Apparat zusammengebaut, um Stimmgabelschwingungen durch Elektromagnetismus nach Willkür zu dirigiren, Intensität und Phasenunterschiede vollständig zu beherrschen, und will damit Klangfarben zusammensetzen."
58. Helmholtz (1859, p. 284). Translation from Idem., 1860b, p. 84.
59. Helmholtz (1863, pp. 184–185).
60. Helmholtz had commissioned Messrs. J. & P. Schiedmayer of Stuttgart to make this instrument. Helmholtz and Ellis (1954, p. 316). The Museo di Fisica at the University of Rome has a Harmonium built by Anton Appunn of Hanau.
61. Ibid., p. 323.
62. Hiebert and Hiebert (1994, p. 303).
63. Helmholtz (1954, p. 366).

Chapter 3
Transformations in the Workshop

The Koenig sound analyser (CR no. 242) (Fig. 3.1), first conceived in the early 1860s, represented a convergence of contrasting Parisian and German traditions in acoustics. In the German context, Helmholtz and his followers *listened* attentively to resonators with their ears, while in Paris observers *watched* them resonate with

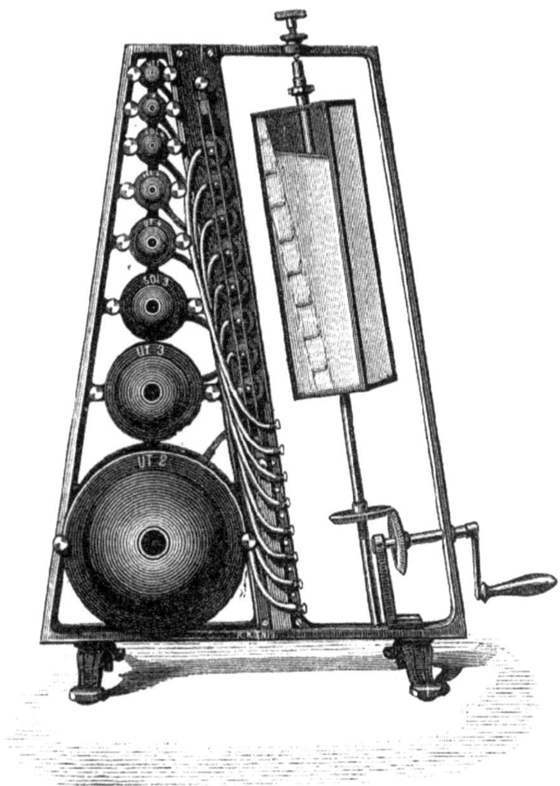

Fig. 3.1 Koenig sound analyser. CR 242a
Source: Koeing (1889, p. 87)

D. Pantalony, *Altered Sensations*, Archimedes 24, DOI 10.1007/978-90-481-2816-7_3,

manometric flames. There were also artisanal influences at work. Helmholtz's first glass resonators were conceived as products of mathematics, while from Koenig's workshop the subsequent spun-brass receptacles were more the products of local tuning solutions. The actual analyser also carries evidence of the manufacturing context in Paris. The cast iron frame on the model at Dartmouth College (purchased c. 1870) has manufacturing marks on the feet revealing construction techniques employed to produce large numbers of these instruments for a growing market. The turned wooden handles found on most surviving models, also found on hundreds of Parisian instruments at the time, represent a vast manufacturing context amidst the growing popularization of acoustics and the significant influence of the booming Parisian precision trade of the 1860s.

Through his combination of mathematics, physics, physiology, psychology and novel experiments, Hermann von Helmholtz provided a comprehensive framework for studying sound. One of the goals of this book, however, is to show that the act of manufacturing and selling instruments contributed equally important features to the scope, methods and content of acoustics. Paris served as a fertile environment for these changes – there was a dense community of instrument makers (musical and scientific), emphasis on making instruments for teaching, strong commercial forces, and traditions in visual culture. Koenig's atelier was a creative engine for the production of instruments; It was also a mediating space between national, intellectual, social and material realms. The practice of acoustics that emerged, and is still with us today in many forms, was as much a product of this Parisian context as much as it was of Helmholtz's studies.

Inside Parisian Workshops

The turned wooden collar found on many of Koenig's tuning forks is a recognizable form from French and English instruments of the late eighteenth century through to the late nineteenth century. One of Koenig's workers or contractors probably made hundreds of tuning fork collars similar to the one pictured in Fig. 3.2. Stylistically, they echoed neo-classical gestures found in buildings, furniture and other instruments. They were the product of a fairly stable and conservative culture of the scientific artisans. Before I explore the question of how Koenig's atelier came to influence acoustics in a larger sense, I shall look closely at the inner workings and context of the more common aspects of the Paris workshops.

It is astonishing to realize that the famous workshops in Paris were so basic in methods and conservative in style. The hand file, a simple tool carrying with it hundreds of years of artisanal tradition, reigned in Parisian workshop shops. Surprisingly, many scientific instruments, even delicate and complex ones, were made with skillful manipulation and simple tools. Parisian apprentices, for example, were given countless exercises using only a file. The simplest exercise was the making of a brass cube. As they advanced they made more complex items such as a telegraph.[1] In fact, when we see the sophisticated nature of many nineteenth-century instruments, it is surprising to learn of the simple construction methods. Horace

Fig. 3.2 Turned wood collar. Photo by author, 2005. Museu de Física, University of Coimbra, Portugal

Darwin, the founder of the Cambridge Scientific Instrument Company, replied to a workshop applicant in 1881 that he did not need specialists for certain instruments, but "what we do want are really first class instrument makers used to lathe, vice, finishing and lacquering."[2]

The lathe was the most common machine found in Parisian workshops and came to represent the changing relation between skills and machines. They were usually hand or foot driven, even late in the nineteenth century. A handful of the larger shops used steam. When Charles Young went to Paris in 1853 he made particular note that some of the more important makers still used hand or foot lathes. He knew of more advanced steam-driven machines in Vermont and New Hampshire and was surprised at the conservative nature of his continental hosts.[3] This simplicity even surprised a hands-on scientist such as the American Henry Rowland. On his first visit to Europe in 1875 he described the workshops as "museums of antiquity." But, he added, "the workmen are so much better than ours that the work turned out is not inferior to ours, and in small instruments or when there are many pieces to be made by hand, they are superior."[4] Standard workshops carried screwdrivers, vices, hammers, abrasive powders, chisels, screw making tools, ovens for melting, and a wood-working bench. Some shops had drilling machines; others may have had a machine for making brass tubes. These tools were managed by an exceptionally skilled and trained workforce which was one of the sources of innovation in nineteenth-century science. Americans did not yet have the density of infrastructure and training to mimic this culture.

Koenig represented the last generation of *artiste-constructeurs*, the kind of master artisan who did not easily relinquish control to an assembly line or outside workers. A more definitive factory style of manufacturing, however, began to creep into instrument making in the last quarter of the nineteenth century. In contrast to a moderate enterprise such as Koenig's, which by the 1870s employed at most 15 workers, the progressive and successful electrical maker Jules Carpentier by the 1880s had built a modern manufacturing facility driven by increasing demand for precision electrical instruments. He divided labour and assembly, expanded into several buildings, invested in machine tools (planning and milling), and adopted the making of standardized, interchangeable parts.[5] At times he had as many as 50 workers. Others had only a few helpers. Most operations, large and small, contracted out jobs to specialists and unskilled laborers. Paris had a large workforce of what were called *travailleurs domiciles* "home workers."[6] These trends were echoed in the making of musical instruments such as pianos, as well.[7]

Instrument making in Paris during the second half of the nineteenth century was at a crossroads between the older master-artisan traditions and mass production. Modern manufacturing techniques, which were being used in making some of Koenig's instruments (CR no. 242a),[8] testify to changes that were already creeping into workshops. Throughout the nineteenth century, however, many of the makers tried to remain firmly and proudly in the artisan tradition. An English visitor to the 1867 Paris Fair took note of growing insecurity in the artisan class. "It is curious," he wrote,

> that most of the French workmen with whom I have spoken are of the opinion that art and handicraft are declining among them. They say that the excessive division of labour has had a tendency to make men more like machines: and the constant breaking up of small workshops has had the effect of disheartening men from attentive study.[9]

All of these changes aside, the second half of the nineteenth century was a relatively prosperous period for artisans in Paris. Skilled workers had access to both training and entertainment. There were rigorous and structured apprenticeships, night schools for drawing and mathematics, public access to mechanical libraries, public courses at the universities and access to the Musée des arts et métiers where they could examine the finest examples of instruments and technology. They were encouraged to travel, learn their trade on the road and diversify their skills. They enjoyed the rich public life in Paris, with its music cafés, gardens and large eating halls. Popular artisan cafés, some of the "grandest working-class coffee houses in the world" were "noble buildings on the outside," with interiors "glittering with gold and decorations."[10]

Paris was a city of artists and artisans with a very public appreciation of things that were skillfully and artistically crafted. There were, of course, the art salons which in the 1860s attracted as many as a million visitors in a six-week run. There were massive international fairs, one in 1855 and another in 1867. Another English observer at the 1867 fair concluded that the superior French displays could be traced to the stimulating surroundings of the Parisian cityscape: "The Parisian cannot walk through the streets without seeing objects of art-workmanship of no common order of merit: the rich carvings on the fronts of the houses, the many statues, columns, fountains, arches, nearly all designed with the purest taste, tend to familiarize him from infancy with the highest standards in art-workmanship, and to make it proportionably difficult to impose on his judgment with inferior productions."[11]

It was within this context that Koenig embarked on the path of a precision instrument maker. His first few years were enormously productive and inventive. He quickly threw himself into making instruments related to two of the most significant developments in acoustics – graphical instruments and the work of Helmholtz.

The Phonautograph and the Origins of Graphical Acoustics

Just as Koenig began business in 1858, he became involved in the construction of the phonautograph, an instrument that recorded sounds directly from the air. It was a precursor to the Edison phonograph (in conception and appearance) and it marked a lasting shift in acoustics from reliance on the ear to the eye. Koenig's involvement in this development was not trivial – both in the workshop and as a commercial agent. The phonautograph emerged from his workshop as a substantially different instrument from its original form; it also found its way into the laboratories of several researchers.

Self-recording instruments were part of a wider trend in mid nineteenth-century instrumentation, with an emphasis on automation (replacing human skills with machines), objectivity in instrumentation (making it possible for data to be viewed and shared by several witnesses at once) and the investigation of previously unobservable patterns and effects (extension of the senses). This trend became part of the fabric of science itself in the twentieth century, where inscriptions came to be viewed as a snapshot of the original material or phenomena under scrutiny. In *Laboratory Life*, Bruno Latour and Steve Wolgar, have argued that when observed from an anthropological perspective, modern laboratory culture resembles "a system of literary inscriptions" whereby the essence of a particular field (material, theoretical and social aspects) become dependent on its visual forms of communication.[12]

When inscription techniques and instruments first emerged in the 1840s, Helmholtz, Emil DuBois Reymond and other German physiologists used them to display previously hidden patterns related to physiological functions.[13] Recording instruments appeared in meteorology as well. At the 1851 London Exhibition, the instrument maker George Dollond displayed a self-registering meteorological instrument.[14] In Paris, Jean-Marie Duhamel invented a method for recording the vibrations of a tuning fork. Later in the 1850s, the young Parisian medical researcher, Etienne-Jules Marey (1830–1904), a friend and contemporary of Koenig, made his reputation by the invention of recording instruments for measuring pulsations of the heart and other physiological phenomena.[15] His instruments, argues Robert Brain, defined practices and conceptions of the French school of linguistics at the turn of the century.[16]

The idea of the phonautograph came from a local inventor, Édouard-Léon Scott de Martinville. He conceived of this apparatus as early as 1853 as a device that could inscribe nature's own language of sound, a universal language unencumbered

Fig. 3.3 The 1857 phonautograph by Scott. As the suspended weight (*left*) lowers, the inscription plate is pulled away from the stylus and collecting drum. Sound waves are recorded on the moving plate. CR 213. Drawing of instrument from Patent
Source: Scott de Martinville (1857)

by conventions, and connected directly to the production of speech.[17] He used the phrase, "speech that writes itself" (*la parole s'écrivant elle-même*) to emphasize the automatic nature of the device. Scott, who had been a typesetter, wanted to reform stenography so as to record thoughts more efficiently in a natural, permanent form. He had been influenced by the idea of photography and wanted to create a similar record for sound. "Sound, just like light," he wrote, "can provide a lasting image at a distance."[18] He was also interested in the physics of sound, had studied with Regnault at the Collège de France, and knew other local scientists who did research in acoustics, such as Duhamel, who had produced a simple apparatus for recording the vibrations of a tuning fork. Yet the key inspiration for Scott's idea came from a description of the mechanism of the ear in a physics textbook he had been editing. He immediately saw that if one wanted to create an "image" of sound, one should build a replica of the inner ear and connect this contraption to some form of writing device. The ear, therefore, became a natural model for transforming sound into a visible signal. God, an inventor and "sublime artist" for whom "nothing is impossible," led Scott to his goal by revealing "the marvel of all marvels, the human ear." (Fig. 3.3).[19]

Scott focused first on making the phonautograph look and operate like the inside of an ear. In his application for a patent submitted in 1857 he described a bowl-shaped sound receiver and a tube with a thin membrane at the end. A writing stylus connected to the membrane rested on blackened recording paper secured on a plate. A weight mechanism pulled the plate at a uniform speed as the stylus recorded the vibrations.[20] Initially he had worked with the instrument maker Gustave Froment. In February 1859 Koenig approached him with ideas to improve the invention. Scott recognized the potential usefulness of the constructor's knowledge of acoustics and craftsmanship (*les connaissances en acoustique et en facture*) as well as his "indispensable experimental skill" (*l'adresse expérimentale indispensable*) that would

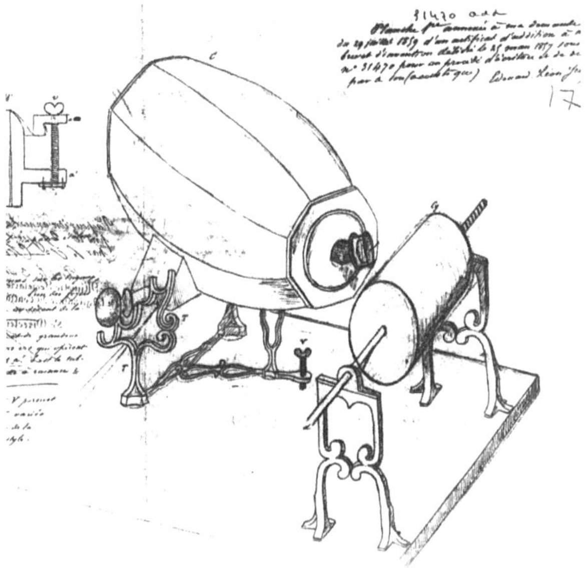

Fig. 3.4 Revised patent for the phonautograph, 1859. CR 213
Source: Scott de Martinville (1859b)

enable the proper scientific and industrial application of the invention.[21] The maker, he wrote, could also help make an instrument that would be more effective for public demonstrations. They signed a contract on 30 April 1859 which gave Koenig "exclusive construction rights" (*droit exclusive de construction*) to the finished product (Fig. 3.4).[22]

In July of that year Scott submitted a significantly revised patent.[23] Koenig had built a rotating cylinder that could record the vibrations in a smooth and more uniform motion and designed an ellipsoid drum to receive the sound more efficiently.[24] He also added a graphic chronometer (tuning fork with stylus) to measure time (see below).[25] Scott, still concerned with modeling the anatomy of the ear, changed the membrane structure at the end of the drum. The membrane, the heart of the apparatus, was a constant source of attention, and they tried different materials – goldbeaters skin, bladder, animal vellum, cellulose, and caoutchouc.[26] Scott provided a preliminary description in *Cosmos* in 1859,[27] which was quickly followed by an additional note from the editor, Abbé François Moigno:

> We are happy to be able to announce that at this moment, M. Léon Scott, aided by the theoretical and practical artfulness [*l'habileté théorique et praticque*] of M. Rudolphe Koenig, has just constructed a new apparatus that registers with the utmost clarity [*la plus grande netteté*] the vibrations of a tuning fork, up to a thousand of them a second.[28]

The reception of the phonautograph owed much to Abbé Moigno, in particular his eager promotion on behalf of Scott and Koenig. Moigno, a colourful and influential figure in Parisian scientific circles, who had played an important role in promoting the career of Koenig's predecessor, Albert Marloye (see Chapter 1), was an ideal champion for the co-inventors. In his early career he trained as a Jesuit, taught mathematics, and gained a reputation as a skilled writer and public speaker. He did original investigations in mathematics, was a pioneer in promoting the magic lantern for education,[29] translated scientific works (from English and Italian), and published popular and scholarly books. But his largest influence came from his publishing activities. In 1852 he founded and became editor of the science periodical *Cosmos*, which renamed became *Les Mondes* in 1862. These bi-weeklies were a mixture of recent findings, translations, scientific reports, news and local scientific gossip. Through his editorship, he actively and enthusiastically promoted the work of scientists and instrument makers.[30]

Moigno, therefore, was one of the first to spread news about the phonautograph to locals and foreigners. In 1859 he described the improved instrument at the Aberdeen meeting of the British Association for the Advancement of Science.[31] Prince Albert was president at the BA meeting that year and delivered the inaugural address. He praised the efforts of scientists to gather facts and build objective knowledge about the world.[32] Baconian style sciences based on collecting data and specimens in the field were predominant at these sessions (in contrast to the presentation of laboratory findings), with many talks on geological, astronomical, and biological activities. Moigno gave talks on his own work and on behalf of several French scientists and makers who either could not attend or could not communicate in English. In the session devoted to instruments he described a photometer to measure the intensity of stars, an electro-medical apparatus by Ruhmkorff, and the phonautograph.[33] He did not have the phonautograph with him, but showed samples of the graphical plates. Giving an account of the Aberdeen meeting in *Cosmos* he wrote: "These tuning forks, pipes, human voices, alone or together, that automatically write hundreds and thousands of vibrations executed per second created a true sensation [*un véritable enthousiasme*]."[34] Nothing like it had ever been seen before, he wrote, and the graphical plates "were judged worthy to be presented to the Queen at Balmoral by her royal consort [Prince Albert]."[35] In the next issue he published his Aberdeen presentation on the phonautograph, reminding readers that they could see it in action at Koenig's shop. The presentation ended optimistically with the declaration that such an instrument could "lift the corner of the veil [*le coin de voile*] that covered the mysteries of the mechanisms of the human voice."[36]

Supported by Moigno's publicity, Scott and Koenig envisioned themselves as modernizers of acoustics. In contrast to Helmholtz, who saw himself as reforming acoustics through mathematics and physiology (Fourier), Scott and Koenig believed they were transforming acoustics through the introduction of a novel instrument. Both sides reflected their own context – Helmholtz was immersed in German academic natural philosophy, while Scott and Koenig worked in an environment of invention, entrepreneurship and workshops. They marketed their instrument, therefore, as *the* instrument that would change a stagnant field. In an 1859 pamphlet entitled "The phonautograph, apparatus for the graphic recording of noises, sounds and the voice, invented by Édouard-Léon Scott and constructed by Rudolph Koenig." (most likely written by Scott, Koenig or Moigno), the author claims that the phonautograph extended the senses and displayed, in a visible permanent record, previously undetectable complexities and patterns of sound:

> Most of the sciences founded on observation and experiment already have in possession a set range of special instruments, suitable to provide a precise and thorough knowledge of certain phenomena, because our senses, it is known, are capable of providing to us only sensations [*car nos sens, on le sait, ne sont aptes à nous fournir que des sensations*], most often defective, irreducible and variable from one individual to another. Astronomy and optics have instruments of great variety that provide a vast extension or an extraordinary subtlety [*subtilité inouïe*] to sight. The natural sciences have their means of observation in chemical analysis and in the microscope that reveals a world that seems intended to elude us through its smallness. These instruments, genuine tools of scientific work, have opened a path of inexhaustible richness to experimentation, and have made progress of unexpected reach in the sciences and the arts.[37]

Acoustics, its author argued, had not followed the lead of the other sciences. The study of sound had been until recently like "astronomy before the invention of the telescope; it languished in waiting for its instruments of observation, measurement and analysis."[38] Acousticians desperately needed a "microscope to see sound, and more, to save the imprint."[39] "The phonautograph fills this gap."[40] The author then described how the inventor laboured for over 6 years developing his device and, at times, even presented some of his experiments on noises, the voice, song, and musical instruments to other scientists, but was not fully satisfied with the results.[41]

> Happily, another person came to him. M. Rudolph Koenig put himself to the task for the complete implementation [*mise en oeuvre*] of the phonautograph. M. Scott owes much to this skilful instrument maker for the proper use of the instrument, the arrangement of the diverse parts with good acoustical conditions, and the ingenious construction [*l'ingénieux agencement*] that allows the apparatus to figure prominently in a physical cabinet.[42]

Unfortunately, praise for the instrument created tension between the co-inventors. Scott was enormously protective of his invention and increasingly became paranoid about anyone who became involved with it; Koenig brought to the partnership a youthful, independent spirit and entrepreneurial drive. He was part of a confident artisanal community, and increasingly adopted the identity of guardian of true scientific pursuits. When Scott suggested making further changes to the membrane, Koenig refused. Scott had been trying to improve the kinds and arrangements of membranes in order to imitate the anatomy of the ear even more faithfully. As well, Scott seemed more interested in using the instrument to document performances in song and theatre.[43] The instrument maker, however, wanted to maintain a simple design that in Scott's words was "destined only for physical cabinets."[44] The two men thus underwent a "complete separation" due to their disagreement over the future of the instrument.[45] Scott's bitterness was still evident in his presentations in the early 1860s, and in his own history of the phonautograph, written to defend priority for his ideas shortly after the invention of Edison's phonograph. He even went so far as to correct the misconception that Koenig had been the maker who helped him with his first invention in 1857: "It was not M. Rudolph Koenig," he wrote, "... but a constructor of the highest order who lived at rue Notre-Dame-des-Champs [Froment]."[46]

Fig. 3.5 Engraving of Koenig's commercial phonautograph. CR 213
Source: Koenig (1889, p. 77)

During the following years, they both made changes to the instrument; Scott continued to focus on the frame and the membrane arrangement, but it was Koenig's design that sold.[47] Koenig transformed the collecting drum from an ellipsoid to a parabolic shape and made the collecting drum of zinc (Scott had originally used a strong form of plaster of Paris). He also worked with various membrane materials, connected the frame directly to the focus of the parabola, and by 1865 had developed an electric tuning fork chronoscope for measuring time intervals on the phonautograph.[48] The form it took at that time remained unchanged throughout his career (Fig. 3.5).

There were a few noteworthy scientists who did research on the phonautograph in the early 1860s. The Viennese physiologist, Adam Politzer, performed pioneering studies with the phonautograph (in Koenig's studio) related to his work on the organs of hearing. (see below). In the Netherlands, Franciscus Donders did research on vowels and reaction-time with the phonautograph.[49] At MIT, Alexander Graham Bell used a Koenig phonautograph for his research on visible speech (Chapter 4). Koenig himself had used it for several experiments, including the detection of combination tones.[50] But as we will see below, he abandoned it for research by 1864 in favour of his manometric instruments.[51]

The phonautograph also entered the scientific imagination as a definitive, objective test of even the subtlest, controversial phenomena. In his attempt to show that science had caught up with even the most farfetched claims, William Crookes wrote: "The spiritualist tells of tapping sounds which are produced in different parts of the room when two or more persons sit quietly round a table. The scientific experimenter is entitled to ask that these taps shall be produced on the stretched membrane of his phonautograph."[52] Koenig never envisioned the phonautograph as a sound reproduction instrument, such as the Edison phonograph developed from the same principles 20 years later. He only saw it as a graphical method. When S.P. Thompson asked him if he or Scott had ever foreseen an instrument like Edison's while working on their invention, he replied: "No, the idea never occurred to either of us; we never thought of anything but recording."[53]

Precision and Graphical Acoustics

Hermann von Helmholtz's theories altered expectations for the precision and purity of tuning forks (Chapter 2). But workshop practices and visual instruments were also changing expectations. In the midst of his workshop activities with the phonautograph, Koenig started to explore sources of error in his tuning forks using graphical methods. In fact, much of his first experiments were directed at investigating the workings and sources of error in his instruments, especially his timing devices. He was not interested in general questions about the nature of sound; instead, his experiments were simply focused on bringing more experimental precision into a rapidly changing the field. This quest had significant implications for the practice of acoustics. Different forms of the refined graphical timing apparatus, for example, would later be used in experiments by Regnault in physics (Chapter 5) and Donders in psychology.[54] Michelson famously used Koenig tuning forks in his experiments to calculate the speed of light.[55] Koenig's narrowly focused struggles in the workshop and laboratory, therefore, led to unexpected changes in practice in other fields.

Aside from making and selling instruments in these first years of business, Koenig had also begun experimenting. His first experiments in 1858 were psychophysical. With a series of high-pitched tuning forks he mapped out the range of frequencies in which humans could distinguish musical intervals (such as the third, fourth or fifth) and discovered that even in the higher octaves of the piano, the best musicians failed to "judge exactly" the basic intervals.[56] But he devoted most of the first years of his business from 1858 to 1862 "especially to the perfection" of the graphical method.[57] By 1862 he had created an album of his key graphical experiments which he displayed at London along with his apparatus. The album consisted of 64 tracings, "les phonogrammes," on blackened paper. He displayed this "masterpiece of patience and skill" at the Exhibition winning the medal of distinction.[58] In fact, the album was so popular that he eventually sold reproductions for 400 fr.[59] Clearly, these attractive graphics created even more appetite for making sound visible. He even complained in 1882 that several people continually reproduced several of these tracings in "a large number of physical texts, most often without any indication of their origin."[60] The instruments, mostly consisting of sets of tuning forks with a special graphical stylus attached to the end of one prong, sold by the hundreds and spread throughout the scientific world (Fig. 3.6).

The striking series of illustrations in the black album included several sections that focused on precision methods: counting vibrations and timing; Lissajous-type combinations of two different "parallel" vibratory movements (e.g. two tuning forks

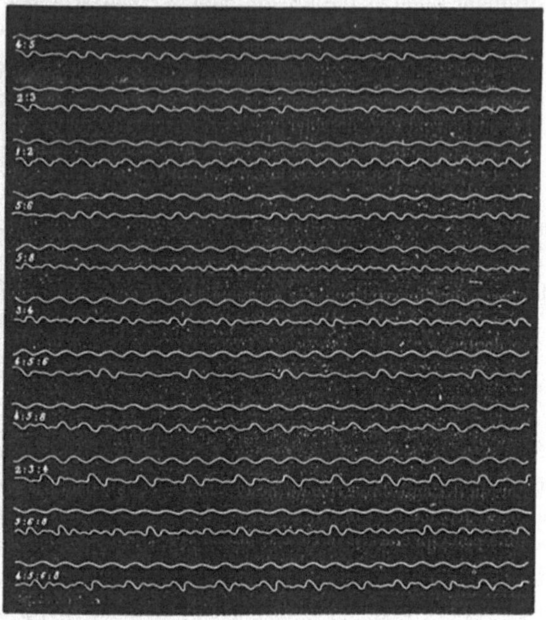

Vibrations de la membrane du phonautographe sous l'influence des sons d'une série de tuyaux d'orgue, inscrites à côté des vibrations d'un diapason chronographe de 512 v. s.

Fig. 3.6 Traces from the graphical album
Source: Koenig (1882c, p. 26)

at different pitches forming a musical interval); vibratory motions (harmonics) within one body (e.g. a stringed instrument, organ pipe or slender tuning fork); displays of rectangular patterns related to musical intervals (Lissajous combinations but with the forks placed perpendicular to each other); comparisons of tuning forks put into vibration by different means – the stroke of a violin bow or sympathetic vibration; recordings of phonautograph experiments; phonograms produced with Adam Politzer (see below) using real animal parts. The tracings served as an advertisement for a corresponding instrument for sale.

One series of studies led to improvements in the tuning-fork chronograph, one of the most significant instruments for precision timing in the nineteenth century. In his initial work for Scott, Koenig developed an apparatus that consisted of a smoothly rotating drum, a tuning fork and stylus for recording and counting vibrations. The chronograph already existed in various forms in physiology and astronomy,[61] but in Koenig's workshop the instrument itself became a subject of systematic scrutiny. He investigated different-sized inscribing forks, changing rates of drum rotation, friction of styluses, and modifications of vibration rates due to changes in current in electrically driven forks. When it came to counting vibrations, he discovered that his original method, using a small escapement chronometer that marked the roller every six seconds, retarded the rolling movement causing unwanted variation.

During his work on the phonautograph he developed a special tuning-fork chronoscope that displayed vibrations of a tuning fork of known frequency on a paper roller for comparison. He also developed a method whereby an electrical signal, marking the beginning or end of an event, recorded marks beside known vibrations in order to measure the exact timing of an event. This allowed him to see and compare, without the awkward disruptions of the chronometer, the number of vibrations in a given time interval.[62] He intended this timer to be used with the phonautograph, but it was also sold separately for precision-timing experiments.

A related series of experiments contributed to the development of another well known precision-timing apparatus, the Regnault chronograph. Through his graphical studies, Koenig became aware of tiny, cyclical perturbations in the frequency of the electrically driven fork. He tested different-sized forks (graphically) and noted their patterns of deviations. This problem created another source of possible error, namely, that one would have to operate a second chronometer to keep track of the irregularities of the chronograph. His discovery of these problems led to Regnault's invention which used a seconds-pendulum to calibrate directly the number of vibrations per second and thus reduce the error of counting successive seconds over a long period of time.[63]

Koenig turned his attention to almost every possible source of error. He performed graphical studies, for example, to show that tuning forks excited by violin bows (a common technique in the laboratory) revealed noticeable variations of amplitude from those excited by sympathy (using another vibrating fork to stimulate a fork). Such a demonstration showed the ability of the graphical method for seeing something that would have been imperceptible to the ear.[64] Membranes, the heart of the phonautograph, also became a subject of scrutiny. He studied the tracings of combined sounds from different sources (e.g. two organ pipes played together) and discovered that the membrane of the phonautograph did not represent equally the intensity of various simple tones that had been played with equal intensity.[65] This finding would lead to his main complaint with the phonautograph (in the context of Helmholtz's studies), namely that it failed to represent intensities of various simple tones.[66]

There was also a personal aspect to these meticulous studies. Koenig was determined to avoid any potential criticism of his instruments that would reflect on his ability as an instrument maker, experimenter and, ultimately, his business reputation. Many passages in his book read like a dialogue with potential critics.[67] If someone suggested that the method for counting vibrations was fundamentally flawed because of the weight and friction of the stylus, he responded: "But this is not serious, because nothing is easier than to determine with precision this little alteration that intervention of the stylus causes to the vibrations."[68] And yet he did take it seriously. As a remedy he compared the free fork with the writing fork using both beats and a Lissajous comparator in order to calculate the precise amount of deviation. Future experiments could simply factor in this variation. To take another example, if anyone was worried about the effect of the rolling drum on counting the vibrations, he conducted a series of graphical tests to show that changes in drum speed could be monitored and taken into account.[69] This defensiveness about the quality and reliability of his products surfaced notably in his disputes over combination tones (Chapter 7).

Some of these experiments were done in his studio with members of the local physics community. In the experiments on the phonautograph, Koenig teamed up with Jules Lissajous (the inventor of the famous optical tuning method), recorded the vibrations of a tuning fork placed in front of a phonautograph membrane, and compared these tracings to the tracings of an identical tuning fork inscribed directly (with stylus) on a rotating drum. They found that the tracings were in fact very similar.[70] In line with the studies mentioned above, the focus was primarily on honing instruments and methods, and not on studying the nature of sound.

Surviving examples of Koenig's graphical apparatus demonstrate his concern with controlling these variables. They consist of a large and very heavy cast-iron frame (1 m in length) with two adjustable steel mounts for tuning forks. One fork holds a blackened glass plate on its prong with a counter balance on the other prong; the other fork has a small writer on the end of the prong that moves slowly and smoothly backwards as it rests on the vibrating glass plate of the adjacent fork. The combined movements create distinctive graphical curves on the glass plate. For more elaborate geometric patterns, the writing fork is placed at different angles to the fork in holding the glass plate. The apparatus comes with two electrical mountings for maintaining the vibrations of the forks. Above all, the bulky, sturdy construction of this apparatus reflects Koenig's determination to cushion unwanted vibrations (CR no. 233).

The "Plaque tournante" at Rue Hautefeuille: Transforming Helmholtz's Acoustics

Koenig's atelier during the 1860s was like a "plaque tournante" (train roundhouse)[71] for ideas, instruments and people. We find him involved with education developing a number of attractive demonstrations such as wave machines.[72] He collaborated with physicians and invented two diagnostic instruments, the "dynascopic tuning fork" and a stethoscope with rubber tubes.[73] He worked with artists to illustrate his catalogue and help record his manometric experiments (see below). He was also around the corner from the medical faculty and Louis Auzoux's anatomical model shop and he sold Auzoux's models as early as 1859.[74] We find him also doing his own scientific research and experimenting with scientists. In 1862 a local physicist, Prof. Faye, collaborated with Koenig to create a instrument to measure the speed of sound. He performed original studies with Chladni plates and designed a special set of rectangular plates that produced theoretical figures predicted by a theory of Charles Wheatstone.[75] And, of course, he had connections with musical instrument making; in 1862, he invented a device that could measure the homogeneity of a violin string.[76]

It was "chez Koenig"[77] that developments were promoted. The Reis telephone, invented by Philipp Reis in 1863, and a source of later controversy related to Bell's patent fights, was one of the more high-profile instruments made at this time in

Koenig's workshop.[78] He was one of the first to produce it commercially, and probably one of the few people who could make it work. A surviving example at Harvard reveals a fairly complex assembly of elements: The transmitter consists of a mahogany box, mica membrane, brass horn, key and coils on the side. There is a delicate strip of coiled flat metal (probably platinum) under the mica diaphragm. The receiver contains an electrical constriction coil covered by a hinged pine resonator, which in turn rests on a pine support box, similar in quality to the resonating boxes made for tuning forks. Both the cover and the box have small resonating holes. The sides are mahogany, while the magnetic spool is boxwood and the bridges maple. Each part had its own story – the skills and knowledge surrounding mica sheets, for example, probably came from an electrical shop such as Ruhmkorff's or his successor Carpentier, who was known to have "his mica scrupulously clean and well selected."[79] Aside from these interesting elements coming together, the instrument is an informative document about the famous "make or break" technology before Bell (CR no. 166).

Koenig's instruments also represent his interactions with colleagues in the precision trade. His Lissajous comparators carried Huygenian eyepieces likely made by local microscope makers.[80] The same apparatus, as with many others, were driven by electrical coils made of finely insulated woven green thread (CR no. 234i).[81] Examination of insulated coils on electrical instruments from the same period in Paris, made by Breton Frères, Ruhmkorff, Deleuil, Loret, Gaiffe and Chardin reveal similar style, construction, colours and materials.[82] To take a more mundane example, several of Koenig's instruments, such as the analyser, have turned wooden handles. The identical handles appear on the electrical instruments made by contemporaries of Koenig.[83] Through the mediation of this busy meeting place/atelier, therefore, acoustics began to adopt a Parisian look and context.

Inside Koenig's atelier, therefore, we see the convergence of the instrument trade in Paris and the acoustical studies in the German territories. During the early 1860s Koenig began using his atelier to spread (and modify) the ideas and instruments of Helmholtz. In 1862, German émigré, Rudolphe Radau, writing in *Cosmos*, informed Parisians about Helmholtz's theories and instruments, "which," he added, "we have been able to admire at Koenig's place."[84] Following the publication of *Tonempfindungen* in 1863, Koenig's atelier gained even more prominence for the production and improvement of Helmholtz's instruments. Moigno commented in *Les Mondes* on the importance of Koenig's connections to his native land, Prussia. He had been "able to maintain continued relations with the scientists of that country, permitting him to gather and realize, in convenient form, research and demonstration instruments unknown in France before his time."[85]

In most cases he played the role of copying and spreading Helmholtz's devices, but even in these circumstances the construction process was not necessarily straightforward. The double siren, for example, was the first Helmholtz instrument that Koenig made and sold. In fact, it was the subject of their earliest known correspondence in 1859 in which he thanked Helmholtz for a "detailed description" (*ausführliche Beschreibung*) of the recent invention.[86] Helmholtz had commissioned the mechanic E. Sauerwald to make the first double siren in 1855–1856.[87] Sauerwald, who had made electrical instruments for Gustave Magnus, created sirens with elegant brass workmanship (such as a decorative counter face) and an extra rim of brass on the disks for better rotation (and possibly aesthetics).[88] Koenig's version is noticeably simpler in presentation, reflecting a changing market (students) and different volumes (sales throughout North America and Europe). Even though less refined in appearance (e.g. fewer decorative touches with a paler brass), surviving sirens show that even early in his career he had mastered precision metal and brass working (the ultra-smooth rotating disks, alignment of the holes, tightly fitting chambers) and delicate metal work (counting mechanism) (CR no. 27). By 1865 he was selling double sirens for 400 fr, a fairly large sum reflecting the substantial labour of manufacture.[89]

In keeping with a pattern throughout his career (e.g. tonometer, wave siren, wave machine), Koenig built an elaborate, research version of the double siren. This instrument, probably unique from Koenig's collaboration with Alfred Terquem of the University of Strasbourg in the late 1860s,[90] survives in storage at the Natural History Museum of Lilles (CR no. 27). It was meant to test Helmholtz's ideas on a grand scale and volume. The chambers are 40 cm in length, and the entire frame is 3 m in height. It came with traditional double chambers with rings of pierced holes; it also came with a simpler siren attachment with diamond and triangle-shaped holes, showing that even in the late 1860s, Koenig and Terquem were working towards new conceptions of sirens (see Chapter 7 on wave sirens).

At the same time as he worked on the first double siren, Koenig also made a copy of the Helmholtz synthesiser. The original synthesiser had been made for Helmholtz by Friedrich Fessel of Cologne in 1857 using electrically driven tuning forks and resonators. Koenig quickly set out to make a copy. In early 1860, soon after opening his business, he read Helmholtz's article on the timbre of vowels (*Ueber die Klangfarbe der Vocale*). "Your work on the timbre of vowels," he wrote to Helmholtz, "has interested me so much, that I would be very pleased to have some detailed descriptions of the apparatus with the tuning forks. ... It would please me to have a full sketch with a few measurements of the main proportions. I would also be so thankful to you, if you would be so kind as to allow me to manufacture and occasionally distribute it [*mir dieselbe anfertigen zu lassen und gelegentlich zu schricken*]."[91] Having received no reply after a few months (Helmholtz's first wife died in December of that year), he wrote again at the end of May 1860 to ask for instructions to build the synthesiser. He informed Helmholtz that a Russian professor (probably Sechenov), who had also read his article, had inquired about the synthesiser and wanted to know how much it would cost to make.[92] Not long afterward he received the reply and began building the instrument.

As with his graphical studies, Koenig used his workshop as a laboratory and did a number of tests that eventually led to improvements. In May 1861, he reported to Helmholtz that he had finished building the apparatus and had been performing experiments on it but was continually forced to stop on account of his active business:

> For some time now the apparatus has been finished and I ever so much regretted not having the honour of your visit as you came through Paris. I naturally would have very gladly shown you this work, in order to allow you to instruct me on improvements or alterations [*die Verbesserungen oder Abänderungen*] to the apparatus.[93]

He was able to replicate only the basic effects of Helmholtz's original studies and was hoping for more conclusive tests in the future:

> The forks [*die Gabeln*] all sound good with open resonators [*offenen Resonanzröhren*], and they were very little heard when the resonators were closed. One can also regulate the strength of each tone [*die Stärke jedes Tons*] conveniently through the keys [*die Tasten*]. The little time that I could successfully work on the experiments, however, was only sufficient for the production of very incomplete vowels [*die Hervorbringung der Vokale nur sehr unvollständig*].[94]

In other words, he could not get it to work. On a positive note, he added, "It was very interesting to show how the whole complex tone [*gänzlich der Klang*] was altered when one changed a single harmonic [*einen der harmonishcen Töne*], more or less, so I lent the apparatus to the Sorbonne and made there several such experiments in the course of Professor Desains."[95]

As with the sirens, the key parts of the synthesiser were copied with only minor changes. The design of the overall assembly, however, was different. In his first descriptions in *Tonempfindungen*, Helmholtz presented the parts as a loose arrangement of forks, resonators, interrupter and batteries. They were not mounted as one piece. This is not surprising for a non-commercial prototype. Koenig reworked these elements become part of one instrument. After seeing his instruments at the 1862 London Exhibition, Vienna physicist Joseph Pisko featured the synthesiser in his book *Die neuren Apparate der Akustik*.[96] The instrument had square wooden resonators and was different from later versions of the instrument with brass resonators, but it appears as one instrument. The later ones after 1865 comprised brass cylindrical resonators mounted on a wooden base with electrically driven tuning forks. A mercury interpreter drove the circuit in series. The maker needed to construct and tune the brass resonators, construct and tune the tuning forks, test the electromagnet coils, calibrate the positions of the resonators, experiment with the metal covers, and adjust the strength of the tuning forks to electrical stimulation. The ivory keys, arrangement of resonator units in order of size, finished wood base and mountings, brass cylinders with decorative pillars, polished and signed steel forks, and finely insulated green wire represented a multi-faceted construction, testing and workmanship enterprise. Many of the parts were most likely bought from other artisans.

As a whole the synthesiser was a very attractive representation of Helmholtz's ideas. By 1865 Koenig advertised the synthesiser in his second catalogue as "the large apparatus for the composition of different timbre of sounds, notably the timbre of vowels, through the simultaneous production of a series of simple notes that form a progression of harmonics [*notes simples qui forment la suite harmonique*]."[97] It cost 800 fr, making it the third most expensive instrument in the catalogue. It was the only one to be commercially produced at this time.

The spherical resonators were more of a challenge and it is here that we start to see the role of the workshop adding to and even modifying Helmholtz's work. In fact, they were the first instrument commissioned by Helmholtz for Koenig to manufacture.[98] In theory, the maker needed to tune each resonator according to a mathematical formula for resonance (volume, size of opening, height of the neck at opening). However, as Helmholtz came to realize, making a precisely tuned resonator demanded more than building from a formula. In *Tonempfindungen* he claimed to have made his earliest resonators from "any spherical glass vessels" that he could find, such as the collecting chambers of retorts, and then inserted a glass tube adapted for the ear into one of their openings. "Later Herr R. Koenig, (maker of acoustical instruments [*Verfertiger akustischer Instrumente*]), Paris, Place du Lycée Louis Le Grand 5 constructed a tuned series [*abgestimmte Reihe*] of these glass spheres."[99] In fact, it had been a demanding job for Koenig. He commissioned glassblowers who could not make tunable spheres. The necks and openings were too thick, and there were long delays in making the spheres. He had to create his own method for heating and opening the necks for tuning. In this way, he developed a practical, empirical understanding of resonance which led to other innovations. For example, he carried out investigations with variously sized chambers in order to extend their range. In a letter of 1860, Koenig informed Helmholtz that he was making a more complete series of resonators.[100] This series, as with the synthesiser, became an icon of Helmholtz's theory of timbre where the array of spheres took on the appearance of a tapering basilar membrane (CR no. 54). Whereas Helmholtz had articulated a theory of resonance, Koenig turned it into real, working objects with an overall design that served as an instructive concrete expression (or model) of the theory of timbre.

Koenig's introduction of brass by 1865 revealed another change in his workshop practices. The spherical resonators were made of two half pieces spun on a lathe (probably pressed against a wooden mould) and then joined together (Fig. 3.7).

Fig. 3.7 Spherical brass resonators. Close-up of spun brass. CR 54. Physics Department, University of Toronto, Canada

Spinning rings from the rapidly turning machine action encircle the resonators (CR no. 54). Even after developing a reliable manufacturing process in order to produce larger numbers of durable, consistently sized resonators, one can see in surviving examples that he continued to make small tuning adjustments by hand (especially with file marks at the neck). At the same time, he developed his cylindrical resonators that consisted of two brass tubes that slid into each other thus changing the volume and frequency. The surviving examples have a range of four to six notes, with the outer surface of the inner tube graduated and stamped with the frequencies. The series of 14 tubes have an overall range from sol1 to mi5 (G2 to E6). The graduated markings give the appearance of great precision, marking precise points on the sound spectrum. Recreated experiments with the resonators (both spherical and cylindrical) reveal that they are not necessarily as precise as their markings indicate; they respond to a range of notes (sometimes as much as one third of a musical note) above and below the marking, with the strongest response at the marking. The peak resonance intensity, however, could have been more noticeable and distinct to a trained ear.[101] Lack of precision aside, they represented a move towards durable, functional instruments for the classroom that relied on a more standard manufacturing process, with a convenient built-in range that could be used for more than one purpose (CR no. 55). Many of them, for example, were used in the sound analyser (CR no. 242).

Like the steel and wood, the brass in Koenig's shop received special attention. The brass in all the spherical and cylindrical resonators was often coated with shellac to prevent corrosion and to enhance aesthetic appeal. Surviving instruments also reveal that Koenig bought (or used) a few varieties of brass. A Savart bell at Harvard, for example, has a cylindrical resonator made with his standard, darkish, hard brass, while the bowl is made of a thick, higher quality, light-pinkish brass (CR no. 74). These kinds of differences, very subtle to us, were probably quite pronounced for instrument makers, especially if important acoustical effects depended on them.

Koenig's most significant transformation came with precision tuning forks based on Helmholtz's studies and Scheibler's beat methods; he also incorporated local innovations into these instruments. In the late 1850s, spurred by a competition to develop a standard tuning fork for France, Jules Lissajous had developed an optical method for comparing and tuning frequencies.[102] Koenig adopted these techniques in order to refine his tuning procedures. The combination of Scheibler's quantitative method, with Lissajous's optical method, made it possible by 1860 to make high quality, precision tuning forks.

As he did with the synthesiser and the resonators, Koenig created and marketed a complete, attractive set – the first commercial set for both teaching and research. He displayed his tuning-fork tonometer at the London Exhibition in 1862 and it would soon sell for 2,000 fr, which was 20 times the price of the average instrument in his second catalogue (1865). It consisted of 65 tuned forks, covering one octave, successively advancing by only four complete vibrations, each mounted on a beautifully finished pine resonator box. The whole series was an attractive package, and like other Helmholtz apparatus, become a symbol for the analytic, elemental theory of sound. It also represented the unique contribution of Koenig's atelier as a space where the techniques of Helmholtz converged with those of Paris. Shortly following the exhibition, Rodolphe Radau, a Königsberger physicist living in Paris, introduced Koenig's tonometer to readers of *Cosmos*, claiming that Koenig was able to make a major improvement to this instrument by combining the benefits of "modern science," such as the local graphical and optical innovations for tuning, with Scheibler's method.[103] Koenig, Radau claimed, had finally made it possible to popularize Scheibler's invaluable method of tuning.[104] We shall see in Chapter 5 how Koenig extended this apparatus for his *grand tonomètre* of over 670 tuning forks.

Demonstrating Helmholtz: Adam Politzer and Koenig at the Académie des Sciences

Helmholtz did not visit Koenig's studio during this period, but his presence was strongly felt in the form of instruments and an important visitor. In 1861 the Hungarian physiologist, Adam Politzer, visited Koenig's workshop to do a series of graphical experiments on the workings of the inner ear. He had recently worked in Carl Ludwig's laboratory in Vienna and in order to obtain his *privat-docent* training in otology he travelled to several laboratories through out Europe, including Helmholtz's in Heidelberg. He then went to Paris where he worked with Claude Bernard and performed a series of experiments at Koenig's atelier (Fig. 3.8).[105]

For both Politzer and Koenig, this was an extraordinary confluence of ideas and instruments. On 10 June 1861 (at the weekly Monday session of the Academie des Sciences) Claude Bernard presented the work of Politzer, who had just conducted a series of experiments on the workings of the inner ear.[106] Using the heads of freshly killed dogs and chickens and a recently deceased human, Politzer had used a form of manometer (pressure-gauge) to demonstrate the workings of the tympanic membrane and its associated muscles and bones. At the end of the presentation, Bernard displayed, with the help of Koenig, some recent "automatic" graphical traces showing related experiments obtained from Koenig's studio. He had connected a stylus to the tympanic membrane (including intact ossicles) and recorded the responses to various simple tones and combinations of tones. For those in the audience, this display was a stunning graphical display of acoustical phenomena connected to an area of anatomy that had only recently been explored. Abbé Moigno wrote that the tracings recorded in these experiments were "an incomparable perfection; they prove that the membrane of the tympanum is largely superior to the best artificial membranes of the physicists."[107] Two weeks later, after experimenting with Koenig, Politzer himself presented a continuation of his experiments on the inner ear to the *Académie*.[108] This time, Politzer did the anatomical preparations and Koenig did the graphical work for a live demonstration of the automatic inscriptions.[109] They used a form of the phonautograph with a blackened roller, stylus and, again, fresh parts of the inner ear, bones and membrane to demonstrate similar findings with beats and simple tones.

Fig. 3.8 Demonstration of early graphical experiment. The person on the left is possibly Rudolph Koenig
Source: Guillemin (1881, p. 655)

The *Académie* was a prominent scientific venue for both Koenig and Politzer, where they learned a valuable lesson about the social codes of presenting original work to a high-profile audience. In the first report of June 14, Moigno stated that "M. Helmholtz, the eminently skilful physicist had tried, but in vain, the experiment that worked so well in Paris."[110] In the next report he reported that "M. Politzer believes that we were mistaken in claiming that the illustrious German doctor M. Helmholtz had tried these experiments without success where he had succeeded; we willingly believe him, and retract what we said in this regard."[111] Politzer, who was returning to the German scientific world, did not want to be seen as too ambitious and spreading incorrect stories about Helmholtz in Paris. Helmholtz, however, was aware through Koenig that they were working together and sited their graphical experiments in his 1863 treatise.[112]

The demonstrations must have caused a stir because priority was now in the air. Léon Scott, clearly upset with Koenig, claimed to be the true originator for the method used in these experiments and was unhappy at not being credited. A few weeks later Moigno reported that Scott reminded the *Académie* that this recent series of experiments by Koenig and Politzer resembled ones he had already done. To prove his point, Scott asked that a package detailing these studies sealed in 1857 be opened and read to the *Académie*.[113] The commissioners at Scott's presentation included Victor Regnault, Claude Bernard and the physicist, Claude Pouillet.

In introducing his sealed package, Scott commented that recently a "foreign savant with the aid of an instrument maker" had presented to the *Académie* a series of inscription-type experiments with the inner ear of a decapitated animal. Aside from pointing out that he had already done these experiments, he also reminded the audience of his senior credentials. He thanked Duhamel[114] for his inspiration relating to his graphical research, and his "former master," Regnault, for his support at the Collège de France. He went on to detail his research related to recording graphically the responses from parts of the inner ear and his attempts to create an imitation of these functions in his invention, the phonautograph. However, Scott failed to appreciate the entirely new context that had opened up for graphical studies. Politzer and Koenig may have been doing experiments similar to Scott's in method, but they were testing Helmholtz's theories, something that Scott had not considered in 1857. Acoustics had changed; it had become a hybrid of Helmholtz's work, artisan influences and Parisian graphical work.

Manometric Flame Capsule and Optical Acoustics

It was not long after this controversy that Koenig abandoned the phonautograph. He turned to what was called an "optical" device that made sound visible with vibrating gas flames. "The most curious of all his inventions, one that we already all know here [Paris]," declared the Société d'Encouragement when they awarded him a gold medal in 1865, "is without a doubt the one that uses gas flames as a means for revealing the vibratory movements of air."[115] *Cosmos* first reported this invention in August 1862, describing how Koenig "showed us an attractive apparatus that had a lot of success at London, and that is destined to demonstrate one of the laws that follow the vibrations for a column of air contained in an organ pipe."[116] The manometric capsule consisted of a small chamber sealed by a membrane. When the membrane was activated by sound vibrations, pressure changed rapidly and gas in the sealed chamber expanded and contracted. A lit flame connected to this chamber flickered rhythmically up and down in time with the vibrating membrane. A rotating mirror displayed this flickering flame as a saw-tooth band of vibrations. Between 1862 and 1866, Koenig applied this technology to several purposes – studying organ pipe vibrations, beats, and the study of musical instruments and voices (Figs. 3.9 and 3.10).

The appearance of optical instruments coincided with Koenig's frustrations with the phonautograph. He claimed that the problems stemmed from limitations of membranes in general. Were they able to translate all sounds with equal intensity? *Cosmos* reported in May of 1862 that although Koenig had verified Helmholtz's general claim about timbre (that compound notes contain other harmonics) with "several phonographic tests (*épreuves phonographiques*) that we have seen at his place," he was beginning to question the reliability of the phonautograph membrane.[117]

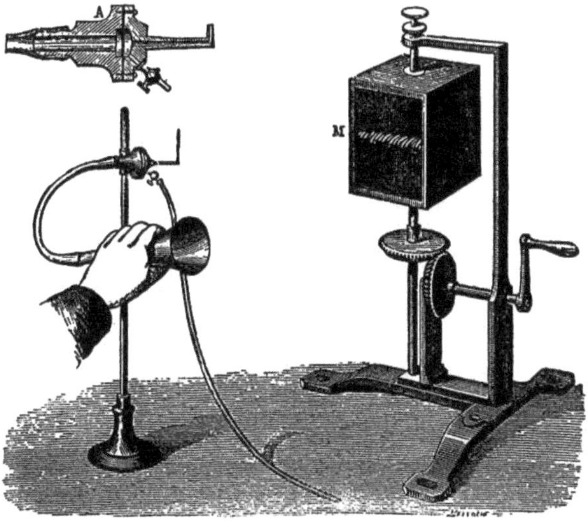

Fig. 3.9 Manometric capsule and rotating mirror
Source: Koenig (1882c, p. 57)

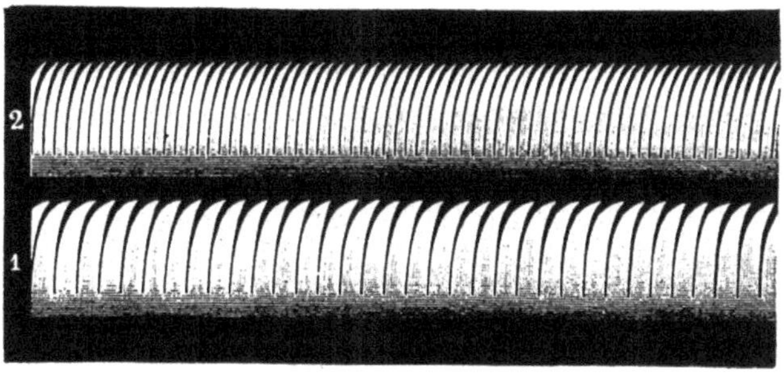

Fig. 3.10 Manometric flame patterns from two different organ pipes
Source: Koenig (1882c, p. 52) (used with instrument CR 239)

> It is unquestionable that the researches of Helmholtz have opened a sure path from which the timbre of sounds must be engaged. For some time, it was hoped that the phonautograph would serve to clarify this question; but M. Koenig has arrived, through his experience, at the definitive conviction that the vibrating membranes fitted with styluses will never give anything but the *number of vibrations* of notes... and that it is impossible to receive from them [the phonautographic studies] a profit for studying the *quality* of sound [*les* qualités *des sons*].[118]

Two weeks later Radau explained that Koenig had one key problem with the use of a membrane for analyzing complex vibrations – he believed that the membrane actually favoured certain notes, thereby making it very difficult to assert any claims about amplitude or intensity. For example, one could not adequately claim that the timbre of a certain compound note was due to the strength of a particular harmonic, if this effect was merely the result of the membrane being more inclined to respond to that tone. Koenig seems to have come to this conclusion after finding in his own series of graphical tests.[119] Two other local scientists, Bouget and F. Bernard tested Koenig membranes in 1860 and came to the same conclusion.[120]

Social pressures came to bear on this situation as well. Léon Scott complained that Koenig and others had appropriated his graphical ideas. Koenig, already a fiercely independent artisan, clearly wanted to remove himself from this perception. In addition, the phonautograph was not as well suited as it could be to the demonstration market. It was finicky and not dramatic enough for large audiences. The manometric capsule could deliver an entertaining, more reliable demonstration for a large audience.

One of Koenig's more famous instruments emerged from this conflict – the manometric sound analyser. The education market, as we will see later, was particularly ripe for such an attractive demonstration instrument. At the 1862 London Exposition, he demonstrated the simpler manometric capsule with an organ pipe.[121] Following the introduction of this device, he developed an "apparatus designed to decompose in a visible manner the timbre of sound into its elementary notes by means of manometric flames."[122] He claimed that using the analyser, one could gauge the intensity of a harmonic through the brightness (sharpness) of the flame.[123] Furthermore, the capsules, when hooked up to resonators, responded to one note only, eliminating the concern of using one membrane to simultaneously detect several harmonics. As he had done with other Helmholtz instruments, Koenig assembled the units of this apparatus into an attractive whole. The prismatic frame resembled the structure of basilar membrane, creating a concrete expression of Helmholtz's ideas for the front of the lecture hall and textbook illustrations (CR no. 242).

He quickly found other uses for the manometric capsule. With a series of organ pipes, he used it to display the relations of musical intervals based on the optical-tuning methods of Lissajous. With a stethoscope or speaking tube, he used the capsule for studying violin and vocal sounds. With a resonator and calibrated tubing, he measured phase and interference effects. Each was seen as research and demonstration apparatus. The manometric technology ushered in a cinematic quality for making sound visible, but was still not a method for precise quantification, nor was it an easy path for recording results. It did, however, prove very successful as an attractive teaching instrument. As the market continued to open for acoustics, it would be Koenig's manometric instruments that spread the ideas of Helmholtz from Europe to North America.

Notes

1. Oudinet (1882, pp. 21–29).
2. Cattermole and Wolfe (1987, p. 23).
3. Pantalony (2004a, pp. 23–27).
4. Henry Rowland to President Gilman, undated, summer 1875. MELSC.
5. Brenni (1994c, pp. 13–14).

6. Paulin (1938, pp. 192–225).
7. Lieberman (1995).
8. Instrument historian, William Andrewes, pointed out several manufacturer marks on the manometric analyser at Dartmouth College. See CR no. 242a.
9. Coningsby (1867, p. 447). Also see Gourden (1992).
10. Coningsby (1867, pp. 433–434). Also see Child (1889).
11. Whiteing (1867, p. 466).
12. Latour (1979, p. 52).
13. Holmes and Olesko (1995).
14. Middleton (1941) and Multhauf (1961, pp. 102–103).
15. Braun (1992). In 1859 Marey introduced the sphygmograph or "pulse writer" for inscribing the pulsation of the heart. He then worked with Auguste Chauveau to produce a new form of cardiograph in 1861. Koenig and Marey were friends and cited each other in later works. Marey's classic treatise of 1878 displayed several of Koenig's drawings and traces, see Marey, *La Méthode Graphique*. They also shared the same artist for their texts, a man named Perot, who assisted Koenig with his manometric drawings.
16. Brain (1998b).
17. For further context on Scott's work, see Hankins and Silverman (1995, pp. 133–135), Silverman (1992, pp. 120–122), and Charbon (1981, pp. 11–15).
18. Scott de Martinville (1859a, p. 314).
19. Ibid., p. 315.
20. Ibid., 1857.
21. Idem., 1878, p. 50.
22. Ibid., pp. 50, 70.
23. Idem., 1859b.
24. David Giovannoni and Patrick Feaster of "First Sounds" have recently discovered that Scott was working with a rotating cylinder as early as 1857. This is based on a description by Scott found in the archives of the Société d'Encouragement à l'Industrie Nationale folder 8/54. Their research as well as their reproduced sounds from the first phonoautograph recordings can be found at http://www.firstsounds.org/
25. Ibid. Moigno (1859a,b) and Schmidgen (2007).
26. Scott de Martinville (1878, p. 58).
27. Idem., 1859a.
28. Moigno (1859a, p. 320).
29. Moigno (1872).
30. Obit. *New York Times*, July 16, 1884. Catholic Encyclopedia (New Advent). "François-Napoléon-Marie Moigno".
31. Moigno (1859b, p. 677).
32. British Association (1860), pp. lix–lxix.
33. The title of his own talk was, "Supplement to Newton's Method of resolving Equations" in Ibid., pp. 9, 62.
34. Moigno (1859a, p. 417).
35. Ibid.
36. Idem., 1859b, p. 679.
37. "Le phonautographe, appareil pour la fixation graphique des bruits" in Koenig (1859, p. 1).
38. Ibid., p. 2.
39. Ibid.
40. Ibid.
41. Ibid., pp. 3–4.
42. Ibid., p. 2.
43. Brock-Nannestad (2007, 2008). Also listen to Scott's (1861) recording of "au clair de la lune" at www.firstsounds.org
44. Scott de Martinville (1878, p. 70).

45. Ibid., p. 69.
46. Ibid.
47. Ibid., pp. 64–67. There are no known examples of Scott's later instruments. It is not known with whom he worked.
48. Pisko (1865, pp. 71–77) and Thompson (1901, p. 630).
49. Donders (1864) and Schmidgen (2007).
50. Koenig (1882c, p. 199).
51. Radau (1862a, p. 623).
52. Crookes (1870).
53. Thompson (1901, p. 630).
54. Schmidgen (2007).
55. Miller (1935, p. 75).
56. Koenig (1901, Deuxième partie, p. VIII).
57. Rudolph Koenig (1901, Première partie, p. III).
58. Radau (1862a, p. 659).
59. Koenig (1865, p. 42).
60. Koenig (1882c, p. 1). Copies of Koenig's tracings can be found in Loudon and McLennan (1895, pp. 109–110) and Pisko (1865, pp. 55–93). Auerbach in a Winkelmann (1909, *Akustik*, pp. 150–152) and Zahm (1900, pp. 421–422).
61. Bud and Warner (1998, pp. 110–112).
62. Koenig (1882c, pp. 2–6).
63. Ibid., p. 11.
64. Ibid., pp. 21–22.
65. Ibid., pp. 23–24.
66. Radau (1862a, p. 623).
67. Koenig's chapter on the development of the standard tuning fork is the best example of this from of presentation. Koenig (1882c, pp. 172–183).
68. Ibid., p. 6.
69. Ibid., p. 8.
70. Ibid., pp. 24–25.
71. I borrow this phrase (turning plate in a train round house) from the French historian, Christine Blondel, who has made a similar characterisation of the electrical instrument makers in nineteenth-century Paris. Blondel (1997).
72. For more on the wave machines, see Holland (2000, part 1, pp. 99–101; part 2, pp. 30–32).
73. Moigno (1862a). Idem., 1864, p. 319.
74. Koenig (1859, p. 31).
75. Koenig (1862), Moigno (1862c), and Koenig (1864d). Idem., 1864a. Charles Wheatstone (1833).
76. Moigno (1862b, p. 700).
77. Radau (1862a, p. 623).
78. Evenson (2000), Koenig (1865, p. 5), Pisko (1865, pp. 94–103), Shulman (2008), and Thompson (1883).
79. Threlfall (1898, p. 261).
80. Turner, G.L'E. (1996, pp. 127–128).
81. For coils, finely insulated with green thread, see Ibid. and CR nos. 56, 166, 189, and 248.
82. One of the finest collections of Parisian electrical instruments is at the Bakken Library and Museum in Minneapolis. See for example, cat nos. 74.13.053, 87.15.004, 91.9, 101.1, 81.137, and 81.52.
83. Eugène Ducretet, for example, used an identical handle on his electrical machines. See The Bakken Library and Museum, cat. no. 82.409. This issue was pointed out to me by instrument historian, Will Andrewes, following his examination of the Dartmouth analyser in 2002.
84. Radau (1862a, p. 623).
85. Moigno (1865, p. 534).

86. Koenig to Helmholtz (Dec. 2, 1859) in Hörz (1997, p. 358).
87. Helmholtz (1863, pp. 241–242).
88. There is very little information on Sauerwald. A handful of his electrical instruments exist in collections in Europe. These finely crafted instruments, however, with their connections to people like Helmholtz and Magnus, tell us a good deal about what kind of workshop he ran and the kind of skills he brought to bear on his commissions. A few museums have examples of his double siren, see CR no. 27. For more on Sauerwald see, Brenni (2004).
89. Koenig (1865, pp. 9, 33–37).
90. Terquem (1870, p. 280).
91. Koenig to Helmholtz, (Feb. 29, 1860) in Hörz (1997, p. 361).
92. Koenig to Helmholtz (May 27, 1860), Ibid., p. 362.
93. Koenig to Helmholtz (May 18, 1861), Ibid., pp. 362–363.
94. Ibid., pp. 362–363.
95. Ibid., p. 363.
96. Pisko (1865, pp. 22–26).
97. Koenig (1865, pp. 10–11).
98. Koenig to Helmholtz, (Dec. 2, 1859 and Feb. 29, 1860) in Hörz (1997, pp. 358–361).
99. Helmholtz (1863, p. 561).
100. Koenig to Helmholtz (Dec. 2, 1859 and Feb. 29, 1860) in Hörz (1997, pp. 358–361).
101. Based on experiments done in 1999, 2000 and 2001 at the University of Toronto. Pantalony (2001).
102. Turner, S. (1996, pp. 33–51).
103. Radau (1862b, p. 111).
104. Ibid., p. 109.
105. Politzer (1864, p. 61). Idem., 1913, 1883. Also see M. Murdy and M. Kraft, "How Adam Politzer (1835–1920) became an Otologist" at the http://www. politzersociety.org
106. Moigno (1861a).
107. Ibid., p. 669.
108. Idem., 1861b, p. 780.
109. For Koenig's brief description of these experiments see Koenig (1882c, p. 29).
110. Moigno (1861a, p. 669).
111. Moigno (1861b, p. 780).
112. Koenig to Helmholtz, Jun. 8, 1861, in Hörz (1997, 363–364) and Helmholtz (1863, p. 248).
113. Moigno (1861c) and Scott de Martinville (1861).
114. Jean Marie Constant Duhamel was a senior professor at the École Polytechnique. In the 1830s he invented a novel graphical instrument called a vibroscope. It consisted of a rotating cylinder and a vibrating writing stylus, see Dostrovsky (1970).
115. Moigno (1865, p. 535).
116. Radau (1862b, p. 147).
117. Radau (1862a, p. 623).
118. Ibid., p. 624.
119. Koenig (1882c, pp. 23–24, 28).
120. Bouget and Bernard conducted experiments on Koenig's improved membranes (using his phonautograph), independent of the ideas of Helmholtz, in order to test Savart's hypothesis that membranes would *respond to all tones equally*. Savart had claimed that membranes responded to all harmonics equally, instead of favouring certain notes (revealing a natural frequency of vibration). Bouget and Bernard concluded that membranes did in fact distort the recording of sounds, and did not, as Savart claimed, respond equally to all tones in the surroundings Moigno (1860). After viewing a particular graphical tracing of a complex tone one could not, for example, claim that the octave of a complex note was indeed stronger than the fifth, if the membrane naturally favoured the octave.
121. Brooke (1863, p. 33).
122. Koenig (1865, p. 46).
123. Detailed descriptions of these experiments can be found in Koenig (1873c, p. 106).

Chapter 4
The Market and Its Influences

The booming market for scientific instruments in the 1860s developed from two streams of science education – augmenting and updating of physical cabinets across Europe and North America, and the growth of student laboratories. Two Koenig artifacts clearly illustrate these contexts. One, a Galton whistle from the Massachusetts Institute of Technology (MIT), comes from the first attempts in the United States to build student laboratories in the late 1860s. Like other Koenig surviving apparatus at MIT, it shows heavy use and is painted with the original inventory number from the shelves of the laboratory. Another apparatus from the University of Coimbra in Portugal, the manometric apparatus for comparing two sounds using manometric organ pipes, tells the story of a different teaching environment. Like almost all the Koenig apparatus at the University of Coimbra (one of the largest collections in Europe), the pine pipes appear unmarked and unused.[1] At the University of Coimbra, where the historic physical cabinets (made of Brazilian wood) remained (and still remain!) in the same condition and place since the late eighteenth century, there was an engrained tradition of classical teaching – professors lectured, demonstrators demonstrated and students took notes. The evidence (lack of use) on the instruments show that professors still adhered to this style of teaching well into the twentieth century. The surviving instruments at MIT, on the other hand, long ago lost their original sheen. Both sets of instruments, which illustrate two sides of the teaching market during this period, were bought at the 1867 Exhibition in Paris, a high point of Koenig's career as an instrument maker and businessman. They also represent two other directions in Koenig's work – deskilling instruments for student laboratories and creating novel visual instruments for attractive lectures (Figs. 4.1 and 4.2).

Even within a strong market, scientific instruments do not sell themselves. The manufacturer chooses the instruments to make, how to promote them, and to whom to sell them. These decisions have a large impact on the transmission, direction and scope of a field. In the second half of the nineteenth century, the commercial aspects of scientific instrument making took root in several ways: There were hugely popular international fairs where instrument makers displayed and demonstrated their wares; there was a growing laboratory movement for teaching and research which stimulated the manufacture of thousands of instruments; there was competition

D. Pantalony, *Altered Sensations*, Archimedes 24, DOI 10.1007/978-90-481-2816-7_4,

Fig. 4.1 Galton whistle. Photo by author, 2005. Physics Department, MIT, USA

Fig. 4.2 Manometric organ pipes (CR 239). Photo by author, 2005. Museu de Física, University of Coimbra, Portugal. FIS.406

among schools to build their collections; and there were periods of rapid economic and industrial growth that created favourable conditions for the instrument trade.

The 1860s were Koenig's most profitable and productive years. He won awards at the 1862 Exhibition in London and the 1867 Exposition Universelle in Paris. In 1865 he won the Médaille d'Or from the Société d'Encouragement pour l'industrie nationale. In April 1868, at the age of 35, he received an honourary doctorate from the University of Königsberg. By the end of the decade, on the eve of the 1870 Franco-Prussian War (Chapter 6), he had successfully spread his instruments through Europe and North America and gained international recognition. Instruments related to Helmholtz's studies and graphical acoustics had rapidly become part of the standard demonstrations and student exercises.

The First Year of Business – from the Workshop to the Classroom

The Musée de la Civilisation in Quebec City is the only known institution that holds instruments from Koenig's first year in business, 1858–1859. They once rested on the shelves of the grand cabinet de physique of the Séminaire de Québec, which in its day was unparalleled in Canada. In fact, between 1800 and 1860, the Séminaire (founded in 1663), developed one of the more impressive physical cabinets in North America. The instruments survive today, in a storage facility in the suburbs of Quebec City, silent witnesses to a dynamic period in Quebec history, the growth of science education in North America, and Koenig's early business.[2]

In late 1858, a young teacher from Quebec City named Thomas Etienne Hamel visited Rudolph Koenig's newly opened atelier near the Lycée Louis-le-Grand. He had spent the last 4 years studying physics and mathematics in Paris in order to bring this training and education home to Québec. He also wanted to bring the latest instruments and books back to his native schools, and therefore went on a buying spree at the end of his stay.[3] He visited the young instrument maker and was treated to a series of attractive demonstrations. Upon returning to Québec, Hamel received a letter from the maker describing the demonstrations he had witnessed and reminding him that he could buy these instruments to reproduce the same demonstrations at the Séminaire.[4] He wrote of Hamel's particular pleasure at seeing the multi-task sonometer invented by I.J. Silbermann (the physics demonstrator at the Collège de France)[5] and Guillaume Wertheim's electrically vibrating iron bar. He also singled out an instrument from Dresden, a siren based on the studies by August Seebeck.[6] This was a series of siren disks with striking geometric arrays of pierced holes that formed the basis of several experiments.

Even in his first months of production Koenig was eager to adopt and develop instruments. The 27-year-old wanted to project an image of advancing the field and going beyond his predecessors. Accompanying his letter to Hamel, he sent a four-page hand-written list of instruments "which are not found in Marloye's catalogue."[7] (The Séminaire already had Marloye instruments). He advertised a handful of models showing the movement of molecules (*moléculaire*) in a wave, demonstrations of polarization and interference, a series of organ pipes, the Seebeck siren and the *grand sonomètre d'après Mr. Silbermann*. The latter object, for demonstrating several experiments with vibrating strings and musical scales, was quite expensive at 500 fr (on average 20 times more than other instruments, or approximately 100 times more than the cost of a book). There were no graphical instruments listed (the Scott-Koenig collaboration was still proceeding at this point), and no instruments related to Helmholtz, two areas that soon became central to the business. However, the wave models that Hamel bought represented recent trends in England[8] and a move towards visually appealing instruments.

One thing that stands out from the surviving Hamel purchases is the simplicity and self-sufficiency of Koenig's early workshop. The cardboard disks for the Seebeck siren, for example, evoke a small construction enterprise. They are made of dense cardboard and have detailed instructions handwritten by Koenig. Later models were made of brass or zinc, and came in more elaborate tambour-style supports. The patterns of the holes, like the Opelt siren disks, were a work of art themselves and appealing as a display of the visual beauty underlying acoustical harmonies (CR nos. 28, 28a, 28b, 28c and 29).[9]

In the Quebec collection there are also a few small, black wave models with simple woodwork inscribed with the title of the instrument in white paint, (e.g. "mouvement moléculaire d'une onde aérienne/produite par une choc simple") and signed by the maker. A train of white beads moves along the path of an axle in the interior of the box. A small wooden crank sets the beads in motion.[10] But even though they appear simple to us today, they were visual and instructive marvels for a young class not used to such demonstrations.

The simple enterprise, however, did not last long. In 1859, only a year into his business, Koenig issued a full catalogue to advertise his products, most important being the latest development, the phonautograph. Ninety-five of the 190 instruments came directly from Albert Marloye's catalogue of 1851. He used the same descriptions, word for word, with identical prices, with only four minor exceptions.[11] But, he also clearly moved beyond Marloye. In the introduction to the catalogue, he emphasized that he had created instruments "nearly indispensable for the study of acoustics" which should be part of any collection or "physical cabinet."[12] He paid particular attention to the phonautograph, of which he was the "sole constructor," because of the "very interesting" graphical experiments that could be repeated and because it provided a means for doing scientific research that had been previously "inaccessible."[13] He expanded the basic demonstration repertoire by listing a wider range of organ pipes and vibrating plates, and for classrooms he advertised over forty oil paintings (1.5 by 1 m) illustrating basic acoustical phenomena, well known experiments, apparatus, and the workings of musical instruments.[14] Even before Helmholtz's work became widely known in Paris, Koenig promoted the idea that acoustics was changing.

1862 Exhibition at London

By 1862 Koenig was ready for a major public display of his products. The International Exhibition at South Kensington in London was his first exposure to a large scientific and public audience. It was a sequel to the hugely successful Great Exhibition held in London in 1851, and the Paris Exhibition of 1855. France was the leader in precision instruments in 1862 with over 60 makers represented in class 13, "philosophical instruments and processes depending on their use." The exhibition established Koenig's reputation in England. The jury, which included William Thomson, Charles Wheatstone, and David Brewster, reported that the "ingenuity in design" and "excellence in construction" of his instruments made his display "one of the gems of the philosophical department."[15] He displayed instruments that would become standard apparatus for acoustical teaching and research: the Seebeck siren, apparatus for inscribing Lissajous figures, manometric organ pipes, tuning forks with mirrors for Lissajous optical demonstrations, and a series of precision tuning forks called the tuning-fork tonometer.[16] Koenig's booth in the French section was between E. Hardy, who sold electrical, magnetic and other philosophical apparatus, and C.J. Columbi, maker of geodesic instruments.[17]

One of the highlights of the booth was an album of recently recorded graphical traces and studies from the phonautograph. "M. Koenig," according to one popular account,

> showed a wonderful collection of instruments applied to the illustration of the theory of the conduction, undulation, and vibration of sound. By a most ingenious but simple instrument – a common glass cylinder, coated with fine lampblack, and applied, turning, to a tuning – key when vibrating – M. Koenig makes sound its own printer. From the impression left on this cylinder all the different vibrations and undulations of sound between A and G are here recorded from their outset to their latest tone, have been made to register themselves, and from the records thus left a most beautiful series of acoustic charts has been drawn out.[18]

The success of his London display attracted positive press in Paris. Following the Exhibition, Rudolphe Radau praised his work in *Cosmos*. He was particularly impressed with the tuning-fork tonometer claiming that this instrument would now make it possible to popularize and spread Scheibler's method of tuning.[19] The 65 fork tonometer, which was displayed in a decorative wood and glass case with a display sign that read "TONOMÈTRE D'APRÉS SCHEIBLER" and with stems of the tuning forks resting in turned wooden handles (CR no. 37), was clearly a novel attraction and one of the reasons that the jury awarded Koenig a medal of distinction, commenting: "By aid of this instrument, and a *practised* ear, very delicate gradations of pitch may be obtained." They held out the hope that "an authoritative establishment of international uniformity [standard pitch] would confer an inestimable public advantage."[20]

The London exhibition also helped spread the word about Koenig's work to the German world. Joseph Pisko, a physicist from Vienna, published a book *Die Neuren Apparate der Akustik* only a few years later (1865) based on instruments that Koenig showed at the London fair (Fig. 4.3).

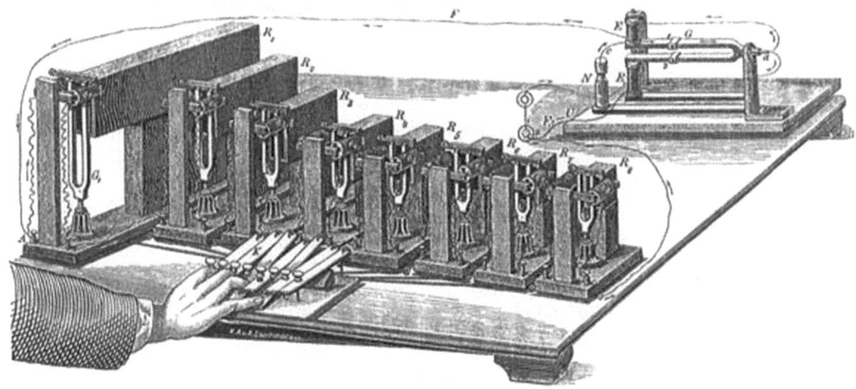

Fig. 4.3 Joseph Pisko's illustration of the Synthesiser. CR 56
Source: Pisko (1865, pp. 22–26)

Selling Helmholtz's Instruments

Just prior to 1865, Koenig moved to a location near the faculty of medicine at 30 rue Hautefeuille in the heart of the Latin Quarter in close proximity to medical students, scientists, artists and artisans. He had just spent 2 productive years developing instruments he could promote as centerpieces of his company – optical, graphical and Helmholtz's apparatus. In an introductory letter to the American physicist, Joseph Henry, he makes the case that acoustics is a reformed science that must take its proper place in the physical cabinet:

> Admirable work has been executed in the last few years and since the illustrious Helmholtz published his admirable treatise on physiological acoustics everyone has been occupied with these researches and acoustics has finally been established as a science. The instruments that serve to demonstrate its phenomena are just as indispensable to a cabinet of physics as all the others that one meets there. Examination of my collection will show you that the causes that previously occasioned the neglect of this science no longer exist. In effect, if the greater part of scientists had recoiled from acoustical researches for fear that their ear, little exercised, would encounter insurmountable obstacles, we are today in possession of such admirable methods that permit the study of sound without the assistance of the ear.[21]

Koenig clearly set out to make Helmholtz's work and visual methods the focus of his activities. His second catalogue of 1865 presented a significantly expanded range of products with detailed woodcuts. Eighty-eight new instruments appeared, 25 of which related to the work of Helmholtz. The largest section "methods for observing the vibration of sounds without assistance from the ear," featured his graphic and manometric instruments. There were novelties like the Reis telephone, a variety of tuning forks (especially the 65-fork tonometer), clever organ pipe demonstrations, a medical stethoscope, electrical tuning-fork interrupters, and more instruments invented by local scientists. The prices remained the same as 1859,[22] with the least expensive instruments near 10 fr, and more expensive items costing hundreds of francs (see Table 4.1). The cover page of the catalogue carried an engraving of his medal of distinction won at the 1862 London Exhibition. With these instruments and

Table 4.1 Prices of instruments from 1865 (labour wages averaged 5–9 fr a day)

Wooden organ pipes averaged between 5 and 10 fr
Brass, spherical resonator, 10 fr
Free reed pipe, 25 fr
Tuning fork, 25 fr
Sound analyser, 250 fr
Double siren, 400 fr
Phonautograph, 500 fr
Lissajous apparatus, 600 fr
Seebeck siren, 800 fr
Vowel Synthesiser, 800 fr
Tonometer, 2,000 fr
Apparatus for production of sound in liquids, 1,000 fr

Fig. 4.4 Koenig's 1862 Medal of Distinction used on the cover of his catalogue
Source: Koenig (1865, title page)

attractive images, the entire catalogue represented a major development and investment in the changing field. That same year he sent a catalogue to Joseph Henry inscribed: "A Smithsonian Institution à Washington D.C. hommage de l'auteur, Rudolph Koenig." (Fig. 4.4).[23]

The cost of labour in Paris provides a useful means for putting the above prices in perspective; it also gives a good estimate of the amount of time involved in making each instrument. According to reports submitted at the 1867 Exposition (2 years later), the average wage in Paris for those who worked by hand or with machines was 5 fr per day. First class hands, however, made as much as 9 fr a day.[24] Prices in the musical instrument trade, in which Koenig had himself been trained, provide another comparison. Musical artisans made between 5 and 11 fr per day, with half of them working by piece. Materials could elevate costs considerably. Wood came from France, but specialized parts such as steel cord and wire came from England and the German territories. Small musical instruments cost from 50 to 200 fr, harmoniums from 100 to 1,500 fr, violins and violincellos from 200 to 500 fr, brass instruments, 80–400 fr, pianos 500–4,000 fr and church organs 2,500–100,000 fr. Similar to the scientific instrument market, half of the Parisian musical instruments were shipped to other countries, especially North and South America.[25]

The 1865 catalogue coincided with an economic boom in Europe and North America that led to more purchases of Koenig instruments. By 1865, the Abbé Hamel in Quebec made another medium-sized purchase totalling 975 fr, representing recent developments in acoustics, in particular the work of Helmholtz.[26] Joseph Henry also bought a large group of instruments. They did this through an agent, who promoted Koenig's work. Not being able to travel to Paris at this time, both Hamel and Henry bought instruments through Hector Bossange, whose offices were located at Quai Voltaire, near the Institut de France and within walking distance of several instrument shops.[27]

Owing to increased sales, acoustics now had a larger presence in physical cabinets and classrooms. Volkert S.M. Van der Willigen, who was in charge of the physical cabinet at the Teyler Museum in Holland, made a large purchase of Koenig instruments at a value of Dfl. 1,065.80. The shipment was delivered in December

1865 and weighed 221 kg. As with the Séminaire, the Teyler Museum had made acoustics a priority and wanted to be on the cutting edge of a growing field. In 1863 the Teyler had already purchased a Wheatstone Kaleidophone from Koenig and in 1864 they bought a series of resonators and a singing flame. The 1865 purchase included, among several items, a Helmholtz Synthesiser and a phonautograph.[28]

Function Replaces Beauty: 1867 Paris Exposition

By 1867, not even a decade into his business, Koenig had issued an illustrated catalogue and shipped instruments throughout Europe and North America. Images of his instruments had appeared in physics and acoustical texts, including those by Jamin, Helmholtz, Ganot, Tyndall, Pisko, and Radau. He had published research and announcements of his inventions and was beginning research on vowels and combination tones. The 1867 Exposition Universelle in Paris was an ideal opportunity to show his work to the world.

The Paris Exposition came at a time of substantial technological and economic change. The entire exposition, the largest since the London exhibition of 1851, was a showcase for advancements under the Second Empire of Napoleon the Third.[29] It took place in an elliptical building on the present-day Champs-de-Mars, with a series of concentric rings, or levels, organized by country and group. The circumference of the building was a staggering mile long. There were over 11,000 exhibitors from 35 countries, represented in ten groups – art, apparatus (science printing, photography, medicine), furniture and domestic objects, clothing and fabrics, raw materials, machinery, food products, agricultural exhibits, horticulture and produce, and articles that improved the "moral and physical condition" of people.[30]

The group "Mathematical Instruments and Apparatus for Teaching Science" had 107 exhibitors, "confined almost exclusively to Paris."[31] Koenig's display stood out for its innovations. According to Augustin Deschanel, the French author of a popular text on natural philosophy, there were items in this section that marked developments of "decidedly enhanced importance," such as telescopes with silvered reflectors, electrical induction apparatus, electromagnetic machines, regulators for electric lights, and optical displays of sounding bodies.[32] Koenig, who won a gold medal, was now considered one of the best in a highly competitive field. Jules Lissajous, the inventor of the optical method for comparing vibrating tuning forks and part of the official jury, stated that Koenig had now attained the same reputation in France and abroad as his predecessor, Marloye. Much of this success, according to Lissajous, derived from the novelty of his collection, being on the cutting edge of science and, as a savant himself, sometimes "ahead of it" – "*Toujours au courant de la science, il la devance parfois en faisant lui-même oeuvre de savant.*"[33] The collection, he stated represented a history of progress in the field. He surveyed the key instruments: the tuning-fork tonometer "comprised all the perceptible sounds of hearing [*toute l'étendue des sons perceptibles*],"[34] Helmholtz resonators, and the Synthesiser for producing vowels. He then described the graphical and manometric

Fig. 4.5 One disk from Crova's projection apparatus, CR 262a. Photo by author, 2005. Museu de Física, University of Coimbra, Portugal. FIS.1282

flame devices that enabled the study of vibrations in a "direct and precise" manner without the "help of the ear."[35] He was particularly impressed by Crova's projection apparatus that demonstrated mechanical vibratory movements on a large screen.[36] In the research realm, he described the chronograph and tuning fork apparatus, with its graphical tracings for precision timing, made possible through collaboration with Victor Regnault (Chapter 5) (Fig. 4.5).

According to Lissajous, however, there was one unusual feature about Koenig's display – his instruments looked surprisingly functional. Even in the second half of the nineteenth century, amidst pressure to produce more instruments for less cost, there was still emphasis and pride among the Parisian artisans to endow instruments with aesthetic appeal and show off their talents. In addition, customers wanted attractive objects for their cabinets and demonstrations. But changing market demands and research-based precision makers such as Ruhmkorff and Koenig had begun to change that emphasis. In his jury report, Lissajous stated that making acoustical instruments required highly specialized knowledge, devotion and skill, so much so, in fact, that other makers avoided acoustics. Marloye, he wrote, had mastered this craft and demonstrated "great skill" and "perfect taste" in working with wood.[37] Koenig, on the other hand, was "less preoccupied by the artistic point of view, than with scientific progress."[38] Aesthetically, the replacement of wood, with its historical importance in musical instruments, by cast-iron parts was one of the more noticeable changes. For Lissajous it was unfortunate that the cast-iron work (*la serrurerie*) was replacing cabinet making (*l'ébénisterie*). The collection seen as whole "did not produce a favourable impression for the eye."[39] Instrument making was therefore at a turning point reflecting broader changes in the scientific market. There was increased demand for precision and function, and, with increased student use, there was growing demand for durability and practicality.

Americans at the Fair

Koenig's display, however functional to French critics, caught the attention of a few influential American visitors, representing what would become one of his best markets. There was no other specialist in the field who made Koenig's range and quality of acoustical instruments; there was certainly no one in the United States who specialized in acoustical instruments. Yet, the demand for acoustics was growing. One source of this appeal in the United States derived from the importance of music in American culture. The highly influential American educator, Frederick A.P. Barnard, the president of Columbia College, who was living in Paris in 1867, wrote of Koenig's instruments in musical terms. The Helmholtz resonators, he noted, facilitated the analysis of musical tones; the tuning-fork tonometer made it possible for those without a musical ear to work with a large range of precision notes; the phonautograph recorded the notes of musical instruments and the human voice; singing flames explored the mysteries of musical notes and their relation to gas flames; the sound analyser permitted a test for the presence of harmonic overtones in all musical notes; and the tonometer, a "striking part of the exposition," embraced 330 forks, covering "the entire range of audible sounds."(CR no. 36).[40] Barnard had a practical interest in Koenig's display as well. Like Alexander Graham Bell (see below), he took great interest in technologies that made sound visible. In fact, he was severely hearing impaired, and wrote at length about Koenig's visual instruments. In addition, Barnard was more familiar than most with the basics of acoustical teaching and practice. Before the American Civil War, he had taught at the University of Mississippi which had an extensive collection of Marloye instruments.[41] His report on acoustics was therefore one of his longest entries on the Fair: "No branch of modern physical investigation," he summarized,

> has been productive of more numerous interesting results than acoustics. And in no branch of experimental inquiry have investigators been dependent for their apparatus upon a smaller number of able constructors. This subject was represented in the Exposition almost exclusively by a single exhibitor, Mr. Rudolph Koenig, of Paris. On the other hand it may be said that Mr. Koenig's exhibition was so complete and admirable as to leave nothing to desire.[42]

Koenig also benefited from being near other French makers, where many other scientific buyers congregated. Barnard wrote a report for the American government on the "exact sciences," claiming that all the precision instruments on display were "miracles of skill," while the French and British sections were "brilliant," and the French displays in particular encompassed "probably a richer collection than any of its kind that was ever before brought together."[43] Similar to Deschanel, Barnard singled out progress in electricity and visual acoustics as examples of significant novelties. As with other American visitors, he had little to compare with at home. He lamented America's poor showing at the exposition: "Americans asked," he wrote,

> where are Ritchie, Green, McAllister, Würdemann, Zentmayer, Grunow, Chamberlain, Pike? Our countrymen could not but feel that, while we were nowhere adequately represented, in this department our representation was so disproportionately inadequate as to be likely to produce very unjust impressions abroad in regard to the state of science among us.[44]

What stands out in Barnard's descriptions, from someone who was keenly reviewing the latest teaching and research equipment, was the manner in which complex and hitherto mysterious acoustical effects could be understood through simple mechanical operations. Barnard described at length, for example, the mechanisms of the Helmholtz double siren, such as the crank that rotated the upper disk allowing one to alter pitch, phase and interference effects. He described the workings of the resonators and the analyser and devoted half of his report to graphical and flame devices: "Inasmuch as the laws which govern the production and the mutual influence of sounds are strictly mechanical, they admit of being demonstrated by methods which are not acoustic in the sense of being dependent on the sense of hearing; and accordingly the most striking illustrations of acoustic phenomena which have been recently devised are addressed rather to the eye than to the ear."[45] Graphical instruments allowed one to witness the path of a vibrating point; optical instruments (e.g. Lissajous mirrors) displayed the "visible image of the same path" on a screen; and a combination of optical and mechanical (e.g. gas flames) amplified "the varying conditions of a vibrating body of air by the effect produced by jets of flame."[46]

William B. Rogers, Alexander Graham Bell and MIT

Koenig's exhibit translated into a significant sale. The founder of the Massachusetts Institute of Technology, William B. Rogers, was in Paris as a representative for the Commonwealth of Massachusetts. He made several purchases for his school and laboratory.[47] By November, nine cases had arrived from Europe, with more to come, and special cabinets had been made to store the instruments.[48] The MIT calendar for 1868–1869 reported:

> Through the liberality of a friend of the institute, an extensive collection of acoustic apparatus has been purchased, including sets of organ pipes, tuning forks, resonators, a large Seebeck's sirene, phonautograph, and other instruments used in performing the more recent experiments in sound.[49]

Graphical acoustics and Helmholtz's instruments had now spread across the Atlantic and landed in a fertile technological and entrepreneurial market. A few years after the arrival of the instruments, Alexander Graham Bell, who had recently moved to Boston, wrote to his parents that there was a complete set of "Helmholtz'

Fig. 4.6 Alexander Graham Bell used this phonautograph pictured in the earliest instrument room at MIT. Photo c. 1867 (PH 533). Courtesy MIT Museum

apparatus" at the Institute and Prof. Monroe was going to "repeat Helmholtz' experiments with me shortly."[50] In the next few years, he performed several experiments with Koenig's phonautograph and manometric flames. He also delivered public lectures in Boston with these instruments and used them to test his ideas and developing projects. He had been working on visible speech and was particularly taken with Koenig's graphical and visual innovations, and described details of their workings in letters to his family (Fig. 4.6).[51]

The Rogers purchase was also part of a changing educational context. In the late 1860s, MIT led a reform of science education in America with the introduction of student laboratories. These developments produced significant increases in sales to America for French instrument makers. Following the Civil War and Reconstruction, America experienced economic and industrial growth. The Morrill Act of 1862, in particular, had provided the foundation for developing post-secondary education. MIT gained thirty percent of the Massachusetts share of funds and land, helping raise large sums of money in its first few years.[52] In 1868 and 1869, under the direction of William Rogers, professor or physics, Edward Pickering, and professor of chemistry, Charles Elliot, who became president of Harvard in 1869, MIT claimed to be the first school in the country to build teaching laboratories. The 1868–1869 calendar emphasized hands-on learning in the laboratory and excursions to foundries, mills, iron works, glass works, chemical works, textile mills and bleacheries. Borrowing social and practical elements from the world-wide phenomena of mechanic's institutes started in Britain, it also advertised free courses offered in collaboration with the Lowell Institute (an educational foundation in Boston that supported a public lecture series and classes). These were open to either sex and took place in the evening for people who had to work during the day.[53] They were often attended by female school teachers who wanted to get practical up-to-date knowledge of the sciences, particularly courses in algebra and trigonometry, geology, zoology, and chemistry.

These developments were known to the citizens of Boston. In 1869–1870, for example, Pickering gave ten lectures on sound on Monday and Wednesday evenings starting on December 13.[54] The instruments, of which Koenig's played a significant part, had become a foundation for an entire public education project. In 1869, *Frank Leslie's Illustrated Newspaper* carried a story about the newly opened building for MIT, describing the courses and the instruments in the physical cabinet. In acoustics, the article read, there were sirens, singing flames and the phonautograph. One illustration showed the phonautograph in the centre of one of the laboratories.[55] MIT and its French instruments had become a showcase for science in New England. There was also a small but significant instrument scene in Boston itself, with makers such as Charles Williams, Daniel Davis, N.B. Chamerblain, and E.S. Ritchie.

The Parisian Science Monopoly and a Portuguese Customer

In addition to the newer American schools, the French instruments appealed to older schools in Europe who were updating their physical cabinets and teaching programs. Another visitor at the 1867 Exposition was 30-year old associate professor (*substituto*) of experimental physics from Coimbra, Portugal, António dos Santos Viegas (1837–1914). He was travelling through Europe on a purchasing trip to visit workshops and scientific institutions and stayed in Paris during the Exposition. His largest purchases were at Koenig's and Ruhmkorff's workshops, totalling 5,000 fr at each establishment.[56] (A complete collection of Koenig apparatus at that time cost approximately 16,000 fr).[57]

These instruments became part of the "Gabinete de Fisica" at the University of Coimbra, which had a rich and long history.[58] It was one of the finest in Europe for a short period at the end of the eighteenth century, created during educational reforms in the 1770s. Struggles with the conservative Jesuits, who were eventually expelled from Portugal in 1759, had led to radical changes in education. At the university, professors created a reformed curriculum in natural philosophy, chemistry, and natural history. The physical instruments, including mechanics, optics, pneumatics, electrostatics, magnetism, heat, and hydrostatics, represented the most up-to-date science and craftsmanship from Lisbon, London and the continent. They were stored in tall, elegant cabinets of Brazilian wood in a renovated building that had previously been occupied by the Jesuits. After the turn of the eighteenth century, however, even with periodic purchases, the cabinet went through a period of decline.

By 1867, following the lead of other schools throughout Europe, Portuguese educators proposed ambitious plans for building modern student and research laboratories. The director of the physical cabinet, Jacinto António de Souza, chair (*lente Cathedratico*) of experimental philosophy,[59] sent Santos Viegas, his junior, to other European countries to buy instruments for updating courses. France was still the center of scientific teaching. Both teachers taught from Jules Jamin's *Cours de physique de l'École Polytechnique* using Parisian instruments.[60] In fact, most of the teachers in geology, mineralogy, chemistry, botany, zoology, and agriculture taught from French texts. Santos Viegas, therefore, went to France, the German territories and England, placing orders in each country.

In Paris, Viegas was drawn into a science monopoly, where lectures, instrument workshops, agents and book dealers worked seamlessly together. He attended the 1867 Exposition and public courses at the Sorbonne, Collège de France, l'École Polytecnique and the Conservatiore des arts et metiers. The French professors, he observed, did not perform any experiments during lectures, lest they interrupt their continuous oral presentation. Demonstrators (*les preparateurs*) and assistants did the experiments, and for more delicate operations, the actual instrument makers performed demonstrations. It was not unusual for makers to hover around lecture theatres. On more than one occasion, Santos Viegas saw Koenig, Ruhmkorff and Bianchi operate their own instruments.[61]

The 1868–1869 calendar for the University of Coimbra reported the instruments bought for the physical cabinet, "representing the latest progress of science."[62] In the German territories, Viegas purchased instruments from Sauerwald, Siemens & Halske, Steinheil and Geissler. In Paris he purchased from Breguet, Alverniet, and a large order from Ruhmkorff, especially a giant coil that had been exhibited at the Exposition. The description of the "new acoustical apparatus from Koenig" took up the largest section for a single maker. He described several items: a Helmholtz's double siren, phonautograph, manometric tubes with revolving mirror (for showing combinations of vibrations), sound analyser, Helmholtz's set of resonators and tuning forks for testing vowels, plates for showing Wheatstone's theory of vibrating plates, and Melde's apparatus for demonstrating the vibration of chords.[63]

Viegas used the instruments as part of a 30-year course based on Jamin's physics – mechanics, gravity, molecular forces, pneumatics, acoustics, heat and magnetism. An additional hand-written program added to the published course description included the study of vibrations and propagation, reflection, graphical traces, speed of sound, auditory thresholds, Lissajous figures and sounding bodies – pipes, strings, rods, plates, membranes, and Scott's phonautograph. He then covered the work of Helmholtz on combination effects, and lectured on timbre and the sound analyser (Fig. 4.7).[64]

In conclusion, this chapter focused on the 1860s, which were highly successful years for Rudolph Koenig. I described the instruments he offered, how he promoted them, and the nature of some of his customers. In the first year of his company he made good use of the demonstration skills he had learned from his master, Vuillaume. He also sold new products to keep up with developments in acoustics. In only a short period he participated in the 1862 Exhibition at London, promoting

Fig. 4.7 Apparatus to show the lengthening and shortening of a rod while vibrating longitudinally. CR 144. Photo by author, 2005. Museu de Física, University of Coimbra, Portugal. FIS.0393

his graphical instruments and making Helmholtz's instruments. Only 5 years later, the 1867 Exposition in Paris was one of the high points of his career from which he gained an international reputation and customers. Following 1867 Exposition, with money to invest in his business, laboratory and workshop, Koenig turned toward serious research. With relative commercial comfort and a vibrant precision workshop, he moved into a confident experimental phase. On a few years later, however, the Franco-Prussian war and the beginnings of a decline in the French precision instrument trade, dampened his business. In the next chapter, as various problems encircled his atelier, we see how Koenig defended his livelihood and turned acoustical experiment into a matter of pride, trust and livelihood.

Notes

1. The instruments have been restored in recent years, but the underlying structure is still in excellent condition. There is little evidence that points to heavy use.
2. For more on the history of science in Quebec, see Gingras (1991).
3. Carle (1986, pp. 139–171).
4. Rudolph Koenig to l'Abbé Hamel, Feb. 28, 1859. ASQ, 77, no. 92.
5. Brenni (2007).
6. Turner (1977) and Seebeck (1841). Idem., 1840.
7. List dated circa early 1959. ASQ, M763.
8. Holland (2000).
9. See specifically, artifact MCQ, acc. no. 1993.13295-300.
10. MCQ, acc. no. 1993.13264
11. It is entirely possible, like other makers in Paris, that Koenig was simply filling out his catalogue. He may not have even carried the Marloye instruments. The Latour siren went down in price from 100 to 80 fr, the differential sonometer went from 110 to 100 fr, the standard tuning fork for 256 Hz rose in cost from 20 to 25 fr, and the free reed pipe went from 20 to 25 fr, Marloye (1851, pp. 54, 50, 47, 43) and Koenig (1859, pp. 7, 19, 44, 27).
12. Koenig (1859, pp. 3–4).

13. Ibid.
14. Brenni (2007) describes similar paintings in optics.
15. Brooke (1863, pp. 32–33).
16. Ibid.
17. Her Majesty's Commissioners. *International Exhibition, 1862*, p. 195.
18. Timbs (1863, p. 134). The American press also reported the graphical innovations, *Scientific American*, August 22, 1863, p. 128.
19. Radau (1862b, p. 112).
20. Brooke (1863, p. 33).
21. The text quoted here is from a translation of Koenig's letter that was done for Henry: Rudolph Koenig to Joseph Henry, October 1865. SIA-JHP, *Incoming Correspondence, Record Unit 26, Box 8, Folder 15*.
22. The Seebeck siren changed from 130 to 800 fr, most likely representing a totally new instrument.
23. Koenig catalogue 1865. NMAH Trade Literature Collection.
24. Lecoeuvre (1867, p. 682). The catalogue of the exhibition reveals a wide-ranging commercial environment for machines, materials and skilled labor. In mining, for example, there were several iron and steel makers present, including a handful of specialized makers of cast steel. Ibid., pp. 465–472.
25. Wolff (1867, p. 167).
26. Five tuning forks and resonators for performing Helmholtz's vowel tests, a phonautograph for graphical studies, two manometric organ pipes, a manometric interference apparatus with two pipes, a new stethoscope with five tubes for medical students, and Auzoux's large paper-maché models of the ear and larynx. ASQ, Université, 84, no. 14.
27. See Henry to Joseph Swain, Dec. 9, 1865, SIA-JHP, Outgoing, vol. 2., 287. Facture from Bossange to Hamel. ASQ, Université, 84, no. 14.
28. Turner, G.L'E. (1996, p. 105).
29. Giberti (2002, pp. 7–13, 38–40).
30. Ibid., p. 11.
31. Deschanel (1867, p. 187).
32. Ibid.
33. Lissajous (1868, p. 480).
34. Ibid., p. 481.
35. Ibid. Lissajous was familiar with Koenig's workshop and had collaborated with him in the early 60s on graphical experiments, Koenig (1882c, pp. 24–25).
36. Lissajous (1868, p. 484).
37. Ibid., p. 480.
38. Ibid.
39. Ibid., p. 481.
40. Barnard (1870a, p. 505).
41. These instruments are now housed at the University of Mississippi Museum, Millington Barnard Collection. Chute (1978 and Fulton (1896).
42. Barnard (1870a, p. 499).
43. Ibid., p. 469.
44. Ibid., p. 470. Barnard visited Europe several times in the following years, see Fulton (1896, pp. 362–363).
45. Barnard (1870a, p. 500).
46. Ibid.
47. While in Paris Rogers received a letter from his treasurer informing him that ten donors had just contributed $5,000 each to a subscription. In addition, Nathaniel Thayer donated $25,000 for a chair in Physics. Mr. Endicott, Jr. to William B. Rogers, July 9, 1867, IAMIT, MC1, Box 4. Prescott (1954, pp. 63–65).

48. MIT Corporation. Records of Corporation Meeting. 1862-Series III. Minutes for meeting on November 23, 1867. IAMIT, AC 278.
49. Massachusetts Institute of Technology (1868, p. 23).
50. Alexander Graham Bell to Alexander Melville Bell, Apr. 9, 1871, *The Alexander Graham Bell Papers*, LC. Also see, Wylie (1975, pp. 8–10) and Prescott (1954, pp. 41–43, 61–65).
51. Alexander Graham Bell, Letters to Alexander Melville Bell, Eliza Symonds Bell, Carrie Bell, Charles J. Bell, April (unknown day) and May 6, 1874. *The Alexander Graham Bell Papers*, LC.
52. Wylie (1975, p. 6).
53. Massachusetts Institute of Technology (1868, pp. 29–30).
54. Massachusetts Institute of Technology (1870, p. 40).
55. Spofford (1869, pp. 228, 234–235).
56. Viegas, "Viagem Scientífica," 2974. Rodrigues, *Memoria Professorum*, vol. II, 291–292.
57. Rudolph Koenig to Joseph Henry, Nov. 20, 1866. SIA, Record Unit 26, Incoming Correspondence, Box 8, Folder 15.
58. University of Coimbra (1997), Martins (2001), and Melo (2002).
59. Rodrigues (1992, vol. II, pp. 288–289).
60. Universidade de Coimbra, pp. 168, 171. AUC.
61. Viegas (1867, pp. 2973–2974).
62. Annuario (1868–1869, pp. 159–160). AUC.
63. Ibid.
64. Viegas (1889, p. 10). AUC.

Chapter 5
Constructing a Reputation, 1866–1879

Hundreds of Koenig tuning forks exist around the world (Fig. 5.1). Koenig and/or an assistant tuned and polished each one through a combination of well-trained hands and ears. These activities represent an extremely time-consuming, labour-intensive approach to making and using *the* fundamental instrument of nineteenth-century acoustics. They also represent the values of Koenig's Parisian workshop – perfection, purity, precision, mastery, quality, concealment and dependability. In the 1860 s and 70 s the context of acoustical practice changed rapidly due to developments in Koenig's acoustical workshop. Acousticians, like practitioners in many fields, had long been gravitating towards automation and de-skilling.[1] In post- *Sensations* acoustics, Koenig encouraged these trends through dozens of graphical instruments and the clock fork (CR no. 32). Ironically, as controversies arose, these pressures seemed to place even higher value on Koenig's own listening and artisan abilities. His authority, and therefore entire branches of acoustical study, became dependant on these older skills and values. This chapter focuses on how these experiences and tensions shaped Koenig's experiments and instrument making. Following the

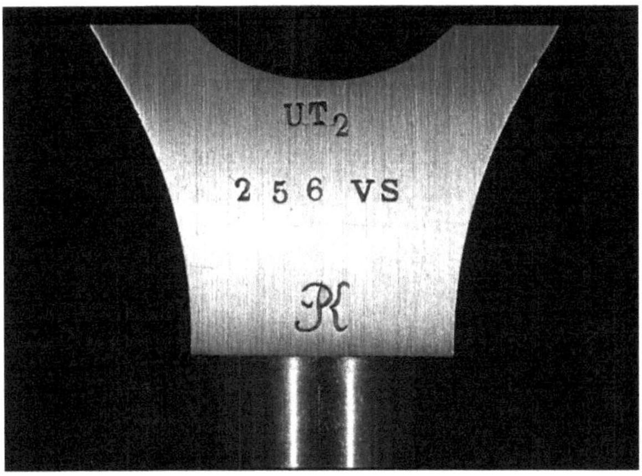

Fig. 5.1 Polished steel surface of a Koenig tuning fork, c. 1880 s. Physics Department, University of Toronto, Canada

D. Pantalony, *Altered Sensations*, Archimedes 24, DOI 10.1007/978-90-481-2816-7_5,

disasters of the Franco-Prussian war of 1870–1871 (Chapter 6), he started a series of experiments that challenged Helmholtz's basic findings. The war brought to the surface the different contexts of their work.

Measuring the Velocity of Sound in the Sewers of Paris

> *I remember Koenig pointing out to me the door by which he used nightly to enter the sewers to make his solitary way through the swarming rats to the place of experiment.*[2]
>
> *James Loudon (1901b)*

In 1866 Victor Regnault asked Koenig to join him as a collaborator in his velocity of sound experiments. Regnault, one of the renowned experimentalists of his age, had started his career in chemistry (for a short time he trained under Justus von Liebig at the University of Giessen) but by the 1840 s had moved to physics making important contributions in thermodynamics. In 1841 he succeeded Savart at the Collège de France. He was a meticulous experimentalist who obsessed over controlling every variable, especially eliminating possible sources of human error.[3] Examining his meteorological experiments, historian Matthias Dörries has shown how Regnault, within a mid-nineteenth-century culture obsessed with objectivity, struggled with "purification" and "expunging impurities."[4] With this perspective, he refined and invented several precision instruments related to the measure of heat. He also had a strong influence on scientists in France and abroad. William Thomson (Lord Kelvin), for example, studied under him during a visit to Paris in 1845.[5] From 1862 Regnault had begun carrying out a series of definitive experiments on sound that would bring all these practices and values into the world of acoustics, and Koenig's atelier.

The velocity of sound had been the subject of experimental trials long before Regnault.[6] In one early attempt, scientists observed the firing of a cannon a great distance away and marked the light flash, effectively almost instantaneous, against the time it took for the sound to reach them.[7] Of course, these sorts of observations relied on the experimenter's reaction time. Regnault was the first to make a serious attempt at removing the "personal equation" from these measurements. He did this with an innovative use of a timing apparatus and the underground sewer system of Paris. "Favourable conditions" for these experiments arrived in the form of Baron Haussmann's renovations of Paris.[8] The laying of gas pipes and water lines provided a perfect opportunity for Regnault to conduct controllable experiments using pipes varying in length from 70 to 4,900 m. He did experiments with a variety of pipes, different modes of producing sound, and differing media. One of his findings, for example, was that the velocity of sound increased in small pipes (11 cm in diameter). In larger pipes, over a metre in diameter, the value reached a limit that was the same as if the waves were traveling in open air.

Aside from establishing a new value for the velocity of sound at 0°C in open air (330.7 m/s), Regnault's memoir is a classic expression of "objective" acoustical research. Above all he wanted to automate these processes. In order to carry out time-measurements, for example, he invented an instrument using Koenig's

Fig. 5.2 Regnault chronograph. The frame is massive and sturdy so as to avoid any unwanted vibrations. CR 216
Source: Koenig (1889, p. 79)

tuning-fork chronograph in conjunction with a series of electric signals. "Regnault's chronograph" (as it came to be known after Koenig began making and selling it)[9] registered the reception of a sound pulse with an electrical signal at the beginning of the sound pulse and again at the end. The tuning-fork chronograph provided a graphical method for precisely measuring the time of travel. It had three rollers for recording and comparing the vibrations of the tuning fork, electrical signals marking the start and finish of the experiment and marks from a seconds-pendulum in order to calibrate the potential errors of the tuning fork. A break in the circuit registered the original report (a trumpet blast) and after travelling through a series of reflections in the pipes (to make distances of up to 20 km) the sound wave activated a membrane that broke the circuit. The chronograph, being connected to the circuit, recorded all of these events on blackened roller paper (Fig. 5.2).[10]

The chronograph represented a trend in acoustical experiment toward the removal of the human observer. However, Regnault could not totally escape the need for a good ear. Near the end of his experiments, he undertook to measure the velocity of sound produced by musical instruments. In particular, he wanted to know if different sounds of the musical scale had the same velocity of propagation. He soon discovered that musical sounds, being composed of a "series of isochronous undulations, whose intensities are unequal,"[11] could not be properly registered by the membranes that triggered the telegraphic circuit. It was often the case, to take one example, that a wave of maximum intensity would follow a succession of weaker waves, making it difficult to determine a single end-point of the signals. He tried to overcome this by making adjustments to the membranes and the circuit, and even using Helmholtz resonators to amplify the sound waves, but these efforts did not work. He was therefore forced to abandon his cherished automatic signal system and hire an "attentive observer" to mark the arrival of the waves. "M. Koenig, our skilful constructor of acoustical instruments eagerly accepted this mission."[12] The instrument maker himself became part of the apparatus.

In the grand pipe of the St. Michel sewer system, which was one metre in diameter and 1,589.5 m long, a trumpet sounded, causing a vibrating membrane to activate the telegraphic signal. At the other end, Koenig placed his ear to the board covering the pipe and broke the signal when he heard the sound arrive. He did several practice trials to gain skill and speed of reflex. Aside from establishing "definitive"[13] results about the differing velocities of high and low notes, Regnault concluded from this experiment that the timbre of a sound, being composed of several simple sounds, was not preserved and actually decomposed during propagation in a very long tube. In the end, Koenig had contributed both instruments and his skill to the series of experiments. It also provided a valuable lesson for future work on timbre about the challenges of experimenting with complex sound sources.

This creative interaction between scientist and instrument maker had a sad ending. Following the horrors of the Franco-Prussian war of 1870–1871 (Chapter 6), when Regnault's son was killed and his laboratory sacked, he removed himself from the scientific community and no longer enlisted Koenig on projects. This series of experiments, however, marked a sharp turn in experimental acoustics towards automation and reduced reliance on the judgement of experimenters.

Creating Vowels Sounds Out of Wood, Brass and Steel

Following his success at the 1867 Exposition, and his work with Regnault, Koenig turned his attention to another enduring problem in acoustics – the human voice. The instruments he created and the studies he undertook embodied the theories of Helmholtz, but also his own emphasis on automation and visualization. The result was the first large-scale study of vowels using an optical method.

In the late 1860 s a theory of vowel sounds emerged based on Helmholtz's analytic framework, which later came to be called the "fixed-pitch theory."[14] Since 1857 Helmholtz's Dutch colleague, Franz Donders, had been conducting extensive tests on vowels (partly with a Koenig phonautograph) and believed, along with Helmholtz, that each vowel sound was dominated by a certain upper partial or small group of upper partials in a fixed region of the musical scale.[15] In short, the cavity of the mouth was specially tuned for specific vowels. Therefore, no matter what note a person sang to produce a vowel sound, the mouth was shaped in a certain way and reinforced the same fixed region of harmonics giving the vowel its timbre or distinctive character. Donders tested this idea with a clever whispering technique. He shaped his mouth cavity in the form of a specific vowel and produced a "windrush"[16] that passed through the mouth and was transformed into a whistling sound. The vocal cords were closed and the windrush came partly from the contracted glottis and partly from the forward contracted passages of the mouth. Helmholtz described the whispering sound as something similar to that of an organ pipe with a defective lip. "A noise of this kind," he wrote, "although not brought up to being a complete musical tone, has nevertheless a tolerably determinate pitch, which can be estimated by a practised ear."[17] Using this method Donders produced seven frequency values for the fixed pitch of the vowels U (French OU), O, A, Ö, Ü, E, I. Helmholtz had also tested Ä and a German Ou.[18]

Koenig was not completely convinced by Helmholtz's and Donder's methods, and brought his own approach to these questions: he had the ability to make, improve, and invent instruments as his research progressed; he also focused on building instruments that would not rely on the judgment of a trained listener. Ironically, his efforts showed how much his methods depended on his own expertise, and how, consciously or not, he was shifting authority to his workshop practice.

On 25 April 1870, after more than 5 years of research and tinkering, Victor Regnault introduced the first series of Koenig's findings on a selected group of vowels at a session of the Academy of Sciences. At that time, Koenig claimed to have obtained his results earlier, but waited to verify them with "eminent physiologists, whose support encouraged me to publish them today."[19] He was particularly concerned, he told the audience, with obtaining exact figures for the *characteristic pitch* of a vowel and more than once delayed publication to seek further verification. At first he had used Helmholtz's tuning fork and resonator method for his experiments on a core set of vowels, but he soon encountered problems. He agreed with Helmholtz's findings for the characteristic pitch of vowels *A*(si4 flat or B5 flat), *O*(si3 flat or B4 flat), and *E*(si5 flat or B6 flat), but for *I* and the French *OU (or U)*, the higher and lower notes, respectively, he came up with different results. For the vowel *I* Helmholtz did not have a tuning fork of high enough pitch and had used Donders's whispering method to determine the characteristic tone to be Re6 (2,304 Hz or D7). Through his own experience, Koenig believed that even this tone was not high enough. By constructing forks of higher pitch he claimed that Si6 flat (3,584 Hz or B7 flat) was the characteristic pitch for the vowel *I*, quite a bit higher than Helmholtz's value.

There were other changes in instruments that embodied concerns about Helmholtz's methods. For the French vowel *OU (English and German U)*, Helmholtz had resorted to a method in which he sensed a peculiar tickling in his throat at the right note to find the value fa2 (F3). In an attempt to reduce Helmholtz's findings to a simpler rule, Koenig believed that each of the vowels seemed to follow a law of jumping one octave in the note of Si flat (B flat), so he hypothesized that *OU* would be characterized by the note si2 flat (B3 flat). "I verified this hypothesis in a meticulous manner with the aid of a tuning fork whose pitch could be varied with sliders [*des curseurs*]."[20] These brass sliders, an invention of Koenig, would become a standard feature of tuning forks. With these results he put forward a law of vowels that conveniently separated the characteristic tones by an octave each.

Fig. 5.3 Resonators and tuning fork for vowel experiments. CR 57. Photo by author, 2005. Physics Department, University of Toronto, Canada

He hoped that it would be possible to discover a physiological cause underlying these simple relationships. Just as one finds the same musical intervals in the music of many peoples, he argued, one finds the same five vowels in different languages (Fig. 5.3).[21]

These studies eventually became a commercial product in Koenig's catalogue and sold throughout the world – the vowel apparatus with five forks and resonators. The first apparatus, which appeared in the 1865 catalog, was modeled after Helmholtz's figures for the vowels (*OU*, fa2 (F3), 175 Hz; *O*, si3 flat (B4 flat), 466 Hz; *A*, si4 flat (B5 flat), 932 Hz; *E*, si5 (B6), 1,976 Hz; and *I*, re6 (D7), 2,349 Hz). As a result of his research in the late 1860 s, Koenig changed these figures to 224,448,896, 1,792, and 3,584 Hz. The hypothetical figures in 1870 were 225 Hz (*OU*), 450 Hz (*O*), 900 Hz (*A*), 1,800 Hz (*E*), and 3,600 Hz (*I*), but Koenig modified his instruments to fit the physicist's scale based on 256 Hz.[22]

Seeing a Voice: Manometric Vowel Studies

While working on the tuning fork and resonator apparatus, Koenig began an even more ambitious attempt to create a visual record for each vowel sung at several different notes. He employed a funnel-shaped speaking tube, a single manometric capsule, and a rotating mirror to record the distinctive character of each vowel in a wide range of pitch conditions. Originally, he wanted to show the "great difference in the appearance of the sound of the five vowels, sung on the same note, as well as to show the manner of the change of the flame-pictures of the same vowel from one note to another."[23] But with some clever control experiments, he discovered a way to measure and assign frequencies to the patterns. He thus produced a quantitative confirmation of his earlier work on vowels and a visual reference, or guide to the vowel spectrum. He made many of these pictures as early as 1867 and first exhibited

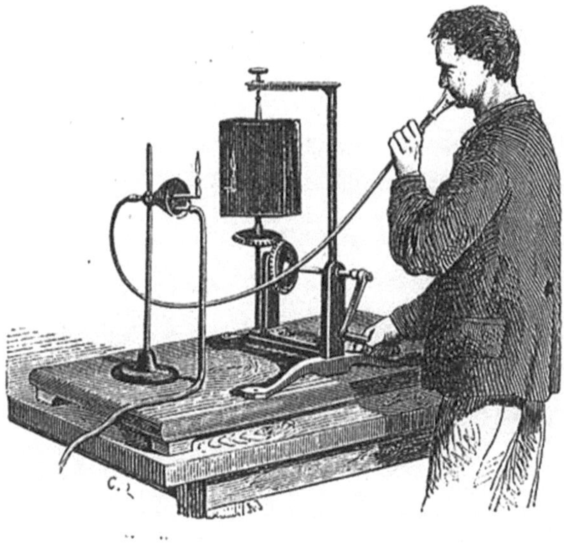

Fig. 5.4 Manometric capsule, funnel and rotating mirror for displaying vowel sounds. Koenig and an artist recorded/drew the sounds on paper
Source: Radau (1870, p. 253)

large drawings of them at the meeting of the Association of Natural Philosophers at Dresden in 1868 (Figs. 5.4 and 5.5).[24]

The final series of images, in fact, emerged from a painstaking series of baseline experiments. He used the drawings of standard combinations to compare with drawings of his vowel sounds in order to analyze the elements of the complex vowel. First, he recorded and studied the manometric effects of organ pipes of known frequency and used them as standards. Second, he recorded and studied combinations of these tones with an apparatus consisting of two manometric organ pipes. From these combinations he was able to record distinctive patterns for specific combinations (musical intervals such as an octave and its lower fundamental, or a third harmonic combined with a fundamental). Third, he used these patterns to compare with recordings of the actual voice or sounds emitted by a violin. For the latter, he placed a rubber hose into the f-hole, connected a stethoscope to the soundboard, and recorded the patterns.

The full process was time consuming. He sang each vowel in 15 different notes ranging from ut1 (64 Hz or C2) to ut3 (256 Hz or C4). "While I sang into the apparatus, an artist drew the picture which he saw in the mirror. I also drew the same picture independently: and if both our drawings were identical they were looked upon as correct; if, however, there were discrepancies, I repeated the experiment until the error was discovered."[25] Once he had recorded a set of drawings for each vowel, he studied each drawing to dissect the combinations. The objective was to find the outlying note, or characteristic pitch. For example, he sang a vowel at a particular fundamental pitch and observed that the resultant pattern was similar to

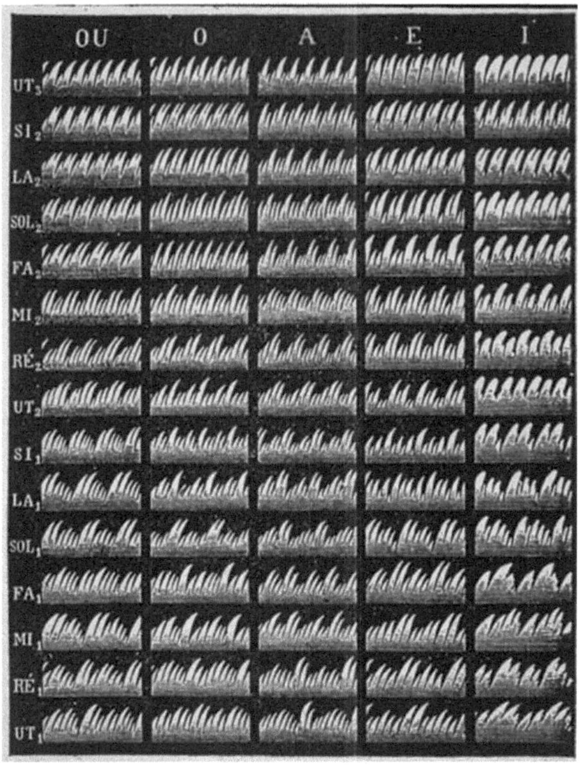

Fig. 5.5 Vowel sounds sung in two octaves of notes
Source: Koenig (1882c, p. 63)

the pattern he had obtained with a fundamental combined with a third harmonic. By obtaining this pattern, he knew that the vowel must consist of the pitch he voiced plus the characteristic tone, which, in this case, was a third harmonic of that tone. After going through the whole range of tones, he determined the location of the characteristic frequency for each vowel. In this way, he came up with a comprehensive table of characteristic pitch figures for each vowel in every possible range.

The drawings had the appearance of being direct, even automatic pictures or snapshots of vowel sounds in action. However, they were events highly mediated by specific human skills, choices and technologies. There was the constant positioning and readjustment of the mouth, efforts to maintain the vowels at one steady pitch, meticulous observation and drawing of the wave patterns, exclusion of visual details not deemed relevant, adjustment of the membrane and gas pressures, and the rotation of the mirror. The pictures also masked the time and effort that went into making them. In fact, they took more than 5 years to complete the whole set of values for publication.

> I delayed their publication until now because I wished to revise them with precision, but was always prevented by the delicate state of my throat, which did not permit me such fatiguing experiments. But now, since I can no longer hope to recover, I have used my best endeavors to make the pictures correct, and give them forth, not indeed as absolutely perfect, but as nearly so as it was possible for me to make them.[26]

Today, sound recordings and spectrograms are common. They are viewed uncritically as a snapshot of sound as it is. Koenig's drawings, the first systemic visual representations of vowel sounds, show the very human, contingent origins of these kinds of representations, with all their inherent culture, choices and limitations. Being a novelty at the time, Koenig made large paintings of these drawings for fairs and lectures and displayed a complete set of them at the International Exhibition of London in 1872. They were copied in several popular textbooks in the late nineteenth century and became a visual icon for vowel studies.[27]

Extending the Tonometer, One File Mark at a Time

In the studies conducted for his treatise on sound, Helmholtz had significantly expanded the range and precision of experimentally produced tones. In the 1870 s, Koenig took this to a new level with the completion of his *grand tonomètre*, first displayed at the 1876 Philadelphia exposition. It consisted of 692 forks, with over 800 tones represented (some of the lower forks had sliders for producing different frequencies), ranging from 16 to 4,096 Hz, making Helmholtz's elements of sound a reality in many different shades of pitch. Such a precision tonometer, it was hoped, would reduce the dependence on skilled listening in musical and experimental practice. The first version of this extended tonometer appeared at the 1867 exhibition in Paris. There Koenig displayed an apparatus with 330 forks which ranged from 16 to 2,048 Hz, with differences between the forks ranging from 0.25 cps to 6 Hz. In 1867, steel rods were used to extend this range up to 32,768 Hz.[28] Afterwards he extended this instrument further to make the *grand tonomètre* for the 1876 exhibition.

The actual process of making these forks served as a foundation for his later experiments and subsequent insights regarding combination and beat tones and his focus on the integrity of instruments. He came to know every possible aspect of his creations from choosing the right steel, making the rough blanks, doing endless comparisons in all ranges, using beats and Lissajous figures, experimenting with minute temperature effects, experimenting with different shapes and fork designs, polishing the surface to look for cracks, and finally fine-tuning by hand with a file (see below). Due to the significance of his tonometer and its place this formative period of modern acoustics, it is important to probe deeper into how the forks were actually made.

Koenig did not record his manufacturing process but there are a few written accounts from the time. In his 1907 book *Die Stimmgabel* Ernst Kielhauser stated that forging was not the preferred method because it would potentially damage the internal structure of the fork and cause distortions in its vibrations. But Kielhauser was most likely referring to a specific form of forging as other accounts contradicted this view. Levi K. Fuller of the Estey Organ Company in Vermont stated in 1892 that "Heretofore fine tuning-forks designed to give absolutely-correct pitch have been invariably made by forging them from a single bar of steel."[29] E.G. Richardson wrote in 1927 that "the size of the fork is determined empirically by the maker, the final tuning adjustment being done by shaving the ends of the prongs."[30] Josef Stefan, professor of physics at the University of Vienna, who was familiar with Koenig's instruments through his colleague Joseph Pisko stated that the forks should sound for a long time after being struck, their stability should result from a balance of mass, form, and elasticity, and they should be made of a material that exhibits both hardness and elasticity. The maker, Stefan argued, needed to insure that the prongs were the same length, width and thickness, they also needed to be perfectly parallel and further apart than was common with earlier forks.[31] One can see from these guidelines that by the late nineteenth-century, tuning forks still relied heavily on artisanal choices and skill. When Koenig was working on his complete universal tonometer in the 1890 s, he wrote to James Loudon of Toronto that he supervised the "purely mechanical work" while doing the fine tuning himself.[32] This claim, although used to assure clients of the care he put into his instruments, was most likely true for Koenig who seemed to trust only his own ear for the final product and whose forks were used extensively by researchers.[33] Even more revealing, one does not find any evidence of shaving or fine tuning on the tuning forks of Max Kohl of Chemnitz, Germany, the firm that became the leading seller of acoustical equipment after Koenig's death. Kohl produced their products in a factory-type setting and institutions used them mostly for classrooms.

It is clear that although guided in general by theoretical knowledge of the fork and its variables, Koenig's activities were highly empirical. D.C. Miller, who visited Koenig's shop, instructed those interested in making forks that, after being machined and finished, the standard fork should be left a "trifle" long, in order to file it down later. He also warned of the heat caused by *any* filing activity and how the maker must wait for the fork to cool in order to proceed, because even small changes in temperature caused changes in pitch. It was Koenig who had performed the fundamental research on the relations between temperature changes and pitch in tuning forks.[34] Kielhauser also described the elaborate efforts to ensure that the prongs were of the same mass. If one prong was longer the efficiency of the oscillating system (duration of vibrations) would be reduced. If a portion of wax added to one prong increased the vibration time, then that prong had too little mass compared to the other. Some evidence for this kind of activity in Koenig's shop can be found on the ends of prongs where one sees beveling of different amounts on the four corners at the top edges. A further consideration of the forks themselves, therefore, would help to clear up some of the mystery concerning the art and science behind precision forks.

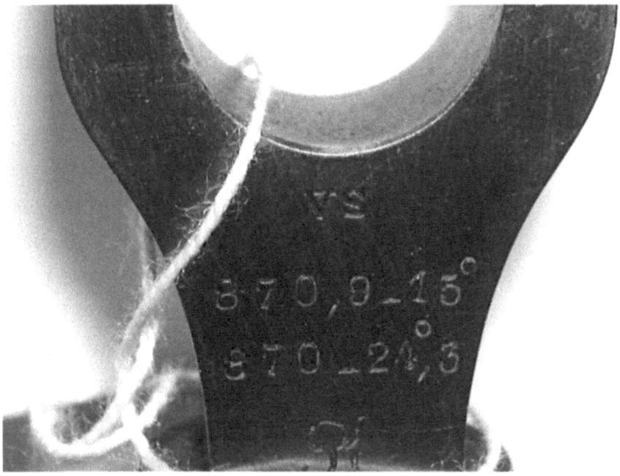

Fig. 5.6 Koenig temperature-adjusted standard fork, la3 (435 Hz or A4). Slight filing at the front edge of the yoke (which lowered the pitch) reveals the fine tuning process. Some forks have a mark three times this size, others have nothing. This one has filing on both sides of the yoke. To raise the pitch, Koenig filed at the top of the prongs. Photo courtesy of the National Museum of American History, Smithsonian Institution, Washington, DC, acc. no. 1989.0306.192. Photo by Steven Turner

Choosing the Right Steel

Forensic historic investigation of artifacts, in addition to consulting written sources, is one of the best means for prying open lost issues related to workshop practice. Archeo-metallurgists such as Dorothy Hosler, Robert Gordon and Martha Goodway have discovered that samples of historic metal can take investigators deep into the conditions and choices of the makers and users of various products.[35] In the case of Koenig, thousands of his forks in museum collections throughout Europe and North America provide evidence of the construction enterprise in his Parisian workshop.

First, what can we find at the surface level? On most forks there are still signs of the delicate fine-tuning procedure through file marks on various parts of the fork (filing on the inside of the yoke minutely lengthens the prongs and thus lowers the pitch; filing on the top of the prongs shortens the prongs and thus raises the pitch slightly). Even on a set of 13 small chrome-coated forks used for musical tuning, one can see the file mark on the back of the yoke. Koenig or his workers had clearly fine-tuned the forks after plating them presumably for both protection and aesthetics (CR no. 44) (Fig. 5.6).

Examination of the tonometer forks has revealed an extensive, three-part manufacturing process: making a rough blank that was approximately the same size for a fairly large range of frequencies (e.g. 100 Hz), rough tuning by changing the length of the prongs, and then fine-tuning by hand using beats (CR no. 36). The walnut rack, with the organization of 677 placements for the forks, also provides interesting evidence in itself of the construction process.[36] It is divided into 4 distinct groups: rows 1–3 have 22 spaces per row, rows 4–6 have 29 spaces per row, rows 7–11 have 35 spaces per row, while rows 12–18 have 50 spaces per row. It appears that the forks in each group came from the same-sized blank and were each fine tuned according to their neighbours (using beats). Wide-ranging measurements were made of individual forks (length of prongs, width of both prongs, width of space inside prongs) and indeed the general proportions are very similar within each group, with significant differences between each group. The one variable that consistently changes in one direction within each group, by very small amounts, is the length of the prongs. As one would expect, as the prongs become minutely smaller, the pitch becomes higher (shorter prongs result in more rapid oscillation). Rough file markings across the plane of the top of the prongs reveal how Koenig or his workers shortened the prongs to obtain a rough estimate of the desired pitch. But not all successive forks were shorter, showing that it was not the only variable involved in the tuning of the final pitch (other key variables included the thickness of prongs, the equal mass of prongs, the distance prongs were apart, and the kind of steel). Furthermore, the forks within each group are slightly different in overall shape, making it difficult to tune them only by shortening. The inside base of the prongs shows small amounts of filing which lengthens the prongs, slightly lowering the pitch. Finally, the width of the prongs near the base is sometimes thinner than at the middle or top. This filing would have weakened the base creating longer vibrations and lowering the pitch. Koenig's fork, therefore, took on a distinctive shape and style due to the tuning methods. This style came to be emulated in places such as the Physikalisch-technische Reichsanstalt, the national institute for standards and metrology in Germany. The director of the technical section overseeing the production tuning forks, Leopold Loewenherz, recommended Koenig's tuning forks as standards.[37]

The microstructure of the steel of a Koenig tuning fork from this period take sus further into this process, and shows that Koenig was as selective with steel as he was with wood, and that his choices reveal clear priorities in the manufacturing process. In fact, the sample studied (from an 1878 tonometer at the University of Toronto), showed that the final sound was as much a product of ideal vibration theory as it was about workshop considerations, culture and available resources. The steel, for example, is surprisingly soft measuring an average Vickers Hardness result of 133 w/25 g, which is far below high quality tool steel (HV 400 and up). With this choice Koenig balanced the length of efficiency of vibrations (revolving around what today we call the Q-factor), with the ability to file and work with the metal. Judging by the microstructure, the full process probably involved selecting a bar stock, forging or cold working it into rough shape and then applying a heat treatment while annealing and slow cooling (not quenching) over a long period of time (so as not to cause any cracks). The slow cooling seems to have been the key to this process, making it fairly hard, but still soft enough with which to work. The fork is about 0.55% annealed carbon steel (hypoeutectoid) which fits this manufacturing process (CR no. 37).[38] The surprisingly soft steel also reveals an emphasis put on regulating or modifying the purity of the instrument. Even the purest tuning forks display weak, yet complex sets of partials that depend on the quality and hardness of the metal. By choosing softer steel, Koenig was not only choosing ease of filing, but a specific harmonic structure and quality of tone (Fig. 5.7).

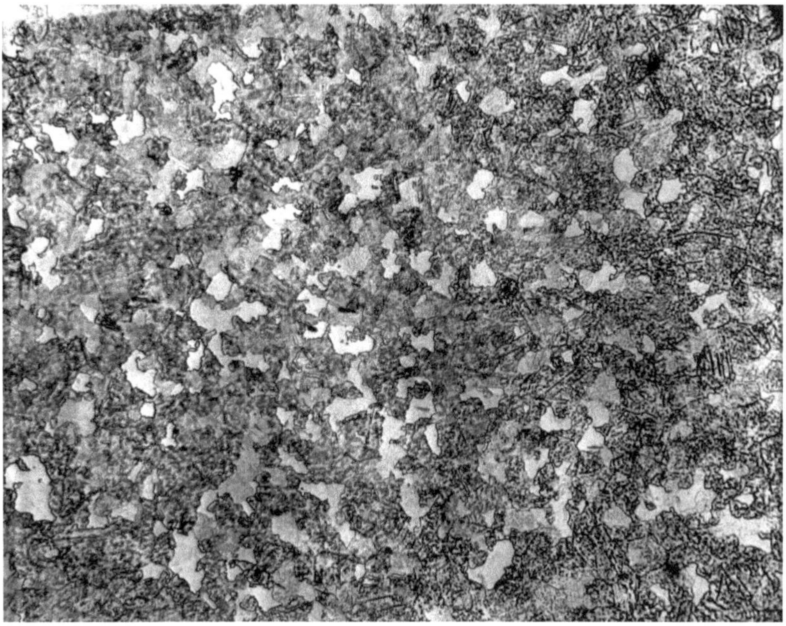

Fig. 5.7 Microstructure-analysis of the surface steel of a Koenig fork (magnification = 135), 0.55% annealed carbon steel (hypoeutectoid). (UT3 512 v.s. from U of T tonometer, dated 1878, CR 37). Photo by Yinlin Xie, Olympus optical microscope, Department of Material Science and Engineering, MIT, USA

The thorough polish and finish was another trait of Koenig forks and was not just cosmetic. Kielhauser stated that good polish was essential not only for preserving pitch (even a little oil from a finger print could throw off the pitch), but for checking the object for cracks, which could potentially dampen vibrations of the fork.[39] Such a statement, combined with examination of extant artifacts, reinforces the idea that Koenig's forks were largely intended for serious use in the laboratory, or as standards of pitch. This evidence is echoed by Miller who claimed that "a tuning fork for scientific purposes should be made of one piece of cast steel, not hardened."[40] Yet even with high-quality steel, one could not assume that it would behave similarly in standard dimensions; Kielhauser cautioned that each fork should be treated on an individual basis without assuming homogeneity of material. In fact, he wrote, the high cost of Koenig's forks derived from the fact that he made them "individually by hand."[41] This underscores the point that no matter how high the quality of steel, the maker still had to intervene through his own tests.

It is not known where Koenig obtained his steel. Varieties of steel were used in Paris by piano- and string-makers, armaments, tools and utensils.[42] Steel also came from specialty markets such as Sheffield in England, where makers bought high quality steel for blades, piano wire, cutlery and surgical instruments.[43] In fact, an example of one of Koenig's largest forks, no. 48 in the 1889 catalogue, is marked, "BEST WARRAN[T]E[D] CAST STEEL SHEFFIELD." (CR no. 48). This implied that, at least for large precision forks, Koenig had blanks cast in Sheffield to be tuned in Paris.

Variations in shape and design over his 40-year career also suggest his attempts to grapple with precision, loudness and purity.[44] Whereas Helmholtz had chosen to amplify specific frequencies with pasteboard resonators, Koenig put more emphasis on working with the sound source itself. He made the sound source a source of intense scrutiny.

Bringing the Workshop into Combination-Tone Studies

Between 1866 and 1876, in the context of the construction of the *grand tonomètre*, and his work on graphical and optical acoustics, Koenig began an extensive series of experiments to study the combination-tone effect. He arrived at different results from those of Helmholtz that forced him to go against the person who had propelled acoustics into a new era for study and teaching. One powerful source of Koenig's different perspective came from his evolving reliance on graphical methods for the study of sound and combination effects. Indeed, historian Robert Silverman has suggested that Koenig's pictorial approach shaped his conception of sound, while Helmholtz adopted a more "analytic" (mathematical) approach.[45] As we will see below, there were other factors that caused the two scientists to oppose each other over combination tones.

In his paper submitted in December 1875, Koenig presented an array of results that contradicted those of Helmholtz.[46] He later claimed that he had originally set out to confirm Helmholtz's findings and theory.[47] Instead, he found observed effects that agreed with Young's older beat theory, where combination effects were viewed as beats or beats that blended into a tone. This was a significant claim and in private Koenig went further, suggesting that he was overthrowing Helmholtz's theory.[48]

From his initial experiments, Koenig developed rules for the appearance of beats and beat tones. He discovered simple patterns of observations that at times agreed with the findings of Helmholtz and previous investigators, and sometimes were quite different. In his studies of "primary beats" he activated separate tones of 74 and 40 Hz simultaneously and discovered the appearance of 34 beats per second, he called this series the *inferior beat*. He also claimed, however, that such a combination produced 6 beats per second, which came from the "negative remainder," from the fact that 40 times 2 equals 80, and 80 minus 74 leaves 6, a series he called the *superior beat*. He then claimed to confirm these rules in a wider range of notes. By keeping one note low and extending the other one through several octaves, he discovered levels of quickening and slowing of beats within different periods. In these situations he found that the primary beats always appeared within one octave of each other, so that, for example, the two tones 100 and 512 Hz created two series of beats, 12 and 88; 100 Hz goes into 512 five times with 12 left over (inferior beat); and 100 times 6 equals 600, and 600 minus 512 leaves 88 (superior beat). Not all of these beats were audible, and he discovered a further rule to describe which set of beats could be heard.[49]

The crucial and controversial step came when Koenig extended these rules to cover what he called "beat tones," which at times occurred in the same location as Helmholtz's combination tones. This is where Koenig applied Young's suggestion that beat tones were simply a rapid succession of beats that blended into a tone.[50] For example, Koenig discovered a beat tone of 256 Hz when he struck two forks of 2,048 and 2,304 Hz. As with the lower-frequency forks, there were two sets of beat tones, inferior and superior. In effect, these beat tones were a substitute for the system of combination tones conceived by Helmholtz. In the first period, the inferior beat (the mathematical difference of the two tones), was identical to what was then termed a "difference tone." The beat tones were also capable of creating "secondary" beats when they interacted with other tones, similar to the way that Helmholtz's combination tones could beat with themselves and harmonics. The two systems, however, predicted different tones in certain cases. Helmholtz predicted and observed summation tones (the mathematical sum of the two prime tones) and several higher-order combination tones deriving from his combination tones; Koenig predicted what he called "superior beat tones" and many different tones from what he termed the "higher periods" of beat phenomena.

Most combination and beat effects were in fact quite difficult to replicate objectively, leaving their existence and explanation up for debate. Beats especially were seen to have a problematic physical status (CR nos. 205-207).[51] This is where the terms subjective and objective enter the discussion. Did combination effects derive from an objective effect, or were they merely products of the mind? Helmholtz, arguing against the older beat theory in 1863, went to great lengths to demonstrate that combination-tone phenomena were true simple tones, not just interference effects, and had an objective existence,[52] meaning that they could be detected independent of the listener, using resonators or tuned membranes.[53] They could also be created in the external parts of the ear (tympanic membrane, hammer and anvil), hence within the listener; but these, he stated, were still objective tones, with their own physical existence that were created by asymmetries in the movements of the external parts of the ear. He felt his detection experiments had shown this to be true, and his non-linear mathematical model provided a plausible description of combination behaviour.

Koenig argued that Helmholtz's combination tones did not exist. In fact, his view was that with proper equipment and experiments it was impossible to prove the existence of combination tones, especially summation tones, which were central to Helmholtz's argument against Young's beat theory. Very rarely, Koenig claimed, could Helmholtz's summation tones be detected using objective methods, and these were either extremely weak effects (sometimes objective effects in the ear itself) or "beat tones" from unwanted harmonics of impure instruments.[54] He supported the beat-tone theory, even though they too did not produce objective effects. In doing this he could not offer an alternative mechanism to Helmholtz's theory. His view therefore came to be seen as "subjective," even though he did not use this term himself, nor delve into mind and matter debates.[55]

Koenig presented himself as someone simply presenting new facts and bringing into question Helmholtz's theory and findings. He did this mainly by attacking

Helmholtz's claims about "objectivity" and bringing into question the purity and precision of his instruments. He moved the debate away from physical, physiological and psychological theory, where he was not comfortable, believing that one could eventually solve these questions and anomalies with better instruments and clearer demonstrations. The instruments used by Helmholtz, he argued, contained too many partial tones or harmonics, thereby creating false combination tones. These tones mixed with the fundamental to create unwanted auditory artifacts, or what Helmholtz thought to be objective, combination tones. In the first paragraph of his 1876 paper Koenig stated that he had been careful to select sources of sound that only produced the purest possible notes. Later in the article, he emphasized again that if one wanted to be certain of dealing with combination tones produced from simple primary tones, one must "set aside both the many-voiced siren [*la sirène polyphone*] and the reed-pipes" and only make use of tuning forks.[56] Even with these precautions, he argued, unwanted harmonics could develop by uncontrollable sympathies (CR no. 159).

Furthermore, according to Koenig, Helmholtz's theory was unnecessarily complex and too far removed from what was really being observed. He gave an example of two notes (ut1 and ut4; or C2 and C5) that produced a beat of 2 Hz. According to Helmholtz's framework, such an outcome resulted from a succession of combination tones which, in Koenig's view, could not even be heard. How, he asked, could they produce a beat at the end of such a complex chain of weak or non-existent effects? "It is far more simple," he concluded, "to presume that the beats... are produced directly from the formation of sound waves."[57] In other words, he supported his beat explanation with observations based on pure, precise instruments that matched his intricate diagrams of beat phenomena. He did not offer an explanation of the mechanism through which these effects came to be heard, but only that they were clearly heard in experiments and seen in graphical form. The diagrams became the explanation. When two forks recorded their vibrations simultaneously on paper, one could see the "beat frequency" in the wave diagrams. For Koenig these beats and beat tones were somehow perceived by the observer (Fig. 5.8).

Whereas Helmholtz had provided a theoretical and experimental basis for objectively establishing combination tones, Koenig offered a series of observations and graphical demonstrations that raised more questions and stoked even more fundamental uncertainty about the nature of sensations and perception. Koenig's workshop and instruments played a key role guiding this debate and revealing significant fissures in the fundamentals of late nineteenth-century acoustics and psychophysics (Chapter 7).

Above all, the 1876 paper was remarkable for its rhetorical use of novel, beautiful and intricate graphical displays, combined with an extensive exposition of cutting-edge instruments. Koenig presented an armada of 56 tuning forks made just for this experiment ranging from 64 to 4,096 Hz (almost the whole range of the piano), many of them being impressively large. The prongs of the lowest forks, for example, were 75 cm long. Precision markings on the sides of the prongs showed where the sliding weights were to be placed for specific frequencies. Previously, people working with tuning forks used pellets of wax to change the frequencies. Massive adjustable,

Fig. 5.8 Graphical diagrams of beat effects
Source: Koenig (1882c, p. 97)

cylindrical brass resonators, over one meter in length and 30 cm in diameter, amplified the weaker low notes. The larger forks stood in heavy cast iron stands and the lowest five forks alone, without their stands and sliding weights, weighed almost 130 kg. The higher-pitch forks (512–2,048 Hz) had stout bottoms, which tapered near the tops of the prongs. This shape, designed to reduce unintended harmonics, later became standard for many of his forks. Aside from considerations of purity and precision, these forks were clearly meant to produce a powerful and convincing lecture-hall demonstration. They were brought to Toronto in 1882 specifically for a series of public lectures at the Canadian Institute (Fig. 6.8)[58]

In his effort to purify beat-tone experiments further, Koenig also designed a tuning-fork instrument to mimic the convenient variability of pitch of the double siren. He invented "tuning forks of variable pitch" to conduct beat and combination-tone experiments with pure, simple tones, while having the capability to make easy, fine-tuned adjustments. It consisted of two massive tuning forks (128 Hz) set in a cast-iron stand, each placed before large brass resonators. An electromagnet drove the forks, and the prongs of one of the forks were hollow and filled with mercury, the level of which could be adjusted "at will [*à volonté*]"[59] by a screw to vary the pitch. Each fork had as well two simple weight mechanisms, which altered the phase of the vibrating forks. Lissajous mirrors on the tops of the prongs permitted precise frequency and phase adjustments to be made. Koenig first showed this apparatus "to several scientists [*à divers savants*]" in 1874.[60] McGill University has a well preserved example of this apparatus made by Koenig. It probably went to Canada in 1882 when Koenig visited Montreal (CR no. 189).

Precision and Livelihood Under Attack: The Koenig Clock Fork

Disputes for Koenig were ultimately about instruments and not necessarily about theory. He treated his instruments like works of art. The beauty was not just in their appearance, but in the way they reflected years of disciplined labour and the unrelenting thoroughness of his experimental routines. He therefore took any criticism of them very seriously. During the period of his intensive work on the combination-tone dispute, Koenig's whole life and identity became intertwined with tuning forks. He had spent a number of years purifying his forks, and he had spent even more time creating his masterpiece, the tonometer of 670 forks. He had contributed more than anyone else to transforming the wide tuning fork into a reliable instrument for scientists.

Precision and accuracy in all fields had a moral dimension at this time. Graeme Gooday has shown how Victorian values entered electrical measurement and instruments in the late nineteenth century.[61] As distributed skills and technologies increasingly contributed to the making and using of new instruments, trustworthiness became an important factor that influenced debates and practices, especially for standards on which many people depended. In addition, as instruments became more automated (e.g. direct reading meters) issues of work ethic and self-reliance surfaced. Good students, for example, should first learn to measure from first principles and rely on themselves more than their instruments.[62] In American physics classrooms as well,[63] laboratory exercises taught students the value of observing and measuring "to train the mind in right modes of thought by constantly bringing it into contact with absolute truth."[64]

It therefore came as a great personal insult in 1877 when Alexander Ellis questioned the reliability of the pitch number for Koenig's standard fork.[65] Ellis had come across his result while undertaking an intensive study on the standardization of pitch.[66] For most of his early tests, he used a tonometer of reed pipes made by Georg Appunn, an instrument maker in Hanau. In an article on standard pitch in *Nature* Ellis announced that he used Appunn's tonometer to measure Koenig's standard fork of la3 (435 Hz; A4) which produced a figure of 439 Hz.[67]

Koenig responded with the full force of his workshop; he invented a novel means for tuning and calibrating tuning forks (clock fork), and wrote a letter to *Nature* defending his forks.[68] He charged that Ellis had attacked the exactitude of the French tuning fork with "too great a haste," too easily ignoring the work of others – Lissajous, Helmholtz, Despretz and Mayer – who had confirmed its accuracy. He quoted an earlier letter from Helmholtz to Appunn in which Helmholtz had praised the reed tonometer of Appunn, and using another example of the same instrument, specifically found the French standard fork to be 435.01 Hz.[69]

Ellis realized that he had erred. After further investigation, he discovered that there had been a constant drift in the frequencies of his reed pipes and in 1880 he retracted his claim. "I feel I owe an apology to Herr Koenig, for my having been unfortunately misled by the unknown error of Appunn's instrument to attribute that error to him, and I make this apology most sincerely, for no one deserves more thanks from acousticians than Herr Koenig, both for the excellence of his worksmanship, and the ingenuity of his contrivances."[70]

The affair showed the authority that Koenig commanded at the time. It also showed how seriously and personally he took criticism. The apology, when it came, was too late. In the midst of his work on beat tones and the role of phase in timbre, Koenig set time aside to make his tuning forks of a precision that was beyond criticism. By 1879, after 20 years of experience making tuning forks, he started a series of experiments that would lead to original findings on the properties of tuning forks, the invention of the most precise instrument to date for determining pitch (clock fork), and the creation of a new international standard tuning fork (the first since Lissajous's in 1859). Such developments had a wide impact on music, acoustics and other areas of physics. For the next 50 years, the tuning fork became the standard carrier of frequency for music, electrical studies and timing apparatus.[71]

In the summer of 1879, in the wake of his disagreement with Alexander Ellis regarding the precision of his standard forks, Koenig started experimenting with a new instrument for determining pitch. He got the idea for this clock-like instrument (clock fork), in which the seconds are produced by vibrations of a tuning fork, from Niaudet, who had presented his invention to the Academy of Sciences on 10 December in 1866, and subsequently displayed it at the Paris and Vienna Exhibitions (1867 and 1873). Koenig, however, was not interested in making a precision clock. He wanted to use it as a comparison tool, along with an actual chronometer, for counting the number of vibrations of a tuning fork (Fig. 5.9).

Testing the true vibrations of an unknown fork involved a comparison between a chronometer and the tuning-fork chronometer. For the latter, Koenig attached the tuning fork of unknown frequency to the escapement of a clock that moved 1/60th of a division (1 s) on the clock dial for every 128 vibrations. The number of hours, minutes and seconds would then be translated into vibrations by multiplying the total (in seconds) by 128. One hour on the dial of the clock fork would be the equivalent of 460,800 vibrations (3,600 by 128). A reading of 1 h and 28 s compared with 1 h on the actual chronometer, however, would mean that a faster fork had been employed, producing more vibrations. In such a situation there would have been

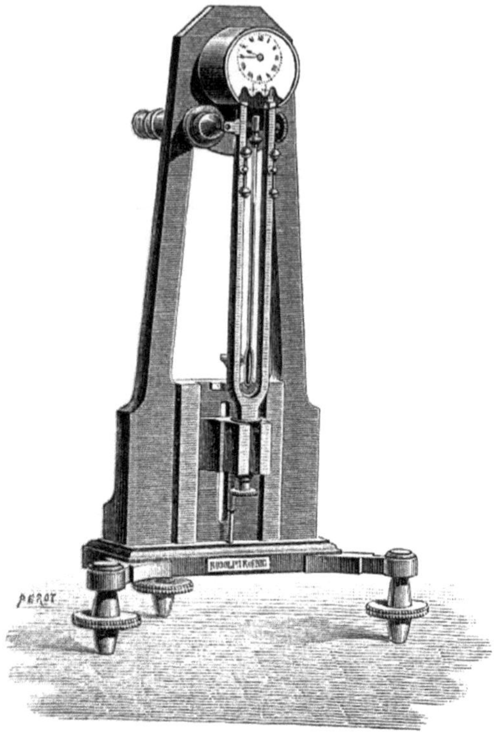

Fig. 5.9. Clock fork with clock mechanism, tuning fork and Lissajous objective lens. CR 32
Source: Koenig (1889, p. 19)

464,384 total vibrations (3,628 times 128) during a period of 1 h on the chronometer, which would mean the unknown fork was vibrating at 129 v.s. (64.5 Hz) or (464,384 v/3,600 s). Therefore, with the simple comparison of clock fork time to real time, the exact pitch could be determined.[72]

Koenig envisioned an apparatus that it was almost completely automatic and thus free of human error. He attached special micrometer screws to the prongs in order to adjust the frequency to the exact number needed for calibrations. The fork could be adjusted until it was at the exact pitch of 128 v.s. (64 Hz) After setting this standard by using comparisons with the chronometer, he employed the Lissajous optical method (with Lissajous microscopes and mirrors) to compare and tune unknown forks. He boasted that this apparatus was not only remarkable for its "extraordinary precision" but it also operated with "little complication or difficult manipulation."[73]

He obsessively studied his tuning forks and isolated the key variables that affected pitch. Temperature became one of the key variables in his study. At the start of his business 20 years earlier, Koenig had made a standard fork of 512 v.s. (ut3; 256 Hz; C4) "without indication of temperature."[74] He discovered in the course of further research that this standard was likely a fraction above 512 v.s. at 20°C (frequency inversely proportional to temperature). Koenig came to the realization that researchers almost never worked at the temperature at which the forks were made, so it was necessary to develop a way of knowing with certainty the variation in frequency of a tuning fork for each degree of temperature change. He first placed a thermometer between the branches of the tuning fork so that the reservoir reached down to the heel of the fork where the influence of heat was at a maximum and the movement of the fork at a minimum. He then performed a series of control tests. He calculated the time required for tuning forks to adjust their internal temperature to their surroundings. He discovered, for example, that it took an average of 45 min for a fork to recover its natural frequency after being held in a warm hand. In another trial it took over 4 h for the fork to recover from a slight drop in temperature overnight.[75] He studied the effects of prolonged use of the fork (and therefore, its internal rise in temperature) and discovered that he could only run an experiment for approximately eight and a half hours before the internal temperature of the steel changed significantly due to overusing the fork.[76] He even studied the influence of the resonator cases on vibrations of a tuning fork and discovered that resonators that were slightly different from the intended frequency prolonged the vibrations of the fork (80–90 s), and at the same time, altered its frequency.[77] On a related point, he found that resonators which were exactly in tune with the forks, caused the forks to vibrate for only 10 s (the correspondence of resonant frequencies was so well aligned that all the energy was dissipated rapidly.), making it difficult to do comparisons. He therefore had to find the right balance that would allow the fork to vibrate for a long period so as to compare it to the clock fork using beats or Lissajous figures, without altering the accuracy of his frequency measurements.

Koenig had to work around the weather and room conditions of Paris. The caves of subterranean Paris had the most stable temperature, but turned out to be too cold at 12°C.[78] He therefore used a room with high ceilings, shut on all sides, in which the temperature varied slightly "especially during overcast and somber days, which there had been a lot of during the year 1879 in Paris."[79] He also developed an oven that was manually regulated to adjust temperature. Using this appliance, he was able to extend his tests into higher temperatures, to determine whether the vibrations of the forks changed at different rates at higher temperature ranges. He also performed a series of tests on forks of the same tone but with different shapes and thickness.[80]

From these studies Koenig gained a remarkable control over the experimental variables and determined that for temperatures below 50°C, a change in temperature of one degree resulted in a change of 0.0143 v.s. for the fork ut1 (128 v.s.; 64 Hz; C2), and 0.0572 per one degree for his ut3 fork (512 v.s.; 256 Hz; C4).[81] In total he conducted more than 300 experiments between July and December 1879.[82] He now had at his disposal a highly precise instrument and combination of techniques for re-evaluating previous standards. He started with his own standard ut3 (256 Hz; C4) fork from 1859. He used the method of beats to obtain the exact number of vibrations by which the old standard differed from the true ut3, which had been established using the clock fork. He found that the old ut3 was actually 512.3548 v.s. at 20°C.[83] Using his conversion figures, Koenig calculated that the old standard would be 512 v.s. at 26.2°C. In order to avoid these calculations in the future he added a small weight-calibrated adjustment device to the prong that allowed the fork to be set at 512 v.s. for any temperature.[84] He then verified these results by placing his old standard fork, equipped with Lissajous mirrors, in his oven and watched as the heated fork came into unison with the new standard. This occurred for a rise in temperature between 6 and 6.5°C as he had measured. Ellis, who had done several experiments (with his tonometer) to determine the pitch number of Koenig's standard fork of 1859, measured a very similar figure for the old standard.[85] In 1880 Ellis stated that with the clock fork Koenig had made it possible to measure changes of vibration "absolutely inappreciable by ordinary methods of observation."[86]

The pinnacle of Koenig's achievement was that he was able to test the actual French Standard (435 Hz; A4) that had been determined by Lissajous in 1859.[87] Through a combination of trials he developed a fork that was exactly 870 v.s. (435 Hz; A4) at 15°C (calculated from a measurement at 20°C).[88] He then went to the Conservatory of Music in Paris and deposited this fork beside Lissajous's fork for 2 days to equalize their temperatures. Using the method of beats he found that Lissajous's fork was actually 870.9 v.s. (435.45 Hz; A4) at 15°C (as above, calculated from a measurement at a different temperature).[89] But he could not determine the number more precisely because the fork vibrated for only 20 s making optical comparisons difficult.

This massive effort to perfect the tuning fork had a considerable impact on acoustics, music and physics. The clock fork appeared in his catalogues of 1882 and 1889 at a cost of 2,000 fr.[90] To emphasize its prominence, he put a picture of it on the cover of his 1889 catalogue. It transmitted the standard of the physicist's scale, ut3 (256 Hz; C4), and the French standard for musicians, la3 (435 Hz; A4), to many institutions throughout the world. The premium standard forks were gilded to prevent rust and came with a brass resonator and stand. In fact, Josef Stefan, president of the 1885 International Conference on standards for musical pitch, stated that standard forks should be gold plated to protect against oxidation. He also stated that standards should not be used frequently so as to prevent unnecessary damage to their elasticity. He recommended Koenig's forks.[91] Some of these forks even came equipped with a small aluminum dial on one of the prongs to be used to adjust the pitch for varying temperature, between 5 and 35°C. (CR nos. 34 and 43).[92] He delivered clock forks and standards to institutions in Italy, Russia, Austria, Canada, the United States and Germany.[93]

Although Koenig had created the means to resolve technical issues surrounding standards, by 1888 there was still a heated debate about what standard(s) to adopt.[94] There were many standards and traditions even within single countries. Koenig proposed the adoption of two standards, one for physics, ut3, 512 v.s. (256 Hz; C4), and one for music, la3, 870 v.s. (435 Hz; A4). Both standards, he argued, were close enough to each other that adopting them simultaneously would not cause any undue confusion.[95] Although not a key participant at the conferences on standards,[96] Koenig became influential in this debate by transmitting his methods and standards throughout the world. In Italy, the King officially adopted his standard fork.[97] In the United States D.C. Miller at Case School provided a "tuning" service to companies and institutions around the country based on Koenig's forks and methods. He used

Koenig's forks and clock fork and later one made by Max Kohl of Germany.[98] In effect, Miller's operation "tuned" America well into the twentieth century. Among others, he certified forks for Steinway and Sons and the scientific instrument maker, William Gaertner and Company.[99]

Koenig's forks also had a wide influence on the practice of science, especially as electrical and timing standards. Between 1882 and 1884 Albert A. Michelson (1852–1931) used a Koenig tuning fork to determine the speed of a revolving mirror through a comparison with a standard clock, as part of his experiments to measure the velocity of light. Although these experiments did not relate directly to acoustics, Michelson, an avid musician, had a keen interest in Koenig's instruments and saw the potential of his precision forks as a standard frequency.[100]

In conclusion, the period between the Paris Exposition of 1867 and the Philadelphia Centennial Exhibition in 1876 was one of the most prolific of Koenig's career as an instrument maker and experimenter. The two activities seemed to blend in studies for the velocity of sound, the timbre of vowels and combination tones. Each of these required the refinement of existing instruments and the creation of new ones. Above all, Koenig's experiments and resulting instruments came to reflect the values of his Parsian context and the growing separation with Helmholtz and his work in Germany. As Koenig's business and personal life became tumultuous, his different background seemed to find expression in striving for flawless instrument manufacturing and experimental work. His obsession with perfected instruments, along with an approach shaped by the kinds of instruments he employed, began to open up deep fissures in psycho-physics. In the next chapter, we shall see how this work coincided with attempts to build a larger market in North America.

Notes

1. See Jackson (2006) on Scheibler, Chapter 6.
2. Loudon (1901b, p. 7).
3. Dörries (1998, p. 256) and Fox (1971, pp. 295–302).
4. Dörries (2001, pp. 233 and 243).
5. Smith and Wise (1989, p. 107).
6. Beyer (1998, pp. 4–7, 32–37).
7. In his history of acoustics, Koenig describes the earlier attempts by Mersenne, Laplace, Humboldt, Bouvard, Mathieu, Prony, Arago, and Gay-Lussac. Koenig (1901), Deuxième partie, I. Also see Beyer (1998, pp. 4–7, 33–37) and Miller (1935, pp. 65–66).
8. Regnault (1868, p. 5). For more on Baron Haussmann's reshaping of Paris, see Jordan (1995).
9. Koenig (1873a, p. 11).
10. For the development and use of the Regnault chronograph, see Koenig (1882c, pp. 11–12) and Loudon and McLennan (1895, pp. 117–118).
11. Regnault (1868, p. 425).
12. Ibid., p. 429.
13. Ibid.
14. Boring (1942, pp. 367–375).
15. Donders (1864).
16. "*Luftgeräusches.*" In Helmholtz (1863, p. 171). Translation from Idem., 1954, p. 108.
17. Ibid.

18. Idem., 1863, pp. 167–173 and Idem., 1954, pp. 105–110.
19. Koenig (1870, p. 933).
20. Ibid., p. 932.
21. Ibid., p. 933.
22. Idem., 1872, p. 179.
23. Ibid., p. 178. Translation from Idem., 1873c, p. 12.
24. Ibid., p. 176.
25. Koenig (1870, p. 176). Translation from Idem., (1873c, p. 11).
26. Ibid., pp. 176–177. Translation from Idem., 1873c, p. 12.
27. Zahm (1900, pp. 53–65).
28. Barnard (1870a, p. 505).
29. Kielhauser (1907, p. 18). Levi K. Fuller, "Method of Making Tuning-Forks," United States Patent Office, Patent Number 483, 513, September 27, 1892. Fuller, who was working with much smaller forks used exclusively for musical tuning, developed a method whereby he bent a bar of steel into a U shape. There is no evidence that Koenig did this.
30. Richardson (1927, p. 113).
31. Jackson (2006, pp. 225–226).
32. Rudolph Koenig to James Loudon, October 4, 1891, UTA-JLP.
33. A.A. Michelson used a Koenig fork to calibrate his rotating mirror in the velocity of light experiments, 1882–1884, see Miller (1935, p. 75).
34. Miller (1916, pp. 29–30) and Koenig (1882c, pp. 182–189).
35. Goodway (1987), Gordon (1994, 1996)), Hosler (1994), and Smith (1981).
36. Over 30 forks were measured from each group, entailing six measurements of different dimensions from each fork. I gratefully acknowledge the help and insights of Roger Sherman in the examinations of the tonometer at the NMAH on May 8 and 9, 2003.
37. Jackson (2006, pp. 226–229) and Loewenherz (1888, p. 265).
38. I would like to thank Professor Sam Allen of the Department of Material Science and Engineering at MIT for providing laboratory time and equipment for this study. Throughout the summer of 2004, I prepared the sample (512 v.s. fork from U of T tonometer) and Allen's laboratory technician, Yinlin Xie, took the micrographs and performed the hardness tests. Hardness HV for the ferrite area: 134, 117, 117, 122.5 and 112.4 for an average of 120.58 w/25 g. Hardness HV for the pearlite area: 146.3, 147.9, 139.9, 152.2, and 136.1 for an average of 144.48 w/25 g. The sample was micrographed in three areas – at the base of the U, on the length of the prong, and near the corner elbow. The micrographs revealed a sample of 0.55% annealed carbon steel (hypoeutectoid).
39. Some of these processes are described in Kielhauser (1907, pp. 17–19).
40. Miller (1916, p. 29).
41. Kielhauser (1907, p. 19).
42. Jordan (1889, pp. 10–48).
43. "Cutlery" in Rees 1820. Appendix on the manufacturing of cutlery in Edmunson (1997).
44. The 1876 tonometer itself demonstrates two styles of fork – the older U shape and the forks with a thicker yoke, see CR no. 36.
45. Silverman (1992, pp. 127–150). On the role of experiment in shaping Helmholtz's work, see McDonald (2003).
46. The original German version was Koenig (1876b); English 1876a and a revised French version (1882c, pp. 87–148).
47. Loudon (1901b).
48. In 1879 John Tyndall recalled that "some years ago, Koenig was ardently engaged on these questions [combination tones]. . .and he then understood Koenig to be of the opinion that he had overthrown the theory of Helmholtz with regard to combination tones, and established the old theory of Thomas Young." Spottiswoode (1879, p. 125).

49. Koenig heard the inferior beat when it was less than half the lowest tone; the superior beat when the inferior beat was greater than half of the lowest tone. Koenig (1876b, p. 181), Idem., (1882c, pp. 90–91), Idem., (1876a, p. 420).
50. Maley (1990, pp. 65–89).
51. See Lord Rayleigh's comments in Spottiswoode (1879, p. 128); or Ellis's comments in Helmholtz (1954, p. 533).
52. (Helmholtz 1863, pp. 227–236), especially 234.
53. Helmholtz (1863, pp. 227–236).
54. Koenig (1882c, pp. 124–131) (see especially p. 147 no. III and the footnote on p. 130) and Koenig (1876b, p. 221). Idem., (1876a, p. 514).
55. For a review of these debates, see Ellis in Helmholtz (1954, pp. 527–538).
56. Koenig (1876b, p. 219), Idem., (1882c, p. 128), Idem., 1876a, p. 513.
57. Koenig (1876b, p. 186), Idem., 1882c, p. 95, Idem., 1876a, p. 424.
58. These forks now reside at the Physics Department of the University of Toronto. Koenig's lectures appeared in a Toronto newspaper, *The Globe*, Aug. 30, Sept. 13, 16, 19, 21, 23, 26, 28 (1882). Also see Rudolph Koenig to James Loudon, November 25, 1881 and March 21, 1882. UTA-JLP.
59. Koenig (1882c, p. 84).
60. Ibid., pp. 84–86.
61. Gooday (2004)
62. Ibid., p. 71.
63. Warner (1992).
64. Rowland (1902, p. 617).
65. Ellis (1877a).
66. Ellis (1968). Ellis in Helmholtz (1954, pp. 493–513).
67. Ellis (1877a).
68. Koenig (1877).
69. Helmholtz's letters had been published by Appunn in an advertisement in his catalogue. Koenig (1877).
70. Ellis (1968, p. 19).
71. Wood (1964, pp. 121–22).
72. Koenig (1882c, p. 173) and Niaudet-Breguet (1866).
73. Koenig (1882c, p. 173).
74. Ibid., p. 172.
75. Ibid., p. 176.
76. Ibid., pp. 177–178.
77. Ibid., pp. 180–182.
78. Ibid., p. 177.
79. Ibid.
80. Ibid., p. 187.
81. Ibid., p. 189.
82. Ibid., p. 182.
83. Ibid., p. 189.
84. Ibid., pp. 189–90.
85. Ellis (1968, p. 61).
86. Ibid., p. 60.
87. The Lissajous standard fork remains in storage at the Musée de la Musique in Paris. The fork is gilded and marked "Secretan, Paris." It is 5 cm from the top of the prongs to the stem of the fork. It rests upside down, in a wooden frame, connected to a pine resonator box. The whole apparatus is 46 cm in height. The base of the frame reads, "Diapason Normal, 870 vibrations par seconde, à la température de 15°C, Arrêté Ministériel, en date du 16 Février, 1859. Sons Excellence, Monsieur Achille Fould, Ministre d'Etat." There are two piano keys

on either side of the two prongs with felt hammers. One reads, "Etouffoir," (dampener) the other reads "Marteau" (hammer).

88. Koenig (1882c, pp. 190–191).
89. Koenig (1882c, pp. 190–191).
90. Koenig (1889, p. 19).
91. See Jackson (2006, pp. 225–226).
92. The Museo di Fisica at the University of Rome has three such Koenig forks with brass resonators on a cast iron tripod stand: ut3 (512 v.s.), la3 (870 v.s.) and si3 (921.7 v.s.) each with aluminum dials on one of the prongs graduated from 5 to 35°C. CR nos. 34 and 43. Descriptions of these forks can be found in Koenig (1889, pp. 19–20).
93. Rudolph Koenig to James Loudon, November 7, 1888, UTA-JLP.
94. Jackson (2006, pp. 151–181).
95. Rudolph Koenig to James Loudon, November 7, 1888, UTA-JLP.
96. In a letter to Loudon, Koenig ridiculed the International Congress of Standards at Vienna in 1885: "I could not stop myself from finding it quite amusing and perfectly ridiculous the explosion of enthusiasm and warm congratulations on the importance of their work." Ibid.
97. Information and instruments related to the standardisation of pitch in Italy can be found at the Museo di Fisica at the University of Rome.
98. This Kohl clock fork can be found on display at the Physics Department at Case Western Reserve University. It is very similar to Koenig's model.
99. The D.C. Miller papers, Case Western Reserve Archives. One such certificate from July 1, 1927 reads: "Certificate of Accuracy of a Tuning Fork. Submitted by Steinway and Sons, of New York.... Frequency of the Fork. – This fork has been accurately adjusted in the Physical Laboratory of Case School of Applied Science to its nominal frequency, the absolute frequency being determined directly from the Riefler Standard Clock, No. 89, by the method of optical comparison with a Koenig Clock-Fork (Miller 1916, pp. 38–42.) The calibration was carried out in the constant temperature clock-room. The exact temperature of the fork was observed at each measurement, and the observations have all been reduced to the standard temperature coefficient, –0.00011, (Annalen der Physik, 9, 408 (1880)). The final determinations show: FREQUENCY OF FORK NO. 3, C = 261.620 at 20°C (68°F)."
100. Miller (1935, p. 75).

Chapter 6
Expanding the North American Market, 1871–1882

Visitors to the 1876 Philadelphia Centennial Exhibition marvelled at the elements of sound in the form of Rudolph Koenig's *grand tonomètre* of over 692 tuning forks, with 800 tones represented, ranging from 16 to 4096 Hz (Fig. 6.1). Koenig had packaged these elements into orderly rows of individual tuning forks covering roughly the range of the piano. The entire display reflected prevalent ways of organizing

Fig. 6.1 Large tuning-fork tonometer (grand tonomètre). Rack is 36 inches high. CR 36. Photo courtesy of the National Museum of American History, Smithsonian Institution, Washington DC, cat. no. 315716, neg. 70524

D. Pantalony, *Altered Sensations*, Archimedes 24, DOI 10.1007/978-90-481-2816-7_6,

Fig. 6.2 Displaying elements. Comprehensive set of nineteenth-century chemical reagents. MCUL 1185. P. Cintra © Museum of Science, University of Lisbon

knowledge at the time – e.g. the period table of elements – and echoed a similar presentation of samples found in chemistry (Fig. 6.2), pharmacy, biology and geology. Perhaps just as important to the audience at the Centennial Exhibition, the tonometer was an instrument that displayed the high art of acoustic instrument manufacturing and precision tuning (Chapter 5). Koenig built and promoted it in trying conditions, however, and it did not find an easy path to its present location at the National Museum of American History at the Smithsonian Institution in Washington DC. The *grand tonomètre*, as well as other instruments brought to North America in the 1870s and 80s, tell the story of significant challenges involved with selling and promoting expensive, cutting-edge instruments, within a market directed primarily towards educating college boys.

Recovery from the Turmoil of 1870–1871

In the autumn of 1870 war broke out between France and Prussia. In order to escape the turmoil, Koenig moved to Switzerland and then to Prussia. During this time he stayed in Königsberg, Berlin and Magdeburg with family. In the spring of 1871, following Prussia's victory, France endured further chaos through the disastrous Commune that was finally brought to a close by the commencement of the Third Republic. Remarkably, Koenig kept business going (albeit at a very slow pace) from a distance, and was pleased to find upon his return in June 1871 that his workers had kept his studio intact, and nothing had been looted.

Despite the loyalty of his workers, Koenig returned to an inhospitable environment in Paris. He was accused of fighting for the Prussians and almost all of his friends in the scientific community turned their backs on him. Regnault, who had lost his son Victor (the painter) in the last days of the war, was a broken man and withdrew from the Parisian scientific scene altogether. One person who remained a staunch supporter was the Abbé Moigno, editor of the weekly scientific journal *Les Mondes*. After Koenig's return in 1871, he defended him against the "odious slander" that claimed he had served in the Prussian artillery.[1] He stated that Koenig could just as easily have moved permanently to Berlin where he was needed, but instead he had returned to Paris where he became a prisoner in his own atelier. He added that the German instrument makers Ruhmkorf and Hofmann had followed Koenig's example in choosing to continue to work in the "inventive atmosphere of France" and "we appreciate and thank them."[2] Another Frenchman who remained a loyal friend and colleague was the physiologist, Etienne-Jules Marey (1830–1904), who, like Koenig, was also a pioneer in the use of graphical technologies.[3]

Koenig was fortunate in having good friends in the international community during this period, as well.[4] They promoted his interests outside Paris and provided needed encouragement during a difficult period. One colleague in Austria was Franz Joseph Pisko, the professor of physics and mathematics at Vienna who, as mentioned earlier, had published a book on acoustics in which Koenig's instruments figured prominently. The American, Walter Le Conte Stevens was interested in Koenig's work and stayed at his apartment when in Paris.[5] Alfred Marshall Mayer (1836–1897), who left the United States in 1863 (during the Civil War) to study in Paris under Regnault,[6] remained until his death in 1897 one of Koenig's closest collaborators and supporters. Mayer, who became one of the leading acoustical researchers of the later nineteenth century, founded the Stevens Institute of Technology in Hoboken, New Jersey. Like Koenig he was a follower of Regnault and therefore had a love of experiment and a deep appreciation for honing methods and instruments. Before his death he spent two summers living at Koenig's place at 27 Quai d'Anjou performing a series of experiments on the tonometer.[7]

England was especially good to Koenig. Following his isolation from the Parisian scientific scene, the English scientist William Spottiswoode became a close promoter and collaborator. Spottiswoode, a wealthy eccentric with a love of acoustics, was the president of the Royal Society in the late 1870s and one of the founders of the British Musical Association, a group of scientists and musicians devoted to "the investigation of subjects connected with the art and science of music."[8] Spottiswoode spread Koenig's name throughout the English scientific community. Upon his death in 1883, Koenig wrote that "during the long years after the war, he was the only scientist who would show a true interest in my work, and I can say that I possibly never received his visits, without having felt after his departure more courage to continue the path of my labours."[9]

Why, then, did Koenig remain in Paris during this tumultuous period? The simplest explanation was that he had invested in and built up a skilled workforce, connections to local infrastructure, including specialized materials and service. Among other possible reasons – his love of Parisian culture and what Moigno called the "inventive atmosphere." The Parisian instrument scene, even with a gradual decline in the French scientific world, still dominated the market, especially with educational instruments. It was still the main destination for North American purchasers. Above all, Koenig was a fiercely independent person. When asked why he remained in Paris, he told James Loudon that he would rather live among his enemies.[10]

His career, similar to other Parisian instrument makers, faced a turning point after 1870 which entailed a shift across the Atlantic. Even with an economic downturn in the early 1870s, the North American market significantly expanded during the period 1867–1882. In fact, one sees with the explosion of new colleges, physical cabinets and teaching laboratories, that North American schools played a major role supporting the French precision instrument trade.[11] Several colleges and universities bought from Parisian makers. Even as the German instrument trade became an increasingly important supplier of precision instruments (especially in electricity and optics), the French traded benefited from large purchases of educational and demonstration instruments. For Koenig there is no doubt this was good for business. He was associated with the good reputation of French instruments – if someone bought a set of Duboscq instruments, they were sure to buy some Koenig pieces as well. But this situation was not always stable. He suffered financially due to fluctuations in the European and world economy, the great costs of making high-end and time-consuming research instruments, the large portion of time he spent devoted to experiments, and the equally large amount of time preparing results for publication.

Besides the growth of the North American market, acoustics was itself growing as a field. It could never compete with electricity and optics, but it had certainly grown in importance for teaching. This surge in interest came from several factors including the work of Helmholtz, other research as well as the expansion of science education in general. In England, William Strutt (Lord Rayleigh) (1842–1919) expanded upon Helmholtz's work by developing a more refined synthesis of the mathematics and physics of sound. John Tyndall (1820–1893) did research on fog horns and sound transmission in the atmosphere and popularized experimental acoustics with his treatise *On Sound* (1867).[12] Alexander Ellis (1814–1890) translated Helmholtz's masterpiece as *The Sensations of Tone* (1875) and performed numerous experiments on the standard-pitch question. William Thomson (later Lord Kelvin) (1824–1907) had a keen interest in experimental acoustics and continued to undertake acoustical research. (In the 1840s he had worked with Regnault for a short time "learning patience and precision" and he also purchased acoustical instruments from Marloye).[13] The physicist Silvanus P. Thompson (1851–1916), who was soon to be involved in work on the telephone, took an active role in supporting Koenig's research and instruments.

The German world as well saw the emergence of a small group of specialized acoustical researchers: August Kundt (1839–1899), Karl Friedrich Sondhaus (1815–1886), Ernst Mach (1836–1916), Gustave Kirchoff (1824–1887) and August Toepler (1836–1912) in physics; W. Preyer (1841–1897), Ludimar Hermann (1838–1914) and Adam Politzer (1835–1920) in physiological acoustics. In addition to the scientists, there were musical instrument manufacturers such as Steinway, who showed an active interest in Helmholtz's work.[14] And then there were the inventors, Bell and Edison in America.

The Third Catalogue, 1873

In 1873, in the wake of the general recovery from the war and the Commune, Koenig published his third, non-illustrated catalogue in order to inform customers of changes in prices and instruments. In a brief forward, he wrote that there had been a large increase in the cost of materials and labour. These changes, however small, were the first significant change in prices since his business began in 1858, and showed the strain that Koenig was under at this time. The catalogue also provided an update on his recent inventions. "The progress of acoustics since 1865," Koenig wrote, "has been considerable enough to demand a new catalogue."[15] He added instruments that would become standard equipment for research and teaching: a set of cylindrical, adjustable, graduated brass resonators; steel cylinders for testing the upper limits of sound; an improved Helmholtz vowel synthesiser with ten tuning forks (instead of eight); a universal Lissajous comparator with two supports, ten adjustable forks, mirrors and inscription devices; Regnault's chronograph with several variations and different forks for precision time keeping; a manometric speaking tube for studying vowels; a manometric apparatus for studying interference effects and the speed of sound; a large tuning-fork apparatus for studying vibratory movements by stroboscopic methods; Crova's rotating-disk apparatus for the mechanical projection of vibratory movements; and a stethoscope with five tubes for multiples users. The optical demonstration instruments and the stethoscope revealed his continuing efforts to make acoustics appeal to large audiences and students. The Lissajous apparatus and the Regnault chronograph were over 1,000 fr (near the top of the catalogue's price range). He offered these instruments with an unequalled range of precision forks. He also continued to modify and emphasize his sirens, electrical interrupters, resonators and vowel analysers and synthesisers (Table 6.1).

In addition to new prices and instruments, the information on the cover, an update of his awards since his last catalogue, was reason alone for publishing the catalogue – Doctor of Philosophy (Königsberg 1868), a medal of distinction at the 1862 Exhibition in London, gold medal from the Société d'Encouragement de Paris (1865), and a gold medal at the 1867 Paris Exposition. Almost all the

Table 6.1 Price changes from 1865 to 1873 (in 1867, before the war, wages averaged 5–9 fr a day)

Reis telephone, 60–65 fr
Double siren, 400–450 fr
Seebeck siren, 800–1,000 fr
Sound synthesiser, 800–840 fr (1,000 fr with two extra resonators)
Single manometric pipe, 30–40 fr
Manometric comparison apparatus, 150–200 fr
Sound analyser, 250–300 fr
Nineteen resonators, 150–165 fr
Standard tuning fork, 25–28 fr
65-fork tonometer remained at 2,000 fr
Phonautograph also stayed at 500 fr

major instrument makers advertised their medals, showing how these awards were a common and necessary part of business.

Joseph Henry and the Smithsonian Institution

One of Koenig's more influential clients in the 1870s was Joseph Henry of the Smithsonian Institution in Washington DC. In 1865 Henry asked his agent in Paris to inquire into the cost of a complete set of acoustical apparatus (Chapter 4). From that time until 1878 the Smithsonian made several orders, including a near complete set of Koenig apparatus. They were given a prominent position in the physical cabinet and promoted them on a national stage. Henry was America's foremost physicist and a pioneer of electromagnetic research. He was also keenly interested in acoustics. He had studied architectural acoustics and performed a series of tests on lighthouse foghorns.[16] After working at Princeton and gaining great international repute for his powerful electromagnets, he was named the Smithsonian's first secretary in 1848. He survived nine US presidents as secretary and became a powerful influence on science policy in the United States.[17]

The instruments of the major Parisian ateliers figured prominently in Henry's vision for science in America.[18] He envisioned a research-based, national institution where scientists, craftsmen and teachers could visit to study and use the latest apparatus of natural philosophy. At the "Castle," the main Smithsonian building on the Mall, there was an apparatus room, or museum of physical instruments, which, in Henry's original vision, "may be used for experimental illustration and original research, and may serve as models to workmen as well as to illustrate the general progress of inventions in this line."[19] He described the role new instruments would play in the future development of American science, in a letter of 1847 to Alexander Dallas Bache, one of the regents of the Smithsonian.

> ... since we are to form a large collection of articles of Foreign and curious research which may serve to excite the love of learning, a collection of Physical instruments should form an essential part of this and be of such a character as to induce a pilgrimage to Washington of all the *quid nunc* professors in our country to enlighten themselves as to the progress of science and to witness the new phenomena.[20]

Unfortunately, his plans had to be put on hold in the early 1870s. A depression hit Britain in 1873 that had a large impact on the world economy. On September 23, 1873, Henry wrote to Koenig about the banking failure in the United States, stating that he would have to countermand one of his large orders for instruments.[21] Koenig responded on 11 October that the Smithsonian could have credit for a year "if you should think it of any advantage for the institution to have the instruments."[22] Perhaps Koenig did not have much stock and had made an order specifically for the Smithsonian. Henry waited a year and ordered a portion of the key instruments again for 3,000 fr. A year later he ordered 50 more instruments for a total of 3,070 fr. In the order of 8 August 1874, for example, he asked in particular for the large bellows for experiments and demonstrations with organ pipes. The next order (January 1875) included a Seebeck siren, the standard series of nineteen spherical resonators, the sonometers of Barbareau and Marloye, the double siren of Helmholtz and several items of demonstration apparatus – plates, membranes, organ pipes, rods and reed pipes.[23] He purchased several more pieces of apparatus before he died in 1878. His last order in 1877 was for high-end research apparatus exhibited at the Centennial Exhibition.[24]

Ordering from abroad could be complex, costly and time consuming. In the United States Queen & Co and N.H. Edgerton of Philadelphia acted as a dealer for Koenig (CR no. 27).[25] British makers as well had agreements with foreign makers, and in the colonies such as Canada and Australia instruments could be bought through the Agent-General in London. One instrument from the University of Sydney touches on the complexity of these dealings. It had ten resonators by Koenig on a wooden stand made by W. Ladd & Co. of London (CR no. 54a).[26] Koenig himself charged 20 percent for packing and shipping.[27] There were also duties. Instruments could be imported into the United States free of duty after 1790, but this fact was not widely known and there was enough confusion on the issue for Queen and Co. to advertise that from June 1874 colleges, schools, literary and scientific societies could import books and instruments free of duty if they were used for educational exercises.[28] Many colleges, however, bought directly from makers or agents in Europe. Union College of New York sent their professor of natural philosophy, John Foster, to Europe in 1875 (his third trip since 1867) to buy apparatus.[29] Landon Garland of Vanderbilt University in Nashville, Tennessee, went to Europe in the summer of 1875.[30] That same summer, Henry Rowland was sent by the newly formed Johns Hopkins University.[31] Foster and Garland bought instruments from Koenig; Rowland, who was not as interested in acoustics, bought a standard teaching set either at this time, or shortly afterwards.[32] In fact, he had some bad luck with his order from Koenig. There was considerable damage to the instruments when they arrived from overseas.[33] Packaging and freight were a major risk and concern for both parties involved. Koenig stated in his catalogue that he did not take responsibility for damage.[34]

Centennial Exhibition, 1876

One thing that gratified me exceedingly was to meet Monsieur Koenig, the inventor of the manometric capsule – (you remember the little instrument with the vibrating flame and the revolving mirrors) – Monsieur Koenig has a splendid exhibit of tuning-forks and scientific apparatus. We had a long talk on scientific subjects in the French Language. He spoke French and I English – and we got on very nicely.

Alexander Graham Bell to Mabel Hubbard Bell, 21 June 1876.[35]

In 1876, ready to capitalize on the American market, Koenig went to the Philadelphia Centennial Exhibition. The president of Johns Hopkins, Daniel C. Gilman, was at the fair and informed Henry Rowland that Koenig was present and "eager to sell."[36] His plans did not fully materialize, however, as the fair turned out to be both a triumph and a great disappointment for the 43-year-old instrument maker. The Centennial Exposition, like all major world exhibitions since London in 1851, was a showcase for the "wealth of the world".[37] One contemporary observer called it "the largest advertising institution the world ever produced."[38] Each aisle represented an unlimited vista of knowledge, consumption and competitive display. Historian Bruno Giberti has written on the task of organizing the layout of the buildings, an enormous classification project in itself.[39] Visitors faced endless variety in all areas – agriculture, horticulture, mining and metallurgy, works of art, manufactured goods, machinery, science, and education. The senses were overwhelmed with beautiful displays, exotic fragrances and sounds. In fact, musical sounds were a prominent feature of the fair experience as America came into its own as a "musical nation."[40] One visitor complained of music from every direction, often at overlapping spaces and times, as if "some melodious yet diabolic influence impels the performers at one and the same moment to rush to their stools."[41] It was also overtly commercial in appearance. The arrangement of the sections, according to Giberti, reflected the well organized space of a modern city, in particular, the "straight corridor streets" of Haussman's Paris.[42]

There was also a national context at play. In studying the major exhibitions from 1851 through the end of the nineteenth century, historian Robert Brain singled out how they "evolved into sites where nations defined and promoted their profile before the wider world as a means of political and commercial advancement."[43] Science and instruments were considered a measure of national vigor. Decline of the English instrument makers was a concern at the 1851 exhibition;[44] the rise of the French at 1867;[45] and the rise of German makers in the fairs of the 1880s and 90s.[46] At the Centennial exhibition, according to one visitor, the French section was one of the more attractive as they "have the very happy faculty ... of arranging everything so as to produce the most attractive effect."[47]

Koenig found himself in good company. He was part of Group XXV, "instruments of precision, research, experiments, and illustrations, including telegraphy and music," and found himself between P. Goumas & Co., maker of wind instruments and saxophones, and Kriegelstein & Co., maker of pianos.[48] Twenty of the seventy displays in this section were devoted to musical instruments. Alexander Graham Bell unveiled an early version of an "electric telephone" at his booth. Thomas Edison displayed an electric pen. Steinway and Sons exhibited six inventions related to their pianos.[49] Important names from the Parisian trade were there: Alvergniat Brothers, Naudet & Co., L.G. Perraux, Deleuil, Brǵuet & Co., Jules Duboscq, A. Nachet, and Secrétan (Fig. 6.3).

Koenig came to Philadelphia with high expectations, bringing his entire collection, especially the masterpiece tuning-fork tonometer and the other tuning forks that he had recently been used for his research on combination tones (Chapter 5). There were graphical forks, the phonautograph, several manometric devices, framed drawings of his graphical and optical figures, Lissajous apparatus, and a large air bellows for demonstrations. It also featured a large aluminum wave siren apparatus with sixteen simple tones for studying timbre, first developed in 1867–1868, subsequently used by Alfred Terquem at the University of Strasbourg and then exhibited at the London exhibition of 1872.[50]

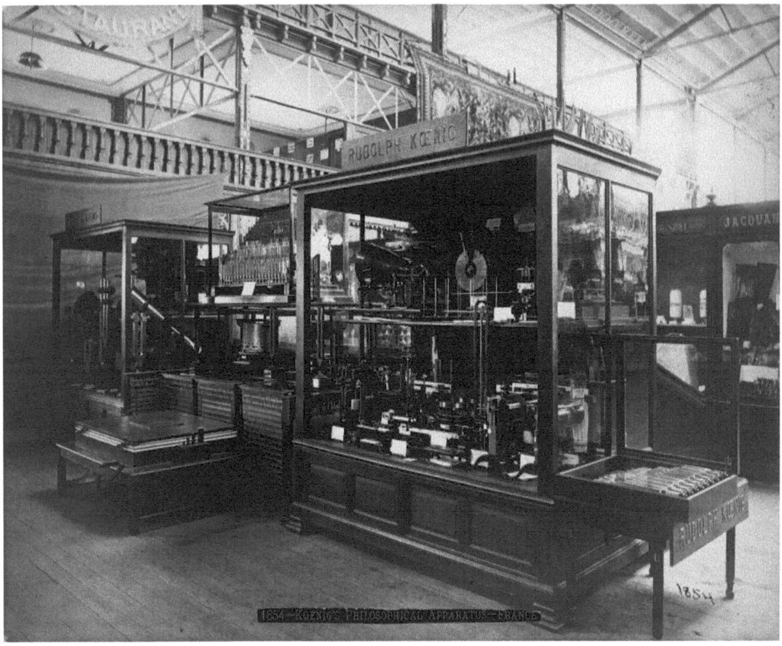

Fig. 6.3 Koenig's display at the 1876 Philadelphia Exhibition. Courtesy of The Print & Picture Collection, Free Library of Philadelphia. #c021854

His display won critical approval with a medal of distinction. The judges included William Thomson of Glasgow, F.A.P Barnard of Columbia College and Joseph Henry. There was also J.E. Hilgard from the coastal survey in Washington DC, Henry K. Oliver, a musician and educator from Salem, George F. Bristow, the composer from New York, J.C. Watson, professor of astronomy at the University of Michigan, Julius Schiedmayer, a piano maker from Germany, and E. Favre Perret, who represented Swiss watch makers.[51] As he had done 9 years earlier, Barnard wrote glowingly of the Parisian maker:

> In the department of acoustics, as represented in the Exhibition, the field was occupied almost wholly by a single exhibitor, Dr. Rudolph Koenig, of Paris. As a constructor, indeed, Dr. Koenig may be said to have monopolized this field before the world almost as exclusively as in the Exhibition, for it is to his skill that most eminent investigators have been accustomed continually to resort for the means of realizing their many ingenious conceptions.[52]

The official report of the judges (also written by Barnard) was far lengthier than any other French recipient and described his tonometer as "giving as many different shades of pitch extending over four complete octaves, and making equal intervals of eight simple vibrations each for the first octave, and of twelve each for the succeeding octaves; the whole forming an absolutely perfect means of testing, by counting beats, the number of vibrations producing any given musical sound, and

Fig. 6.4 Aluminum wave siren shown at the Philadelphia exhibition. This instrument marked the beginning of Koenig's research with wave sirens (Chapter 7). CR 210. Courtesy of The Print & Picture Collection, The Free Library of Philadelphia. #c011530

of accurately tuning any musical instrument."[53] (Fig. 6.1). It also mentioned the manometric flame invention and his instruments for challenging Helmholtz's theories. "Of the exhibit of Dr. Koenig as a whole, it may be said that there is no other in the present International Exhibition which surpasses it in scientific interest."[54]

Koenig's exhibit seemed to attract a disproportionate amount of attention. This was partly due to the novelty of the visual instruments and interest in the large tonometer and vowel apparatus. But it was just as much a measure of his skill at making his products appeal to both serious researchers and wide audiences. The first secretary of the U.S. Society of Science and Mechanism, Samuel Burr, for example, wrote a popular account of the exhibition with highlights from each section. His description of Koenig's booth was the second longest for the entire French section (the largest being for the Parisian carriage makers, Million, Guiet & Co.):

> This gentleman is here personally, and himself attends the several cases that are filled with his various inventions. M. Koenig has made sound a faithful study, and has perfected a great variety of very remarkable appliances. He has tuning forks, from half an ounce to many pounds weight, adjusted to the most delicate influences. Also, instruments by which he can detect sixty-five different tones in a single octave.[55]

Aside from his attractive visual demonstrations, Koenig also made point of promoting his precision research instruments. "He is not here on speculation," Burr wrote, "but to bring his inventions to the notice of scientists. He has given two exhibitions to chosen parties, and before the fair closes, purposes to give still further exhibits of the control he has in detecting the most delicate tones. He has instruments that give all the vowel sounds, and is really far in advance of all who have heretofore studied acoustics."[56] From these observations, it is apparent that Koenig succeeded in building a reputation as a scientist-instrument maker first, businessman second.

James Loudon and the University of Toronto

Koenig, however he wanted to portray himself, was in Philadelphia to sell instruments, and he made a connection that would prove crucial for his future business. Following the exhibition, he traveled to Buffalo for the meeting of the American Association for the Advancement of Science where he met James Loudon of Toronto. Loudon, recently named professor of physics at the University of Toronto, was eager to set up an undergraduate teaching laboratory based on the successful German model of laboratory-based teaching. Following other colleges in North America, he wanted to establish an "object-based" practical, teaching program.[57] Loudon, who had an interest in acoustics, was immediately attracted to Koenig's work. "As it was the year of the centennial Exhibition a large number of foreigners were present [in Buffalo] and amongst them Koenig, who addressed Section A in German, speaking with great animation, and receiving a most enthusiastic reception." Thus began a friendship that would last until Koenig's death in 1901 (Fig. 6.5).

In 1878, after 2 years of administrative jockeying for his vision of science, Loudon secured the large sum of $12,000 from the university to establish the first physical laboratory in Canada. Refusing to use agents he went directly to Europe to buy his equipment. He first stopped at the Cavendish laboratory at Cambridge. "There I met W. Glazebrook, assistant to Professor Clerk-Maxwell who was absent. On mentioning my intention of proceeding to Paris, and consulting with Dr. Rudolph Koenig, the acoustician whom I had met 2 years before. . .Glazebrook said I could do no better."[58] In the late 1870s, in an attempt to reach the English market, Koenig had placed a small advertisements in Nature: "Rudolph Koenig (Dr. Phil.) Manufacturer of Acoustical Instruments to illustrate the laws and produce the phenomena of sound. Paris, 26 rue de Pontoise, price list free." (Fig. 6.6) [59]

Paris was still the centre of the precision instrument trade at the time, and therefore could be a rather complex, intimidating market to navigate for a thirty-seven year-old professor from Toronto. Upon arriving in Paris, Loudon was greeted by a business commissioner (agent) who, somehow learning of his mission, offered help in introducing him to all the reputable instrument makers. He coyly replied that he had merely come to get an idea of the scientific instrument market in Paris and

Fig. 6.5 James Loudon (1841–1916). The University of Toronto and its Colleges, 1827–1906. Toronto: University of Toronto, 1906, p. 120. Photograph by F. Lyondé

would not as yet need help. Proud of his independence, he went alone to Koenig's atelier on 26 rue de Pontoise (in 1877, due to the demolition of his premises for the new medical faculty being erected in Paris, Koenig was forced to move from 30 rue Hautefeuille).[60] "On arriving at Dr. Koenig's place, his first expression was one of delight that I had come unaccompanied by a commissioner (who it appeared generally accompanied purchasers from America) adding that I should get better instruments and full value for my money."[61] Koenig told him of the high percentage that agents often exacted from clients and he coached him on the best way to go about his mission. Over the next 20 years he became Loudon's unofficial agent helping him buy instruments from the leading makers in Paris – Golaz in heat, Lutz and Laurent in optics, Carpentier in electricity, and Froment in mechanics.[62] But the largest number of instruments came from the acoustical instrument maker himself.

Through his connections and guidance Koenig had a large influence on the eventual shape of Canada's first physics laboratory. In fact, Loudon's choice of Koenig was no accident. He saw in Koenig an ideal model for his vision of science education. "Dr. Koenig was not only a most celebrated maker of acoustical instruments but an eminent scientific man who had received an honourary degree from the University of Königsberg for his discoveries in acoustics."[63] Koenig was also a very

Fig. 6.6 The physical laboratory at the University of Toronto, about 1890. University of Toronto Archives, *A1965-0004/1.91*

good promoter of Loudon's mission for bringing practical hands-on science to the University of Toronto, a campaign he had waged against the Toronto establishment for 6 years before securing the money. Loudon attributed part of his later successes in building the laboratories at Toronto to his initial visit to Paris and the atmosphere of Koenig's atelier.

> On learning that our Chancellor (Edward Blake) [who had recently been premier of Ontario] and Vice Chancellor C.T. Thomas Moss [Loudon's neighbour in Toronto] were in Paris at the same time, Dr. Koenig invited us all to a scientific séance at his place where we witnessed many of his most beautiful acoustical experiments, such as Lissajous figures, sympathetic vibrations, interference of sound, sounds of beats etc. The success of this séance was not without its influence in interesting the Chanceller and Vice Chanceller in the development of experimental science in their own university.[64]

The first $12,000 had been hard to get, but in the following 20 years there seemed no limit to University of Toronto's eagerness to spend on laboratories, a measure of Loudon's increasing power at the university (he eventually became the university president from 1892–1906). During his first trip alone, Loudon purchased an almost complete collection from Koenig at a large sum of 21,000 fr. Even Joseph Henry at the Smithsonian had not spent that much in one area of physics. In addition to Loudon, the physiologist Ramsey Wright (1852–1933), another proponent of research-based education, befriended Koenig and purchased instruments for his new laboratory at the University of Toronto. In the 1890s James Mark Baldwin (1861–1934) and August Kirschmann (1860–1932) (a student of Wilhelm Wundt's),

Fig. 6.7 Koenig's brass resonators became an icon of teaching in physics and psychology. The tapering series of resonators echoed the structure of the basilar membrane in the inner ear. CR 54. Photo by author 2005, Psychology Department, University of Toronto, Canada

the founders of the psychology laboratory at Toronto, also made purchases. Even the "Toronto Technical School" (a specialized high school) enquired about his instruments (Fig. 6.7).[65]

Loudon used his purchases for teaching (he did not publish acoustical research), with some interesting local modifications. The near complete series of organ pipes, which represent a fairly old-fashioned Marloye-Savart demonstration of every possible organ pipe effect, reveal elements of a comprehensive acoustical program in Loudon's laboratory (CR nos. 89–116). The synthesiser, a fairly difficult instrument to operate, survives in good condition and was most likely used by skilled demonstrators for lectures, and not trusted to students in their laboratory exercises. The synthesiser that survives at the Science Museum in London (previously part of the South Kensington Museum), in contrast, has adaptations on it (locally added gradations to control and measure intensities) that reveal skilled researchers used it for experiments on vowels and timbre (CR no. 56). The set of eight large tuning forks that Koenig brought to Toronto in 1882 for his public demonstrations of beat tones are still almost new, showing that they were not heavily used and most likely not understood.[66] The surviving collection, therefore, shows a multi-layered teaching program – heavy use of simple, conservative instruments, more rarified use by demonstrators of the flashy instruments. Finally, Koenig's simplest instruments found a voice in provincial Toronto: the pine "dropping sticks" for recreating simple melodies, survive with local music instructions (written in ink) for "How dry I am," "the Maple leaf," "Oh Canada," "Doxology" "Onward Christian soldiers" and "Toronto is our University." (CR no. 1).

The Toronto instruments, aside from telling us about their context of use, also provide a glimpse of Koenig's workshop. The steel cylinders for producing high frequencies just past 30 KHz seem to have been made in the same methodical fashion as his forks (Chapter 5). Each cylinder was cut from a large, mother cylinder, each 2 cm in diameter, including the cylindrical, steel hammer. They were then carefully filed down to a proper size and suspended by thin, silk threads at the nodes. How were the nodes located? Were they calculated before-hand? Even with fairly constant dimensions this would have been difficult. Koenig most likely found the nodes by trial and error, which would have formed an experiment in itself (CR no. 51).[67] The membranes on Koenig's manometric instruments, crucial and delicate elements at the centre of these technologies, were also the subject of much attention. The ones that seem to be original in the Toronto collection (Smithsonian and Dartmouth as well) were made of paper coated with rabbit glue (CR nos. 237, 242, 242a). But there were also ones made of thin rubber or caoutchouc. Regardless of their origins, they were a source of a revealing dynamic between the maker and users. Loudon asked Koenig for details on how to repair his membranes.[68] Joseph Henry preferred to defer the dark art of making the membranes to Koenig. He wrote to Paris asking bluntly: "The membranes attached to the apparatus for showing the vibrations of flames have become broken, please send a supply."[69] Finally, there are large and extremely rare oak pipes that were over a metre in length. They were built for powerful, strong demonstrations of low notes, and would have demanded manufacturing and tuning skills different than those needed for making the smaller, pine pipes (CR no. 112a).

"Cette Ville de Malheur"

Schools such as Toronto purchased several of Koenig's educational instruments, but following the 1876 Exhibition, the economy was still not strong, and serious research instruments, especially those Koenig made specifically for his own experiments on beat tones, were simply too expensive for most colleges and universities. In the 1870s there were still only a handful of serious American researchers in acoustics who could use such instruments – Bell, Henry and Mayer. But even for them, the beat-tone instruments were beyond their own work. There was no one in the United States, for example, who seriously investigated psycho-physiological problems posed by the combination-tone debate. There were also few people who could appreciate the sophisticated mathematics involved in Helmholtz's studies of combination tones (Chapter 7).

Scientists viewed acoustics mainly as a teaching resource. When Wolcott Gibbs asked American physicists to present an inventory of precision instruments in their research laboratories, Harvard did not even list their Koenig apparatus; neither did MIT and the Stevens Institute (where Mayer worked). Columbia listed a few Koenig pieces, and Johns Hopkins (Henry Rowland), which described at length their precision apparatus in optics, electricity and magnetism, simply stated that they had "all the ordinary apparatus by Koenig of Paris, including Helmholtz's double siren, Lissajou's vibrating microscope, Hasting's pendulum comparator, &c."[70] There was always a large market for Koenig's popular demonstration pieces, but there was clearly a much less welcoming practical, intellectual or financial environment for his research instruments.

Even with such a small market for research apparatus, during the excitement of the last days of the Philadelphia Exhibition, George F. Barker of the University of

Pennsylvania decided that he would buy the entire collection that had proved such a spectacle for the local scientific community. Barker did not have the money, so he devised a plan to raise a subscription. For Koenig this was a large sale and a convenience; he had just lost his packaging equipment in a fire at the Centennial making it cheaper and easier to leave his equipment in North America. Peter Munzinger, a Pennsylvania businessman, who, among other enterprises, owned a gas works at Doylestown, became his agent. He and Barker signed an agreement that the former would use "every exertion in his power to raise the sum of 10,000 [US dollars] more or less."[71] In order to encourage prospective buyers, Joseph Henry, F.A.P. Barnard, J.C. Watson and J.E. Hilgard wrote a letter of appeal to American scientists to keep the instruments in the United States. They emphasized the value and scientific interest these instruments would provide for scientific associations and educational institutions. Some of the instruments they argued were built on a scale "without previous example" and with acoustical effects of great power "which must make them invaluable to the instructor or investigator." The tonometer, for example, was said to produce "any given sound for investigation." "These instruments are the perfected result of years of laborious study and great mechanical skill, and they are entirely unique of their kind. The undersigned cannot but feel that it would be a misfortune to the cause of scientific progress in the United States if they should be permitted to leave the country." They underscored that "their costliness puts them beyond the reach of any of our institutions of learning except a few of the most wealthy," but they hoped that "friends of science in the United States ... may purchase this valuable collection and generously present them to an institution in which they may be made useful in promoting the advancement of science."[72] In light of this appeal, the Board of trustees for the university unanimously adopted a resolution that "heartily" commended the citizens of Philadelphia to help raise a subscription for the apparatus. The instruments were then moved to the physical cabinet of the university, and Munzinger had a year to raise the money. In the mean time, the university had bought separately 4,040 fr worth of Koenig's education instruments.[73]

One year later, however, nothing had happened. Koenig received no reports from Munzinger, and he became worried when the latter did not respond to his inquiries. Even without a word on the status of the subscription, however, he still believed that the collection would be sold; Barker had asked several times for Koenig to send more instruments to complete the collection. Unfortunately, the real situation was much more precarious. Lawyers and creditors were chasing Munzinger. One lawyer, in fact, wrote to Koenig that Munzinger had no money, owed $75,000 to creditors and threatened to shoot him when confronted. Koenig, now very worried that his instruments and livelihood were in the hands of a madman, reported this to Barker who promptly took over the account. But it still floundered, giving rise to dozens of frustrated letters back and forth, and eventually a falling out.

This arrangement evolved into a 6-year struggle to control what was at that time a unique and invaluable body of acoustical knowledge in the form of instruments. Barker was caught in a situation beyond his control, selling high-end instruments in an almost non-existent research market, but he also benefited from the situation. For 6 years he had this famous collection for classes and showing to visitors.[74] High-end instruments, even if the owners did not know how to use them, were a status symbol for a growing university. Koenig, on the other hand, had lost a resource for his own experiments, construction activity, workshop demonstrations and promotion. Three times he asked Barker to return a novel aluminum wave siren, which was a new instrument from his research on timbre (Fig. 6.4).He wrote to Barker that he had rushed to finish it before the Exhibition and that it had been damaged by an accident on its way to Philadelphia in 1876. He seemed quite desperate to have it returned, giving detailed instructions for its packing, and offering to pay the shipping.[75] In a letter to the treasurer of the university in September 1882, he claimed to have lost sales and much time making new instruments for his research.[76] The *grand tonomètre*, for example had taken years to make, and he had to start a new one. On the commercial side, he felt he was losing sales. In the late 1870s and early 80s there was considerable research interest in combination tones in London and he must have been frustrated that he did not have the chance to sell these instruments to scientists who could actually do proper research with them (Chapter 7).

The failed subscription not only revealed a struggle to control an elite collection of instruments, but also exposed tensions between English scientists and the Americans. At one point, in early 1880, a rumor circulated in London that the University of Pennsylvania or the Stevens Institute had bought the instruments and not paid for them, refusing to return them to Koenig. Henry Morton of Stevens Institute wrote to *Nature* and *Engineering* to state he had nothing to do with the affair. James Dredge, editor of *Engineering*, then wrote to Barker, "There appears however to be a good deal of bad feeling here in professional circles against the Penna. University on account of a charge made against it, to the effect that the university acquired in 1876, a valuable set of models exhibited by Koenig at the Centennial, and that it has refused to pay for them or give them up. In fact you are accused (not personally of course but the University) of swindling Koenig."[77] Barker's reputation was on the line, so he wrote a lengthy summary of the facts to James Dredge to correct the rumors.[78]

Eventually Koenig was forced to return to America to sell what he could to other institutions and take the remainder of the collection home. He was particularly worried that his reputation had been damaged by the ensuing battles with Barker and the University of Pennsylvania, so he published a pamphlet quoting Barker's letters to clear up "erroneous rumors." "While some said they [the instruments] had been bought by the University, but never paid [for], others thought I had abandoned these instruments to the mercy of the University. So I think it will be well both for the good of the University and for me that the truth should be known."[79] He claimed that he was induced to permit the instruments to go to the University and remain there year after year. When he returned in 1882, he spent almost an entire month scrambling from university to university trying to sell the collection. He managed to sell a number of instruments to Alfred Mayer at the Stevens Institute, to Barnard at Columbia,

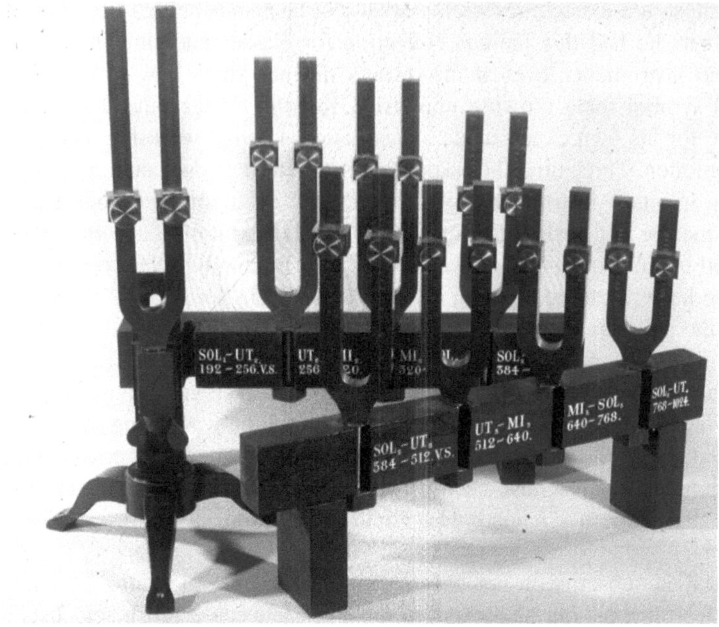

Fig. 6.8 Large tuning forks used in Koenig's 1882 demonstrations in Toronto. Photo by Louisa Yick. Courtesy of the Physics Department, University of Toronto, Canada

and Lafayette College. Before his return, the University of Toronto purchased portions of the combination-tone collection, which consisted of a series of eight tall (up to 75 cm in height), heavy tuning forks for demonstrating beat-tone phenomena to large audiences.[80] Professor Peter S. Mitchie, professor of natural philosophy at the United States Military Academy at West Point, purchased the *grand tonomètre* of 670 tuning forks for over 8,000 fr. But even with this minor success, Koenig had to pack and send eleven crates home at considerable expense. He even had to spend several days dealing with customs. He was devastated by the ordeal and would thereafter refer to Philadelphia as "cette ville de malheur." (Fig. 6.8)[81]

Public Lectures at Toronto

Koenig did make, however, some important sales during this period, notably to Toronto and the Smithsonian, and the other schools mentioned above.[82] Ironically, the combination of ups and downs of his business, the money from Toronto and the loss of his own instruments as a resource, seems to have spurred new directions in construction, invention and research. In fact, the period 1877–1882 was surprisingly productive. In the absence of several key instruments, he took radical new directions in research and invented a whole family of wave-siren instruments, published novel research about timbre and beat tones, and started construction on his final, masterpiece tonometer (Chapter 7). He also used the turmoil to nurture the image of a starving scientist who cared not for material gain, but pure scientific motives. James Loudon and others believed that he sold instruments primarily to fund his own research. S.P. Thompson in Britain, for example, remarked upon his death:

> If by some stroke of luck he sold instruments that brought in a few hundred francs above the regular income of his business he would hail it as the means of constructing some new piece of experimental apparatus that might never find a sale, but would help his investigations. And so with a slender business and a few faithful workmen at his back he maintained a proud independence, sufficient to enable him to continue research.[83]

In the early 1880s, just as the Philadelphia saga was coming to a head, his atelier was again buzzing with new instruments and audiences. During the electrical congress in Paris in 1881, for example, several notable physicists, Helmholtz among them, came one evening to witness his latest experiments (Chapter 7). In fact, as he became more skeptical of fairs and conferences, his atelier became the main place for witnessing his experiments and purchasing his instruments. His challenges never seemed to cease, however. In January 1882, only a few months before his long journey back to America, Koenig was again forced to move from 26 rue de Pontoise to another address, 27 Quai d'Anjou on the Île St. Louis. This move, which would be his last, was quite disruptive as he had to wait until April to move his entire atelier. Eventually he set up the atelier on the ground floor and lived in an apartment above. Quai d'Anjou was centrally located and, being on an island, offered one of the last relatively quiet places in the city. He came to love this apartment and in later years referred to it as his "cher Quai d'Anjou."[84]

He was now in one of his busiest years. He was preparing for his return to Philadelphia, compiling a revised catalogue (1882), planning a lecture series at Toronto, publishing recent research in *Annalen der Physik*, carrying out orders for clients, and producing a book, a collection of his writings since 1858. He also translated his earlier articles from German into French. The catalogue itself was a major undertaking that took almost 2 years to complete. He changed prices (owing to inflation in the early 80s) and added several instruments, due to inventions from himself and the growing field. In 1880, as he was preparing the catalogue, he wrote to Loudon in Toronto: "These publications have become all the more necessary for me, because during recent years, despite the good order from your university, business has been far from good enough to erase, even by a tiny amount, the consequences of my disastrous undertaking [*ma disastreuse entreprise*] at the Philadelphia Exposition."[85]

By August he set sail for North America. On top of the anxieties of resolving the Philadelphia situation (a frantic search for buyers, customs problems, shipping and repackaging instruments) he also agreed to deliver an important set of public lectures in Toronto. Koenig, eager to keep up sales in North America, saw these lectures as a kind of advertisement. "I would be naturally delighted if [the collection] would inspire in some professors the desire to stock their physical cabinets in acoustics."[86] He also planned the trip to coincide with the American Association for the Advancement of Science meeting in Montreal in August, where it "would be important for me to see as many American scientists [*savants Américains*] as

Fig. 6.9 Koenig's double siren (*left*) sound analyser (*middle*) and wave siren (*right*) in the Lecture Theatre of the Macdonald Physics Building, McGill University, Canada. date: 1893. Photo courtesy of the McGill University Archives, PL028671

possible."[87] Most importantly, he saw the lectures as a chance to demonstrate the significance of his research. "I think the delay of a few months [September instead of June] would only serve to increase the interest of these lectures, by permitting you to take into consideration my latest research on timbre that you find in my book, and by giving me the time to prepare more drawings and photographs."[88] For 6 months before the visit, Loudon and Koenig corresponded regularly regarding the appropriate outline and content of the lectures. Koenig suggested that he bring some of his new instruments. Loudon agreed and paid for the shipping and, of course, ended up buying a few of the items (Fig. 6.9).

The six lectures, sponsored by the Canadian Institute, were advertised in Toronto's main paper, the *Globe* : "Canadian Institute Science Lectures. A Course of Lectures on Sound will be delivered by Dr. Rudolph Koenig, Paris, France, inventor of many of the most important instruments used in the study of acoustics, and Prof. Loudon, in the library of the Canadian Institute, Richmond Street."[89] The lectures, delivered twice a week during the month of September, were "illustrated by a series of new and beautiful experiments, commencing at 8 sharp." Tickets were three dollars for the entire course of lectures. The *Globe* published articles summarizing each of the lectures and praised Koenig as "the most profound experimenter in acoustics in the world" whose "his invaluable instruments have a world-wide reputation." The audience was "distinctly one of 'culture rare' ... drawing heavily from the ranks of the *savants* of the city."

The daily reports are a rare, thorough description of an instrument maker delivering a complete lecture series from his catalogue. The University of Toronto was one of the few institutions that could offer such a series. In the crowded library of the Institute, Koenig "performed experiments" while Loudon lectured. The lectures began with a demonstration of Auzoux's anatomical model of the ear and inner ear, followed by a manometric flame demonstration with the "curves illustrating the condensations and rarefactions that occur in the ear."[90] Koenig performed what were called "palpable experiments" to make basic phenomena such as longitudinal vibrations visible. On another day, Loudon explained the phenomena of beats and their uses in tuning while Koenig performed a demonstration with tuning forks that was "rapturously applauded."[91] There were experiments with all his manometric instruments, sirens and projection apparatus. In the context of discussions about timbre, he explained the importance of the selection of woods and materials for musical instruments. He did graphical demonstrations with rotating drums and optical demonstrations with Lissajous forks. He took up the last few days of the lectures explaining and demonstrating his revised work on timbre and beat tones. He sounded the extra-large tuning forks for demonstrating his "beat-tone" effects. He also demonstrated his wave siren on the final evening.

The lectures at Toronto, even amidst the disorder of the Philadelphia affair, marked one of the high points in Koenig's career. Most importantly, he had just completed a series of experiments on timbre and combination tones that he turned into another family of instruments. In the next decade, as his research, instruments and social life became increasingly separate from the mainstream; he became consumed with reforming Helmholtz's acoustics, especially his notions of timbre and combination tones. His atelier, that had been a place of construction, experiment and business, would become a platform for challenging the most celebrated German scientist of the time.

In conclusion, in the mid 1870s Koenig made an ambitious attempt to infiltrate the American market. When one looks through collections of historic scientific instruments at colleges and universities throughout North America, it is clear that these institutions were a large consumer of French instruments in the nineteenth century. In fact, the more we learn about these collections, the more it appears that in the late 1860s and early 1870s, the North American market became a major source of support for the declining French instrument trade. Even as the German instrument trade ascended, North Americans, especially teachers, continued to buy their standard teaching and laboratory instruments from Paris. William Rogers of MIT, Frederick Barnard of Columbia, Joseph Henry of the Smithsonian, John Foster of Union College, Charles Young of Dartmouth College, James Loudon of the University of Toronto, William Dawson of McGill, and Henry Rowland of Johns Hopkins, each customers of Koenig, represented this market for science education. The Smithsonian Institution purchased French instruments to create a national showcase for artisans and teachers; the University of Toronto bought them as a centerpiece for an ambitious teaching laboratory.

Whereas teaching instruments were part of a growing market, high-end research instruments were another matter. Koenig's failure to sell his top research equipment following the 1876 Centennial Exhibition revealed another side of the precision instrument market – the difficulties of balancing the need to mass produce teaching instruments with the considerable financial and labour demands of manufacturing highly-specialized research apparatus. They were difficult to sell, and most customers did not know what to do with them. In effect, they were a nineteenth-century version of today's Haute Couture fashion business in Paris – they drove innovation, promoted the maker, added status to the client, and represented an embodiment of elite scientific and artisanal knowledge. Koenig's post-exhibition ordeal in Philadelphia centered on one of the best collections of acoustical instruments assembled in the nineteenth century. Through his demonstrations at home and abroad, with mixed results, he blended the promotion of business with his controversial views on sound.

Notes

1. Moigno (1871, p. 602).
2. Ibid.
3. Loudon (1901b, p. 10).
4. Ibid., pp. 10–11.
5. Rudolph Koenig to James Loudon, Jul. 24, 1892. Dec. 16, 1892. UTA-JLP.
6. Cohen (1970, DSB).
7. Mayer (1896, p. 84). Also see Rudolph Koenig to James Loudon, Jul. 24, 1892; Jul. 26, 1894. UTA-JLP.
8. *Proceedings of the Musical Association*, vol. 8.
9. Undated letter circa 1883 Rudolph Koenig to James Loudon, UTA-JLP.
10. Loudon (1901b).
11. This observation is based on the prevalence of French instruments in North American collections that date from the nineteenth century. For a background to some of these collections, see Warner (1993). Also see Tom Greenslade's web site, http://physics.kenyon.edu/EarlyApparatus/index.html
12. For more on Rayleigh, see Ku (2005, 2006). For information on Tyndall, see Beyer (1998, pp. 70–79).
13. Smith and Wise (1989, pp. 28, 107–108).
14. Hiebert and Hiebert (1994, pp. 306–307).
15. Koenig (1873a, p. 2).
16. Henry (1856).
17. Moyer (1997). For his early purchasing trips, see Gee (1990).
18. Warner (1993, pp. 17–22).
19. Henry (1850, p. 18).
20. Henry to Alexander Dallas Bache, Mar. 31, 1847, in *The Papers of Joseph Henry*, vol. 7. January 1847–December 1849, the Smithsonian years. ed. Marc Rothenberg (Washington and London: Smithsonian Institution Press, 1972), p. 70.
21. Henry to Koenig, Sept. 23, 1873, SIA-JHP, outgoing, ru 33, vol. 35, reel 55, pp. 635–636.
22. Koenig to Henry, Oct. 11, 1873, SIA-JHP, Incoming, Record Unit 26, vol. 137, p. 286.
23. Henry to Koenig, SIA-JHP, Outgoing, Record Unit 33, vol. 40, reel 61, p. 352; vol. 41, reel 64, p. 733; vol. 44, reel 69, p. 416. This order included, from the 1873 catalogue, nos. 35, 52, 66, 118–124, 126, 128, 129, 131–134, 137, 140, 141, 142, 143, 144, 146, 147–151, 153–158, 162, 193a, 160, 163, 167, 168, 172, 174, 182–185, 186, 188, 189, 193, 196, 199.
24. Koenig to Henry, SIA-JHP, Incoming, Record Unit 26, vol. 166, pp. 269–275. The instruments that Henry purchased in 1877 included Koenig's premier graphical instruments (1873

catalogue): the Regnault Chronograph (205a), tuning forks for graphical composition (208a) and demonstrating Lissajous figures (209a).
25. Queen & Co. 1884, *Notice*. Also see the double siren sold by "N.H. EDGERTON PHILA, PA." to Smith College. CR no. 27.
26. See the series of ten resonators at the University of Sydney, plaque insert on wooden base, not worked on resonators "W. Ladd & Co/11 & 12 Beak St/Regent St W." A draft for 200 Pounds was made out to Koenig in 1880. CR no. 54a.
27. Koenig to Henry, SIA-JHP, Incoming, Record Unit 26, vol. 166, p. 271. Also see, Koenig to Barker, Jan. 4, 1877. UARCUP.
28. Queen & Co. 1884, *Notice*. Drummeter (1989). See Holland (2003) for an example of a customs case involving a scientific instrument imported to Australia.
29. Pilcher and Union College (1994, pp. 62–67).
30. Lagemann (1983, pp. 44–46).
31. Rezneck (1962).
32. It is not known exactly when Rowland bought instruments from Koenig. He may have bought some in 1875 and others following the 1876 exhibition.
33. George F. Barker to Rudolph Koenig, June 30, 1882. UARCUP.
34. Koenig, Catalogue (1889), in the front sections titled, "Advertisement," he wrote, "The greatest care is taken in packing, but the goods are forwarded at the risk of the buyer."
35. Bell papers, LOC.
36. Gilman to Rowland, July 22, 1876. MELSC.
37. Giberti (2002, p. 106). Also see Brain (1993) and Bennett (1983, 1985).
38. Burr (1877, p. iii).
39. Giberti (2002) focused on the complex task of organizing and classifying the 1876 Exhibition.
40. "Characteristics of the International Fair," *Atlantic Monthly*, 38 (Jul. 1876), pp. 85–91; (Aug. 1876), pp. 233–239; (Sept. 1876), pp. 350–359; (Oct. 1876), pp. 492–501; (Dec. 1876), pp. 732–740; 39 (Jan. 1877), pp. 94–100; p. 284
41. Ibid., p. 495.
42. Ibid., p. 109.
43. Brain (1993, p. 151).
44. Bennett (1985, p. 23).
45. Barnard (1870b, p. 469).
46. Brenni (1991, pp. 462–463).
47. "Characteristics of the International Fair," p. 238.
48. United States (1876, p. 345).
49. United States (1880, pp. 516–517).
50. Koenig (1882c, p. 157). Koenig (1901), Première partie, p. IX. Terquem (1870, p. 291).
51. There were also Mr. E. Levasseur, France and P.F. Kupka, Austria, who judged the astronomical, meteorological, and surveying instruments. Ibid., p. 19.
52. United States (1880, pp. 334–335).
53. Ibid., p. 489.
54. Ibid.
55. Burr (1877, p. 400).
56. Ibid.
57. For more on the history of physics at Toronto, see Allin (1981). For more on the development of the German teaching model in Canada, see Gingras (1991). For more on the social and cultural context of these developments in the United States, see Warner (1992).
58. Loudon (1916, p. 40).
59. *Nature* 29 Aug. 1878 and 5 Sept. 1878.
60. Koenig placed a notice in scientific journals at this time advertising the move. Mack (1970, p. 57) also describes the same situation for Koenig's neighbour, Gustave Courbet.
61. Loudon (1916, p. 40).
62. Brenni (1993–1996).

63. Loudon (1916, p. 40).
64. Ibid.
65. Rudolph Koenig to James Loudon, Mar. 10, 1895. UTA-JLP.
66. CR misc. instruments.
67. The frequency is proportional to the inverse square of the length (if diameter remains constant). See CR no. 51.
68. Rudolph Koenig to James Loudon, July 25, 1879. UTA-JLP.
69. Joseph Henry to Rudolph Koenig, August 4, 1875. Also see Pantalony (2001, p. 19).
70. Gibbs (1879, p. 7).
71. Contract between GFB and P. Munzinger, December 30, 1876. UARCUP.
72. F.A.P. Barnard et al., in Koenig (1882b). UTA-JLP.
73. Koenig to Barker, Jan. 4, 1877. UARCUP. Adjustable tuning fork, Toepler and Boltzmann pipe, sensitive flame apparatus after Mr. Govi, 5 frames with 10 tableaux, bow for contra bass, locomotive whistle, Trevelyan rocker, Reis telephone, 22 cylinders for high pitch, lycopodium demonstration, 4 plates, 8 plates to demonstrate Wheatstone's theory, stroboscope demonstration, Wheatstone demonstrations, Helmholtz double siren and a grand bellows.
74. Barker apparently demonstrated some of the instruments for Henry. See, Koenig to Henry, Jun. 22, 1877, SIA-JHP, Incoming, Record Unit 26, vol. 166, 270.
75. Koenig to Barker, Jun. 26, 1877. UARCUP. Also see Jan 4. Aug. 17, May 25, 1880.
76. Koenig to Caldwell Riddle, Sept. 22, 1882. UARCUP.
77. James Dredge to Henry Morton, June 16, 1880. UARCUP.
78. Barker to James Dredge, August 1, 1880. UARCUP.
79. Koenig (1882b) found in UTA-JLP.
80. See Rudolph Koenig to James Loudon, Oct. 17, 1882. UTA-JLP. Also see CR misc. instruments.
81. In October 1882 Koenig wrote daily descriptions of these events to James Loudon in Toronto, Ibid.
82. Henry bought nos. 205a, 208a and 209a for 2,936 fr. SIA-JHP. Contrary to what Miller (1935) claimed, many of the instruments bought for Toronto in 1878 came from Koenig's studio in Europe, and not from Philadelphia. In 1878 Koenig still expected the Philadelphia collection to sell in the United States and therefore did not ship them to Toronto.
83. Thompson (1901, p. 630).
84. Rudolph Koenig to James Loudon, Oct. 3, 1890. UTA-JLP.
85. Ibid., Jun. 24, 1880.
86. Ibid.
87. Ibid., Nov. 25, 1881.
88. Ibid.
89. *The Globe*, Aug. 30, 8; Sept. 13, 8; Sept. 16, 14; Sept. 19, 6; Sept. 21, 8; Sept. 21, 8; Sept. 23, 14; Sept. 26, 6; Sept. 28, 8.
90. Ibid., Sept. 16. The Physics Department at the University of Toronto still has the Auzoux model of the ear.
91. Ibid., Sept. 19.

Chapter 7
The Faraday of Sound

Rudolph Koenig was not formally associated with any school, institute, laboratory, or academy, nor was he even educated past secondary school. He was a scientific instrument maker, earlier trained as a violinmaker, who lived in a workshop/apartment near his products. Since the early 1860s, he had helped refine and spread Hermann von Helmholtz's studies in acoustics through his creations in steel, brass, wood, glass and cast iron. Later in his career, however, he became one of the strongest critics of Helmholtz. In the controversies recounted below, he disputed Helmholtz's theory and experimental findings related to the elusive, yet fundamental acoustical phenomena of combination tones and timbre.

One outcome of the disputes with Helmholtz was that Koenig's instruments became much larger and louder. They were built in the rhetorical spirit of previous "grand appareil" such as the wave siren (CR no. 59), double siren (CR no. 27) and tonometer (CR no. 36). The storage facility of the Science Museum in London has a set of massive stands, tuning forks and resonators that came from a series of demonstrations on beats and beat tones delivered by Koenig in May of 1890 to the Physical Society of London (Fig. 7.1). The cylindrical resonators are over a metre long, with the forks between 67 and 93 cm, and the stands between 73 and 107 cm. The three large forks (as heavy as 50 lbs) are graduated to produce low notes ranging from ut1 to ut2. There are six heavy brass and steel sliding clamps, with a precision lock for fixing the weight on the required note. Sturdy screws attach the forks to the cast iron stands. The instruments are engraved "DS & SK Mus" referring to the South Kensington Museum, the institutions that bought the instruments after the lectures.[1] They derived from Koenig's work at the time on his complete universal tonometer. As we will see below, aside from producing powerful notes for research purposes, they were also meant to impress, overwhelm and convince an audience (Fig. 7.2).

Koenig's vibrant workshop, like these massive demonstration forks, loomed over acoustics during this period, altering the social dynamics of these disputes and the scope of acoustics in general. His research and artisan activities raised questions about the integrity of observations and instruments, and stimulated debates about the nature of acoustic sensations. His reputation as a highly skilled artisan, his reliance on the expert ear, his scepticism about Helmholtz's theories, and his graphical techniques fuelled these questions and shifted the path of debate. Before we look closely

D. Pantalony, *Altered Sensations*, Archimedes 24, DOI 10.1007/978-90-481-2816-7_7,

Fig. 7.1 Remnants of large cylindrical resonators and tuning forks used for Koenig's 1890 demonstrations in London. Science Museum storage facility, Wroughton, UK. Photo by author 2003. acc. no. 1890–53

at the details of these disputes, however, I shall reconstruct the social, commercial and scientific context of the final years at Koenig's atelier.

Life at Quai d'Anjou: 1882–1901

In the 1880s there was continued growth in science education and research market. During this period, the German trade grew, but there was still enough business for the Paris makers. Long delays were the norm as eager professors waited to stock their laboratories. The University of Toronto was typical in ordering instruments to build research and teaching laboratories; but it seemed to have much more money than the average college or university in North America. Following Canadian Confederation in 1867, the Ontario provincial government had gained almost total control over education and one of its goals was to build a centre in English Canada for training in science and technology. Provincial administrators became interested in any form of technical or scientific education that could contribute to economic and industrial expansion. With this in mind, James Loudon created the School of Practical Science (SPS) (engineering) at the University of Toronto.[2] Loudon was therefore in charge of ordering instruments for the SPS and Physics in the 1880s.

Koenig was essential to Loudon's vision for science at the University of Toronto. In effect, he acted as Loudon's agent in Paris. The letters between the instrument maker and teacher provide a rare picture of the fluctuations of the instrument trade

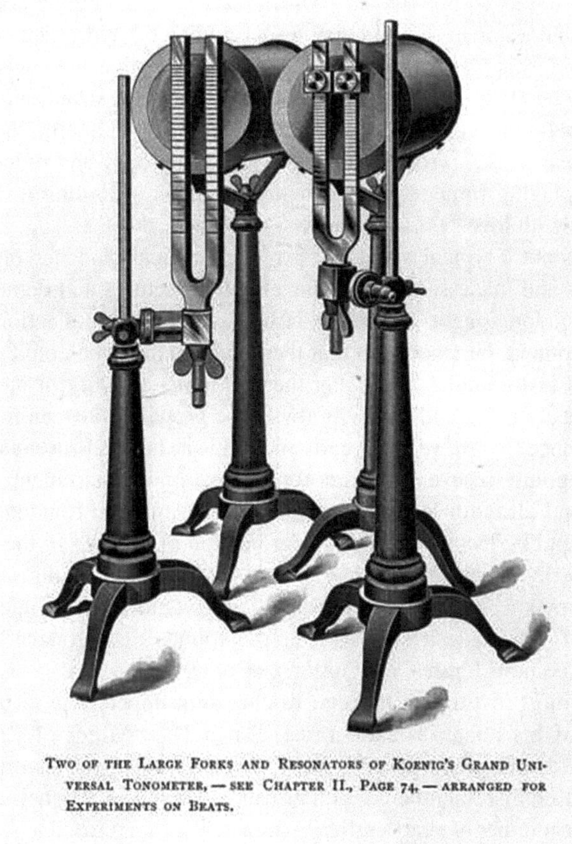

Fig. 7.2 Large forks and resonators from Koenig's complete universal tonometer for experiments on beats
Source: Zahm (1900), frontispiece

in Paris, and the challenges of building a laboratory in North America, far from the centre of action in Europe. Due to the sheer number of instruments being ordered by Toronto and other schools, Koenig spent a lot of time managing Loudon's orders. The famous Brunner brothers who made precision instruments for astronomy, geodesy and meteorology were, according to Koenig, "maniacs, who being very rich only accept work that merely pleases them to do, not wanting generally to deliver, neither for foreigners or through agents" and he was anxious to terminate his business dealings with them.[3] Jules Carpentier, the precision electrical instrument maker that Koenig greatly respected, was also "very rich" and a "constructor by preference rather than to make money [*il est trés riche et constructeur plustôt par gout que pour ganger de l'argent*]."[4] In fact, accept for the Brunner brothers, Koenig valued instrument makers like himself who strove for quality, shunned profit and worked from pure motives. He referred to one unnamed maker, who was constructing a precision lathe for Loudon, as the "mechanician who works a little more like an artist than as a shop-keeper [*commerçant*]."[5] Alvergniat, who made glass apparatus for laboratories (e.g. precision thermometers, hydrometers and mercury vacuum pumps), was totally disorganised and took months, even years to complete orders. Koenig had to go through his books methodically to save Loudon from being cheated; he finally gave up with "this detestable atelier."[6] Laurent, the famous optician, does "great work," wrote Koenig, but he is very busy and ordering from him "will be like playing roulette."[7] The Seguy brothers, according to Koenig, made Geisler tubes "with love."[8]

Koenig was not a typical agent. He provided critical guidance on the purchase of instruments and made suggestions for effective lectures and demonstrations. In 1882 when Loudon sought advice for buying a complete collection of advanced electrical equipment, he responded that there was no one who could possibly make all the desired instruments: "In seeing the enormous quantity of electrical instruments that already exist, and that will always be growing, one can no longer know where true science [*la vrai science*] ends and industry begins [*la simple industrie*]."[9] He suggested going to several makers for a selection of instruments. He provided reading lists and ideas for lectures as well. He forwarded to Loudon several pages of notes for a public lecture on the standardisation of pitch.[10] In the late 1880s he developed a series of projection instruments for Loudon's lectures on sound. On one such improvement he wrote: "I thought you would increase interest in acoustics courses, if you made it easy to hear the sounds that produce the projection phenomena, Lissajous figures, inscriptions, or others."[11]

More than most instrument makers, Koenig kept impeccable records and knew the contents of his clients's laboratories. When D.C. Miller of Case School in Cleveland visited his studio in 1896 and ordered a number of instruments, he was surprised that Koenig remembered each instrument that A.A. Michelson had ordered for the school a number of years earlier.[12] In fact, he often refused to sell instruments if they overlapped with the client's collection.[13] His relationship with his clients did not stop at the sale of his instruments; he constantly made tiny improvements to his instruments and informed them of the changes. He sent Loudon advice on how to keep the instruments in good working condition, how to prevent rust on the tuning forks, and how to replace membranes in certain instruments. He provided long hand-written instructions for the use of the more complex instruments.

Aside from waiting for instrument makers to complete orders, scientists had to endure shipping delays as well. This sometimes meant that a whole course would be cancelled, or important research put on hold while instruments made their way to remote destinations. Much of Koenig's time, as an instrument maker and agent, included packaging orders into crates and organising shipping to his clients. For Toronto Koenig usually used a company called Sherbette, Kane, & Co. He sent the crates by train to Le Havre, by steamer – "the Labrador" – to New York, and then by train – "Merchants Dispatch Transportation Company" – to Toronto. There were often problems with the delivery and he threatened to take his business elsewhere. But he remained with Sherbette and Kane for over twenty years. Part of the shipping routine involved follow-up letters where Loudon listed parts that were broken in transit. Koenig would then replace the broken parts, which entailed more visits to instrument makers, and more waiting, followed by another shipment. To save money he would wait to send these replacement parts with another big shipment months later.

Promotion was another constant concern for an instrument maker. Koenig spent an enormous amount of time and energy on his catalogues. His last catalogue, began in 1887, took almost a year and a half to prepare. All of the descriptions were in French, but the titles of the instruments were also given in English and German, (Loudon helped with the English translations). He was clearly interested in catering to the English-speaking world, where his best supporters and customers were to be found, but he could not afford the costs of printing a second version of the catalogue. With the arrangement of having the titles in three languages, "someone without French can easily go through the catalogue and if an instrument interests him, he can take the time to go through the description in the less familiar language."[14] In addition, the catalogue had engravings on almost every page. He had first intended to take the instruments to a photography studio, but soon realised that was too difficult. He took the photographs himself, fretting over the right angles, and submitted them to a reputable photoengraver for the final images. "In order to have good figures it is most difficult to find the best arrangement of the apparatus and the most convenient size for each engraving, these trial and error sessions take a long time."[15] By early 1889 the catalogue was not yet complete and there were problems with the photoengraver. Koenig wanted the "maleureux catalogue" done before the summer expositions and upcoming conferences, so he was forced to go to a wood engraver to redo all the images. In February he was still arranging apparatus and taking pictures for the engraver and was finished by early summer 1889. "When this catalogue is finished, no one will be able to imagine all the troubles and all the sacrifices that it cost me."[16]

He produced the 1889 catalogue at the start of a less stable phase in his career. A world economic slump hit the instrument market quite suddenly and forced him to contract his business.[17] By the end of the 1880s the Parisian instrument scene, which had dominated the instrument market for over 50 years, was entering a period of decline. Many scientific and technical innovations were now coming from German laboratories, and Germany was emerging as the world leader in precision instrument design and manufacture.[18] Universities in North America and Europe, which had traditionally looked to Paris for guidance, were now looking further east.[19] In addition, most French workshops did not adopt factory techniques for manufacturing instruments more efficiently. Firms that did change their manufacturing techniques, such as Carpentier, thrived in the more competitive market.[20] Firms such as Koenig's, that worked with an artisan ethic, had a harder time keeping up with orders and keeping prices down.[21] To make matters worse, by the end of the 1880s interest in acoustics was declining. Electrical studies were becoming increasingly popular, and more traditional areas of study were not a priority for professors outfitting laboratories.

The first clear indication that times were changing for Koenig and his Parisian colleagues came with the Paris Exposition held in the summer of 1889. Koenig had been working hard on his catalogue, partly in the hope of meeting potential buyers and scientists at this international fair. Unfortunately, the fair turned out to be poorly organised and he ended up not participating in it at all.[22] In fact, he reported to Loudon that the precision instrument section was empty for almost two weeks at the start of the fair. Such a state of affairs reflected the beginning of a decline in the once powerful French instrument trade. It was also a tough year for business and he had to trim back his staff to a "minimum." Even still, he managed to complete his collection and reported to Loudon that it looked "truly excellent."[23]

Throughout his work as an unofficial agent and the fluctuations of his own business, Koenig's happiest hours were spent researching. This seemed to be his escape. In August 1888 after 5 years of straight business, and a few good orders providing "tranquillity" for a while, he was able to get back to his studies of beats and timbre.[24] Amazingly, for a prolific researcher, he had not published a paper since 1882. He wrote to Loudon during this period that "I will nicely supplement the theory of 'timbre' through some new experiments [*je compléterai bien la théorie de la 'Klangfarbe' par des experiences nouvelles.*]"[25] He was still determined to modify some of the fundamental aspects of Helmholtz's theory. As mentioned in Chapter 6, he had demonstrated these experiments in the presence of Helmholtz himself earlier in the decade, but had failed to convince the famous German physicist of his controversial claims. In the mean time, he had gathered more evidence and, in the case of his work on timbre, invented instruments that he believed imitated better what was actually going on in nature.

The Paris Exposition and a disappointing lecture at an 1889 congress at Heidelberg (see below) marked a turning point for Koenig's career. One of his friends, Le Conte Stevens, professor of physics at Washington and Lee University, wrote to him in 1894 that acoustics should be abandoned because it had nothing new to offer. "Under these circumstances," Stevens wrote, "not only do I feel no stimulus to scientific research, but I feel that research which does not relate to questions of industrial importance tends rather to injure me in my relations with those who are my colleagues."[26] This letter upset Koenig greatly and drove him into further isolation. The next year he wrote to Loudon: "Business is as good as it can be at this time where the electrical rage [*la rage électrique*] fills most scientists with contempt for acoustics, which produces neither lighting apparatus nor electric motors, but one lives all the same and make ends meet."[27] On the other hand, as with other difficult periods in his life (early 1870s and early 1880s), he published a flurry of papers in the early 1890s.

Even though Koenig still dominated the acoustical instrument trade in the 1890s (he never had any serious competitors) the ascendance of German science and the German precision instrument trade meant that there were fewer pilgrimages by scientists to Paris, and therefore less business. When one of Loudon's students, J.C. McLennan (1867–1935), visited Quai d'Anjou in 1898, Koenig informed him that Carpentier was the only good maker remaining in Paris, the rest were in Germany.[28] In fact, McLennan was at the Cavendish Laboratory in England at this time and was on his way to Germany to learn German, tour laboratories and buy instruments for Toronto. Koenig refused to leave Paris, even though his family in Königsberg and colleagues suggested that he move back to Prussia. He told Loudon that he "would rather live in Paris amongst his enemies than in any other city in the world among his friends."[29] In reality, Germany may not have been friendlier to him. Through his attempts to dethrone Helmholtz, he had become unpopular in Berlin. During a visit there in 1892, Le Conte Stevens reported that a laboratory assistant warned him not to mention Koenig's experiments in front of Kundt or Helmholtz.[30]

Koenig's best customers and supporters remained in the English-speaking world. He was somewhat of an experimental folk hero in Britain, at times even compared to Faraday.[31] British scientists celebrated the independents and experimentalists who had worked their way to the top using their hands.[32] They admired Koenig's courageous stand against Helmholtz. As well, a deeper national context was at play. In Germany, France and Britain, the interests of industry, science and empire had merged, surfacing at international fairs and conferences.[33] English-speaking scientists, as noted above, were acutely aware of the shift away from London and Paris toward a growing dependence on Germany for knowledge, training and precision instruments.[34] In Germany, rapid industrialization after 1870 stirred national sentiment in different forms. The "mandarins" or educated elite (including scientists), concerned about their declining authority, and threatened by English-style democratic reforms and the advancing modern machine age, saw their role as protecting German values from deterioration.[35] These national overtones were particularly acute in psychology and psychophysics, and it was around this time that different schools of thought began to emerge in both countries.[36]

These tensions worked to Koenig's advantage in Britain. Even though he was German, and even though some of his views on timbre fit closely with the emerging German gestalt thinking, the British scientists ignored these contradictions and celebrated him for his stand against Helmholtz and for his practical, workshop approach to acoustical problems. As we will see below, these differences became especially apparent in the late 1880s and early 1890s when Koenig presented his work to both German and British audiences. The former was a bitter disappointment, the latter a success. In the 1890s, English scientists such as S.P. Thompson continued to visit Koenig's studio to see the latest experiments. Lord Kelvin wrote to Koenig congratulating him on his latest papers on timbre.[37] James Loudon, C.A. Chant, J.C. McLennan and even some administrators from Toronto continued to visit Koenig in Paris, well after the instrument market shifted to Germany. MIT as well continued to buy more Koenig products to update their cabinet. Professor Charles Cross, who joined the staff in 1870, and who had experimented with Bell, continued to develop the acoustics laboratory. In the 1890s he used Koenig's tuning forks in his collaboration with Levi K. Fuller of the Estey Organ Company to standardize pitch in America (Fig. 7.3).[38]

In 1891 Koenig returned to a job he had been working at since 1877, the completion of an even larger tuning-fork tonometer (complete universal tonometer). He had announced in his 1889 catalogue that he had nearly finished the job, but realised that he first had to overcome serious technical difficulties. He completed it in 1894 and believed it to be his masterpiece.[39] It ranged from 32 v.s. to 43,690 v.s. (16–21,845 Hz), which was just above the threshold of human hearing.[40] There were 158 forks with resonators, stands and sliding weights to adjust the frequency. In

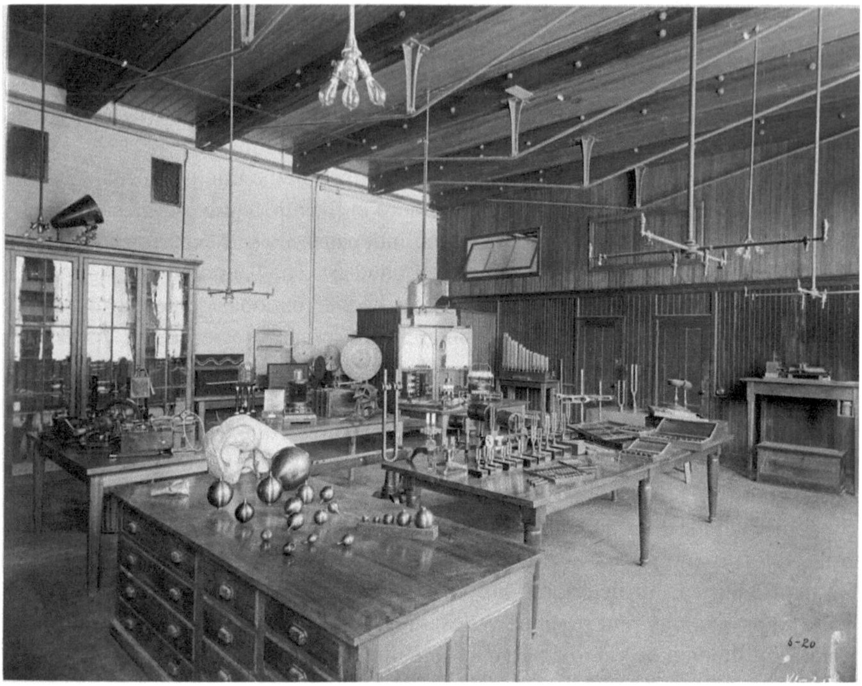

Fig. 7.3 The acoustics laboratory at MIT, about 1890 (PH 552). Courtesy MIT Museum

total the complete universal tonometer produced 1,618 tones (CR no. 36). He had finally succeeded in his lifetime goal – to produce the definitive instrument for precision tuning that offered a full range of sounds in the smallest possible gradations of pitch. He had also intended to create an instrument that could confirm in all ranges of sound (even above normal hearing) his own findings of beats and combination tones as a powerful empirical argument against Helmholtz.

During this period, Koenig built upon his notion of a wave siren as part of his continuing efforts to challenge Helmholtz's theory of timbre. He had first worked with this kind of instrument in the late 1860s (Fig. 6.4). The latest wave siren was intended to imitate more accurately the role of phase in the production of vowel sounds. At this time, physiologists such as Ludimar Hermann of Königsberg were paying more attention to his work. By 1895 Koenig had prepared two articles describing his *grand sirène à ondes* (large wave siren) and his latest series of studies.[41]

Visitors sometimes stayed at Koenig's apartment to conduct experiments on his prized instruments. Mayer spent two summers in the 1890s at Quai d'Anjou performing experiments with the complete universal tonometer. Le Conte Stevens, as well, was a regular visitor. At one point in July 1892, both Mayer and Stevens were staying at his studio performing experiments. A.A. Michelson as well visited his studio that year.[42] When McLennan visited in 1898, he spent a week of afternoons at the studio.[43]

Aside from the focus on science, Koenig was a gracious and entertaining host. Evenings at his apartment included humorous stories, music and literature. He knew Heine's poems by heart (he told Loudon that had always been before him when he worked at Vuillaume's) and Goethe, Schiller and Shakespeare were also favourites.[44] His guests were often treated to special wine from his cellar, and a simple meal from his kitchen.[45] He had a housekeeper and kept a garden in the courtyard of his apartment.[46]

Koenig's apartment was a busy place, but as the precision trade and acoustics declined, he became isolated in certain respects. He no longer went to fairs and did not have the luxury of reporting to an academic institution with its built-in social life. He often waited for scholars who wrote in advance of a pending visit to Paris, even passing up vacations to the Baltic in the hope of experimenting or making a sale. In 1897, for example, he decided not to attend the British Association meeting in Toronto because he felt it would be "ridiculous" to take his large wave siren across the Atlantic for a group of scholars who could see the same experiments in his studio.[47] Unfortunately, the guests did not always come as he planned. One year later, on Dec. 17, 1897, Koenig seemed discouraged and quite alone. He wrote to Loudon with New Year's greetings saying that the "year was less lively, because I did not leave my Quai in order to wait for the visits of foreign scientists who, besides, did not come."[48] Throughout this period, he was continually disappointed when scientists promised big orders and then backed out.[49] Saddened by this state of affairs, he had once written to Loudon that the big donations to American universities were usually not for acoustics, a science that in his words had been "almost entirely abandoned."[50]

In the midst of the difficulties of the 1890s, Koenig won a major order from Moscow in 1895.[51] This order was so large that it kept his studio working following his death in 1901. In the summer of 1897 he also sold his wave siren for 10,000 fr, a large sum for the day. These successes provided the financial security to carry out more research. With some new techniques for constructing high-frequency tuning forks and measuring the vibrations of high frequencies, he created a whole line of inaudible frequencies up to 90,000 Hz. Through this work he added considerable range to his masterpiece complete universal tonometer. His last publication in 1899 on the production of these high frequencies was the first comprehensive study in ultrasonics, destined to become a major part of twentieth-century acoustics.

In 1897 one of his last close friends, Alfred Mayer, died, which saddened him "profoundly."[52] At this time Koenig's own health was starting to slip. He developed Bright's disease, a degeneration of the kidney.[53] For the next four years his stomach was often swollen and sore, he lived off milk alone, and he had surgical procedures to relieve the symptoms. In the midst of this he continued research on ultrasonic frequencies. He also continued to fill orders for Moscow and other institutions.

As the end neared, and he was increasingly bed-ridden, Koenig received a number of orders "as if people know that my end is near and they want to profit."[54] Nevertheless, he was forced to start slowing his business and he asked Loudon for advice on how to sell the remaining instruments in his shop. He was particularly worried about his complete universal tonometer that now included his inaudible

Fig. 7.4 Sketch of Rudolph Koenig by his niece, Helene, in 1901
Source: Neumann (1932b)

tuning forks. He was asking 50,000 fr for the instrument, an unheard-of sum in those days. He was also trying to sell another large wave siren for 6,000 fr.[55] By the spring of 1901, knowing that he was dying, he wrote to Loudon that he would be willing to sell his remaining collection for half price, including the complete universal tonometer and the large siren. Loudon came that summer to try and sort out these sales. He ended up buying Koenig's prized collection of projection instruments that, in Koenig's earlier description to Loudon, would "render very good service to a professor who would like to present an acoustical course before a large public."[56] But the potential buyers for the tonometer, which included the South Kensington Museum in London, and Stanford University in California, hesitated at such a large sum, even at the reduced price (Fig. 7.4).

The difficulties Koenig went through were somewhat relieved by a letter from S.P. Thompson in London informing him that he had been named an honourary fellow of the Physical Society in London. "I was quite surprised by this whole affair," Koenig admitted to Loudon, "because I had thought that the new generation of scientists in England no longer knew me at all."[57] The English had remained his most ardent supporters. Loudon in Canada remained Koenig's most important customer. In the spring of 1901 he asked Koenig to help him draft some notes for a lecture he wanted to give on "progress in acoustics."[58] Even though bedridden, Koenig responded by writing a short, but fairly thorough history of acoustics from his own perspective, with references to himself in the third person.[59] Following his death, Loudon translated this history, gave it as a presidential address to the American Association for the Advancement of Science, and published it in *Scientific American* under his own name, with no mention of that Koneig had written it.[60] Loudon visited Paris in August 1901 to help Koenig sort out his business affairs and to seek buyers for his prized instruments. Perhaps during his visit they agreed to publish Koenig's memoir under Loudon's name, or perhaps Loudon simply stole it from a dead friend. This inexplicable omission is surely a mysterious ending to what had been a fruitful and warm friendship.

Koenig died on 2 Oct. 1901 with his niece Helene, his sister Anna and his brother-in-law, Ernst Neumann at his side. He was cremated and buried at the Parisian cemetery, Père Lachaise.[61] Anna and Helene stayed for a month afterward sorting out his possessions and arranging to sell some of his prized instruments. There were still six workers at the shop and the family decided to keep the firm going until the Russian order was completed. L. Landry, his main collaborator for thirty years, eventually took over the business,[62] but immediately following Koenig's death, the family also wanted to create the impression that the firm was not closing so as to keep bargain hunters from taking advantage of the situation (they kept the apartment until 1903). Abbé Rousselot, a phonetics researcher at the Collège de France, came by later in the month and offered to buy the complete tonometer for 25,000 fr. The family initially held out for more money, but eventually sold it to him at that price.[63] Part of this tonometer (from group 4) is now in storage in the Rousselot collection of instruments at the Mitterrand Branch of the Biblioteque Nationale in Paris, "département de l'audiovisuel." They range from ut7 (4,096 Hz; C8) to fa9 (21,845.3 Hz; F10). Mixed in with the tonometer are forks marked "LL" for L. Landry, Koenig's successor.

In 1901 the family also sold and donated Koenig's large collection of books and journals. He still had over 800 copies of his own book from 1882. Among the personal items left by Koenig which were divided into six parts for his nieces and nephews – over 60,000 fr and a beautiful set of furniture – was his prized violin and a handmade clock, representing two sides of his career as an artisan.

The Combination-Tone Controversy in England

> *But the observations most difficult of reconciliation with the theory of Helmholtz are those recorded by König. . .and these observations, coming from so skilful and so well-equipped an investigator, must carry great weight.*
>
> *Lord Rayleigh (1896, p. 468).*

In the last quarter of the nineteenth century, *Sensations of Tone* received wide acknowledgement and became a standard part of teaching acoustics, and a basis for research in all areas of acoustics. Why, then, would someone like Lord Rayleigh, who understood and appreciated the strength of Helmholtz's studies more than any of his contemporaries, take this challenge so seriously? What does this tell us about the place of instrument makers in nineteenth-century science, and the role of workshops in scientific controversy? And what does it tell us about the tenuous nature of psycho-physics in different national and social contexts?

Koenig spent many of his final years stoking controversy in insecure areas of acoustics. His initial paper of 1876 on combination tones, one of the unresolved areas of acoustics, attracted immediate attention in England. William Spottiswoode, a close friend of Koenig's and the president of the Royal Society, published a translation in the *Philosophical Magazine.*[64] In May 1879, he presented his own version of Koenig's results (with demonstrations of his instruments) to the British Musical Association. John Tyndall, Alexander Ellis (translator of *Sensations*), Lord Rayleigh, and the Oxford physicist R.H.M. Bosanquet, were all present at this meeting.[65]

The British scientists realised the potential significance of Koenig's experiments as a major threat to the foundation of Helmholtz's theories, and subsequently performed several of their own experiments on combination tones. There was a flurry of activity around these questions during the period from 1876 to 1883. Bosanquet, for example, presented an extensive series of studies to the Musical Association and the Physical Society. He developed a special resonator that sealed the ears from anything but the resonator tube. Following observations with this instrument, he claimed to agree with Koenig's findings, but presented his findings in stronger terms, saying that the disputed combination tones and Koenig's beat tones did not have an objective existence (i.e. could not be detected with the resonators) and were therefore "subjective."[66] Nevertheless, he took a middle ground in his interpretation of the results and stated that some of the tones Koenig himself had observed were probably due to impure instruments. He developed his own theory of "transformation" to explain his findings.[67]

Because Koenig and others had shown that some of Helmholtz's combination tones did not objectively exist, their position came to be conflated as the "subjective" argument. But what did such a claim mean amidst a constant stream of findings in sensory anatomy, physiology and psychophysics? In one of his contributions to this debate in 1881, Koenig's friend, S.P. Thompson, took issue with Bosanquet's definition of "subjective":

> If he means by this term that the phenomena of beats and difference-tones only exist in the mind, brain, or nerve structures of the ear, being generated in the sensory apparatus by something which physically has no existence, being in fact only phantoms of the imagination, then I entirely differ from him. But if by *subjective* Mr. Bosanquet means that the existence of these phenomena, though physically and mechanically true, is limited to the receptive mechanism of the ear, then I beg in the first place to disagree with such a perversion of the adjective, and in the second place to deny that any such limitation exists. The beats are *objective*; they can be *seen* in the manometric flames if the primary tones are sufficiently loud.[68]

In these positions, Thompson was close to Koenig, who rarely even used the terms objective and subjective. Although Koenig never described the above manometric test of beats, his graphical display of beat phenomena were real to both men and reflected something that could be clearly heard.

More importantly, the many "subjective" and "objective" studies revealed how the debate came to revolve around Koenig's primary focus on purity and precision in the instruments, while moving away from issues in psychology, physiology or the application of mathematics to sound. During the presentation by Spottiswoode in May 1879, Alexander Ellis reported that W. Preyer (1841–1897), professor of physiology in the University of Jena, had claimed to confirm that many combination tones were subjective, but like Bosanquet, concluded that some of Koenig's observations may have been due to harmonics (impurities) in his tuning forks. He stated that his own tuning forks (made by Appunn in Hanau) were so sensitive that he had to do his experiments alone at night to prevent unwanted vibrations from contaminating the forks.[69] This suggested that even the purest fork could possibly vibrate sympathetically to unwanted contaminant harmonics, thereby undermining the experiments. Koenig's friend in America, Professor Mayer, had also gone to great lengths to prevent unwanted vibrations in his instruments. He performed his experiments at night in an open field several miles out of town, but still "was disturbed by grasshoppers."[70] There seemed to be no end in the attempts to hone instruments and instrumental practice in order to preserve notions of objectivity.

Ultimately, the reliance on single observers and the listener's judgment undermined the quest for purity. Many combination and beat tones were simply not detectable by resonators or the phonautographic or manometric flame technique. Therefore, competing claims about the nature of the disputed tones were almost impossible to settle. One participant at the early meetings of the Musical Association, D.J. Blaikley, who himself was a musical instrument maker and acoustical researcher, stated that "owing to the great difficulty different observers have of really judging what they do hear, it [the matter of beats] has certainly become confused."[71] He then recounted a story of a visit to Koenig's atelier shortly after hearing Spottiswoode's lecture. In this encounter we see Koenig clearly priming his witness:

> My opinion was rather different to his as regards the extent of purity that existed in the tone of his two large forks. He took a pair of forks, a mistuned octave, and there was a beating note heard. He said to me, 'You hear distinctly the octave beating', and I said, 'It is the upper fork beating with the second partial of the lower fork.' He said, 'the second partial does not exist in sufficient strength to be heard.' It is just a question of the difficulty two observers may have, both competent to hear these notes, to observe exactly what does take place.[72]

Workshop as Theatre

At this observational impasse, Koenig's atelier became a tool for persuading others to his side. One of his more noteworthy demonstrations came in October 1881, when scientists from around Europe had gathered in Paris for an electrical congress. He used this opportunity to invite visiting scientists to witness his new series of experiments. As described in Chapter 6, he had been corresponding with James Loudon in Toronto about organising a similar demonstration/lecture series in Toronto and Montreal for the summer of 1882, and in suggesting a possible course of events for their series, he informed Loudon that he had just hosted some of the top scientists of the day, including Helmholtz, Kirchhoff, Lord Kelvin, Clausius, Mach, Hittorf, Kundt, Du Bois Reymond, and Quincke,[73] showing them his latest experiments (Fig. 7.5).

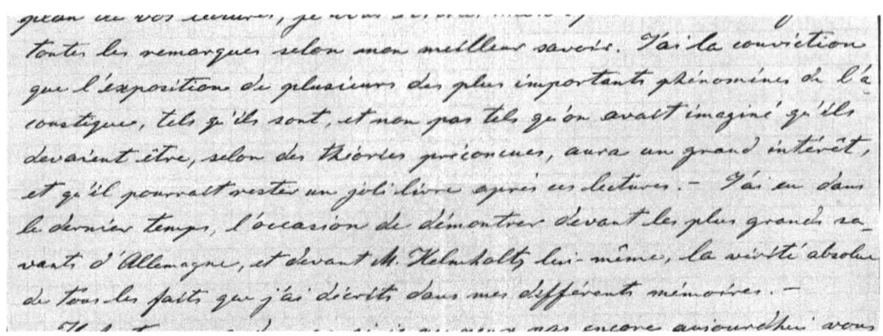
toutes les remarques selon mon meilleur savoir. J'ai la conviction
que l'exposition de plusieurs des plus importants phénomènes de l'a-
coustique, tels qu'ils sont, et non pas tels qu'on avait imaginé qu'ils
devaient être, selon des théories préconçues, aura un grand intérêt,
et qu'il pourrait rester un joli livre après ces lectures. — J'ai eu dans
le dernier temps, l'occasion de démontrer devant les plus grands sa-
vants d'Allemagne, et devant M. Helmholtz lui-même, la vérité absolue
de tous les faits que j'ai décrits dans mes différents mémoires. —
... pas encore aujourd'hui avons

Fig. 7.5 Letter from Rudolph Koenig to James Loudon, Nov. 25, 1881. UTA-JLP (B72-0031/004). Courtesy of the University of Toronto Archives

> I am certain that the exposition of several of the most important phenomena of acoustics, such as they are, and not as we have imagined them to be according to preconceived theories, will be of great interest, and could result in a very fine book after these lectures. I recently had the occasion to demonstrate, before the most important German scientists, and before Helmholtz himself, the absolute truth of all the facts that I described in my different articles.[74]

The mentioning of "imagined" and "preconceived theories" was clearly directed at Helmholtz, showing Koenig's faith that his demonstrations spoke for themselves, thus providing a direct view of nature.

The most revealing part of the debate, however, came with the responses from who were not doing research on these questions (only a small number of physicists were actually doing in-depth research on sound). These were the teachers, popular lecturers and researchers in other fields who had frequented Koenig's atelier and were converted by his demonstrations and arguments. Koenig's strongest allies, for example, promoted his findings with effusive rhetoric about his skill as a maker and experimenter. In fact, this appeared to be one of the main reasons why the dispute carried on for so long. In the English-speaking world, especially with the Americans, his supporters were ardent Baconians suspicious of abstract theory and mathematics. Alfred Mayer, of the Stevens Institute in New Jersey, who had spent the American Civil War in France working under the famed experimentalist Regnault, claimed that Koenig knew his forks so well he could read the rhythms of the beats to attain more precision in tuning.[75] Other enthusiasts like August Zahm, who published a popular textbook on acoustics, simply concluded that Koenig's findings were "in the main, correct" and "generally accepted."[76] He gave away his bias in the preface of his book when he wrote that,

> the only apparatus that can be depended upon for exactness and never-failing operation are those made by the learned and painstaking Dr. Koenig. The making of a perfect instrument is for him a labour of love. It is for this reason that the tuning-forks which bear his stamp are so universally sought, and, when secured, are so highly prized.[77]

D.C. Miller stated as late as 1935 that Koenig was "in the main, correct."[78] W. Le Conte Stevens, who had visited Koenig's studio, compared his mechanical genius to

Newton in optics, Herschel in astronomy, Ruhmkorff in electricity and Wheatstone with his numerous inventions. The framework of Stevens's argument was similar to Koenig's standard presentation.

> Helmholtz discussed "differential tones" and "summation tones," whose existence was inferred from mathematical analysis; and certain phenomena seemed for a time to confirm the conclusions of the great German physicist. But Koenig subsequently applied the most patient care and consummate skill in the experimental examination of these phenomena. Without detracting at all from the credit due to Helmholtz for his splendid researches, it may now be safely said that Koenig's experiments have shown that differential and summation tones are due exclusively to the beats which the ear perceives when impressed simultaneously by systems of waves differing in length. The effect is physiological, and such combination tones are not at all re-enforced by resonators like the separate primaries that enter into combination.[79]

The British were especially supportive of Koenig's position. In the introduction to his review of the controversy, Alexander Ellis stated that much of the debate had revolved around Koenig's results: "We must distinguish the phenomena from any theoretical explanation of them that may be proposed. The phenomena described by such an acoustician as Koenig, so careful in experiments, so amply provided with the most exact instruments, will, I presume, be generally accepted."[80] As late as 1891 one of Koenig's most loyal devotees and visitor to his atelier, S.P. Thompson, told members of the Physical Society of London that Koenig "lives and works in seclusion, surrounded by his instruments, even as our own Faraday lived and worked amongst his electric and magnetic apparatus."

> It is not surprising that one who lives amongst the instruments of his own creation, and who is familiar with their every detail, should discover amongst their properties things which others whose acquaintance with them is less intimate have either overlooked or only imperfectly discerned. If he has in his researches advanced propositions which contradict, or seem to contradict, the accepted doctrines of the professors of natural philosophy, it is not that he deems himself one whit more able than they to offer mathematical or philosophical explanations of them: it is because, with his unique opportunities of ascertaining the facts by daily observation and usage, he is impelled to state what those facts are, and to propound generalised statements of them, even though those facts and generalised statements differ from those at present commonly received and supposed to be true.[81]

In Thompson's view the specialized craft knowledge from the workshop gave privileged status to Koenig in this dispute:

> No living soul has had a tithe of the experience of Dr. Koenig in handling tuning forks. Tens of thousands of them have passed through his hands. He is accustomed to tune them himself, making use of the phenomena of beats to test their accuracy. He has traced out the phenomena of beats through every possible degree of pitch, even beyond the ordinary limits of audibility, with a thoroughness utterly impossible to surpass or equal. Hence, when he states the results of his experience, it is idle to contest the facts gathered on such a unique basis.[82]

Thompson made his comments during his lectures to the Physical Society of London in May of 1890. He had invited Koenig to present his case against Helmholtz's and make the key demonstrations while Thompson read the paper.[83] It was a success and Lord Rayleigh invited him to make the same presentation before the Royal Institution a month later.[84] By this time, Koenig had created a whole line of instruments for proving his case with authority. One particularly impressive set of three forks that were used during these demonstrations survive today in remote storage rooms of the Science Museum in England (Fig. 7.1). They are thick steel forks that rest on a large cast-iron stand at well over 1.5 m. The giant brass cylindrical resonators, which also stand on cast-iron stands, are over a meter in length and 37 cm in diameter. The forks have large, brass sliding weights for precision adjustments (CR no. 194).[85] They must have been an effective form of persuasion in their own right. Another instrument "Apparatus for the continuous sounds of beats" amplified in a clear, dramatic manner, the weaker tones from Koenig's studies.[86] It consisted of two, tuned-glass rods in a tall iron frame with a wheel between them covered with felt that made contact with the glass tubes. As the clothed wheel rubbed against the rods, the resultant friction caused the emission of powerful and pure simple tones via longitudinal vibrations. These powerful tones combined to produce strong beat tones. As with other instruments of Koenig, this one served as a good teaching device and a source of information itself on the mechanics underlying beats tones. The instruments also illustrated Koenig's theatrical strategies for persuading large audiences. A surviving example at the University of Coimbra is tall (over a metre in height) with decorative features and a large display sign that reads "RUDOLPH KOENIG À PARIS." (CR no. 202). These features indicated that it was made for exhibition or travelling demonstrations. Koenig also developed a pair of massive electrically driven tuning forks with large brass resonators for demonstrating combination effects to large audiences. One of the forks had mercury in its prongs allowing the pitch to be adjusted at will. These forks were a cheaper version of the above instruments but could produce very powerful notes (CR no. 189).[87]

Heidelberg 1889: the German Response

The Germans reacted differently. They did not share the same enthusiasm for the German artisan-researcher who lived in Paris. In 1889, during a tough, yet demanding year in business, he prepared to go to the sixty-second congress of physicians and naturalists at Heidelberg to present his demonstrations and make his case to German scientists. He had not attended this congress in 21 years and wanted to show his experiments to a new generation.[88] His description of the congress to his Canadian friend, James Loudon, revealed his insecurity surrounding the reception of his presentation. Helmholtz, he wrote, spoke too long leaving less time for the next speaker, himself. Nevertheless, "in going very fast I succeeded in giving all the needed explanations and the most important experiments on the co-existence of two sounds [*la coexistence de deux sons*]."[89] He was put off by Helmholtz's aloofness and felt that he had responded only out of obligation, "because I doubted that he himself thought that what he had to say was very strong."[90] Helmholtz's rebuttal concerned the notion that Koenig's beats and beat-tones could not be strictly classified as sounds, and that his theory showed how important it was to treat the generating tones as independent. "It appeared difficult for him to admit," Koenig wrote, "that beats can become sounds because the theory that he had given of combination tones had been found confirmed in other branches of science, as in the theory of tides, where two equally different forces, the sun and the moon act at the same time on the water."[91] In other words, Helmholtz saw combination tones as two objective phenomena coming from two distinct sources, not from one single vibratory motion or waveform.

> Then he again gave the description of. . .experiments that had absolutely nothing to do with my talk, and finished in excusing himself from having to speak, but having been directly attacked in his position, he could not do otherwise. He thanked me again in the name of the section, as he had already done before his observations, and we passed on to other things.[92]

This anti-climactic confrontation revealed the disadvantage of having to rely on a travelling demonstration studio to reinforce one's arguments. It also showed how issues of status and class were more keenly felt by Koenig in Germany. By the end of the conference, he was thoroughly deflated and worn out by all the packing and installing of instruments: "I had to work for almost the whole trip to Heidelberg as a true labourer (*un vrai manoeuvre*) by unpacking, mounting, and transporting from one auditorium to the other and back again, to repack and resend my 700 kg of instruments."[93] It was Heinrich Hertz, Helmholtz's student, who stole the show with his groundbreaking announcements on electromagnetism, a topic that had become exceedingly more popular than acoustics.[94]

As uncertainty remained, however, Koenig's findings continued to raise questions. In text-books of the 1890s, there was no consensus, and scientists who devoted much time to acoustics such as Lord Rayleigh regarded the dispute as "open" as late as 1896. He was reluctant to question the investigations of someone "so skilful and well-equipped" as Koenig, and yet he sided with Helmholtz's attempts to keep a strict definition of simple pendular tones in order to preserve Ohm's law (that sound could be analysed into distinct elements). "Experiment may compel us to abandon this law," he wrote, "but it is well to remember that there is nothing to take its place."[95] Indeed, there was nothing from the perspective of a physicist or someone with a physical view of hearing, but as new assumptions and theories about sensations came into being in the late nineteenth and early twentieth century, Koenig's findings stubborning remained.[96]

In the main, however, the obsession with developing objective methods continued to dominate the debate. Sir Arthur W. Rücker and E. Edser created a completely isolated system for detecting some of Helmholtz's combination tones on an objective basis.[97] In 1909, Felix Auerbach, who had worked under Helmholtz on acoustical questions, wrote a lengthy summary of Koenig's research on beats and beats tones where he concluded that there was still no resolution to the two sides.[98] In the twentieth century, researchers have discovered a multiplicity of combination-tone effects created at the sound source, in the transmission medium and in the ear, as well as complex cognitive effects dependent on several variables that were not part of the framework in the nineteenth century (and still not fully agreed upon even today).[99]

The Debate over Timbre

Alexander Graham Bell's invention of the telephone in 1876 sparked a dispute on the nature of timbre which became tied to the combination-tone debate and influenced the direction of Koenig's work. After the introduction of the telephone, Helmholtz's colleague Emil du Bois-Reymond used the invention as a way of illustrating the nature of timbre.[100] He claimed that the partial tones (i.e. harmonics) of a complex sound travelled through the wires as electrical vibrations and maintained their amplitude and frequency. There was a slight phase displacement, but the timbre remained the same. The Königsberg physiologist Ludimar Hermann set out immediately to test this theory with an experimental current-producing telephone. Based on the arrangement of the coils, and taking account of the laws of electrodynamic induction, Hermann hypothesised that *there were* changes in amplitude of partial tones, but he found that the timbre remained the same. He concluded that these results were incompatible with Helmholtz's theory of timbre (where amplitude changes should change timbre). Helmholtz responded with a lecture in 1878 showing that Hermann had neglected to take into account the properties of the entire circuit he was using; in addition, he added that the differences of phase were negligible.[101] Koenig replied with his own article in 1879. Reflecting on the disagreement between Helmholtz and Hermann, Koenig thought it was "important to offer an experimental method that permitted exact verification" of the effect (or lack) of phase changes.[102] He constructed a telephone that operated with tuning forks in order to transmit simple tones of different frequencies. He compared the phase of the input frequency with the outgoing frequency using the Lissajous method and found there to be a sizeable displacement of phase (approximately one quarter of a wavelength).[103]

Historian Julia Kursell has argued that the phase debate revealed a basic assumption of Helmholtz's theory of resonance, namely how it conceptualized hearing in terms of one ear acting as a Fourier analyser.[104] In 1878, S.P. Thompson published findings on binaural phenomena that sited the work of Ernst Mach and showed that phase played a role in sound localization.[105] Lord Rayleigh as well raised this issue of sound discrimination in his presidential address at the British Association meeting in Montreal in 1884. Pure sounds, he said, are hard to locate, whereas, "any other sound" such as those from musical instruments are "easy and instinctive" to place due to phase differences.[106] Koenig himself noted that the dropping sticks (CR no. 1) gave off short bursts of sound with complex timbre that enabled the listener to discriminate the note of each stick.[107]

These questions confirmed Koenig's instinct that timbre was more complicated than portrayed in Helmholtz theory. In *Tonempfindungen*, Helmholtz had used his synthesiser to test for the influence of phase on timbre and concluded that it did not play a role. He adhered to his analytic conception of sound, whereby the number and strength of harmonics determined the timbre of a compound sound. He did, however, note "an apparent exception."[108] He noticed that when he played the fundamental of the synthesiser with the next note (the octave) but slightly out of tune, "an attentive ear [*ein aufmerksames Ohr*] will observe very weak beats which appear like small changes in the strength of the tone and its quality."[109] He concluded that these beats were associated with changes in phase. Furthermore, he stated, the "apparent" changes were merely due to combination-tone effects where "slight variations of quality are referable to changes in the strength of one of the simple tones."[110] When doing his own experiments, Koenig heard these slight differences of quality, but interpreted them quite differently from Helmholtz: "But if timbre [*le timbre*] depends precisely on the existence of harmonics [*des harmoniques*] and their relative intensity, and if this relative intensity is modified by the difference of phase, it is clear that the influence of the latter is not only apparent, but very real."[111]

The questions about timbre touched on a number of issues that marked Koenig's career and the birth of this new science: the role of instruments and craft knowledge in influencing this debate; the manner in which pictorial representation determined what was deemed "real"; the importance of demonstrations for arguing a case; and the uncertain boundaries between physics, physiology and psychology in the latter part of the nineteenth century. As we will see, the dispute also related to Koenig's other controversy with combination tones. He wanted to replace Helmholtz's theory of resonance with a coherent alternative based on his pictorial perspective and perfected instruments. In his own history of acoustics, Koenig remarked: "In the sense of timbre being understood as an assemblage of coexistent sounds [*un assemblage de sons coexistants*], the study of phenomena that are products of the joining of two or more sounds becomes inextricably linked to the study of timbre itself."[112]

Koenig's training as a violinmaker was important in this dispute. His artisanal knowledge of musical instruments gave him a different perspective on the nature of complex tones and timbre. In his first full-length discussion of phase and timbre in 1881, he made a distinction between harmonics and partial tones. Harmonics, he wrote, represented the ideal mathematical series of tones related to the fundamental; partial tones, on the other hand, were the actual sounds that approached, more or less, the theoretical values. These slight enharmonic deviations, Koenig argued, were quite apparent to anyone who listened carefully to organ pipes and vibrating strings and plates. Imperfections in stringed instruments such as violins, produced partial tones that were not pure. This different timbre was especially apparent in violins of different quality.[113] In his article he reproduced one of his earliest graphical inscriptions of a vibrating string producing a fundamental and its octave (Fig. 7.6). By studying the inscriptions, Koenig found a continually changing waveform. The first partial (octave) was not exactly an octave, thereby creating a slight difference in phase, and therefore continual changes in the waveform. These slight changes, according to Koenig, were detected as slight changes in timbre. Homogeneity of the violin string had been a source of concern to him as a former violinmaker. He now made use of this knowledge in challenging Helmholtz.

His work on the influence of phase, he argued, showed that Helmholtz's understanding of sound needed modification. "If this influence exists," he stated in his article on timbre in 1881, "the hypothesis that had existed before the work of Helmholtz on this subject, where timbre depended on the form of vibrations, should be conserved."[114] Koenig's argument in favour of the primacy of the

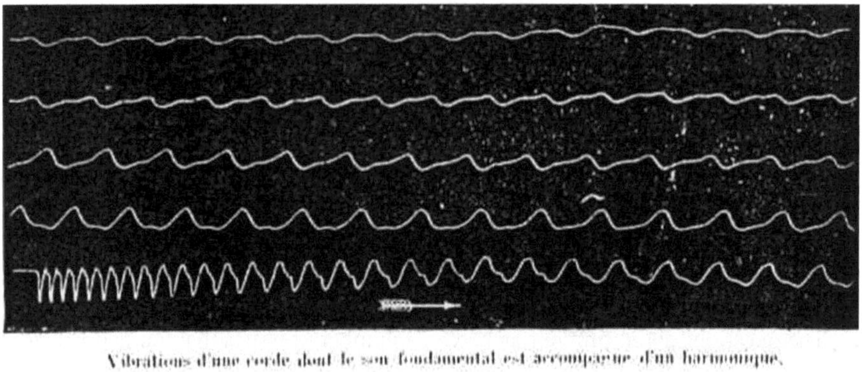

Fig. 7.6 Phonautograph tracing of a string producing a slightly mistuned octave
Source: Koenig (1882c, pp. 16, 221)

waveform constituted a direct challenge to Helmholtz's analytic conception of sound. Helmholtz's physiology, for instance, depended on a linear, one-to-one relationship between simple tones in the outside world and simple-tone receivers in one ear.[115] Accepting Koenig's experiments would have demanded a complete reconceptualization of this mechanical, piano model of the inner ear. Helmholtz's early biographer strongly echoed Helmholtz's position in his assessment of the dispute: "By establishing ... that the difference of phase does not come into the question, Helmholtz confirmed his previous assumption that our sensation of different qualities of tone [*die Empfindung vershiedener Klangfarben*] is reduced to the fact that other nerve-fibres [*andere Nerven fasern*], corresponding with the partials [*den Nebentönen*], are simultaneously excited along with the fibres [*der Faser*] that respond to the fundamental tone. This simple explanation would not suffice, if the difference in phase [*die Phasenunterschiede*] of the deeper harmonics had to be considered."[116] In other words, for Helmholtz the notion of a one-to-one correspondence had to be preserved.

As with the other disputes, the focus from Koenig's point of view was on the level of instruments. He argued that the Helmholtz synthesiser had many shortfalls, namely that it produced compound sounds of "doubtful clarity" and that the tuning forks and resonators interacted thereby complicating the internal phase relations and harmonic structure of the instrument.[117] It also did not fit well with Koenig's emerging view of timbre and partial tones in real conditions, where it displayed continually changing intensities and phases.[118]

Wave Sirens

Koenig saw the changing waves and the different waveforms produced by phase changes as constituting a real difference in timbre. Just as with beats and beat tones, he resorted to his wave-siren technique in order to recreate artificially the natural timbre from actual waveforms. He first constructed waveforms for complex tones

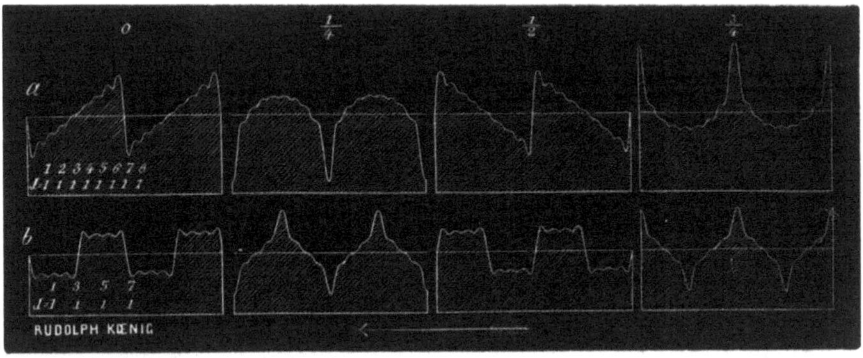

Courbes résultant de la superposition de sinusoïdes qui représentent des sons harmoniques de même intensité.

Fig. 7.7 Compound waveforms resulting from harmonics of equal intensity with phase shifts 0, ¼, ½, and ¾
Source: Koenig (1882c, p. 227)

consisted of a series of harmonics of equal intensity. Using his graphical inscriptions and photography to reduce some of the curves, he drew the resultant compound waveforms under four different phase conditions, with shifts of 0, ¼, ½, and ¾ of a wavelength (Fig. 7.7).

He then traced and cut these figures on the circumference of a cylindrical band of thin brass. Like the combination-tone siren, these cylindrical bands wrapped around a central rotating axle. Wind slits, connected to a large wind bellows, were positioned beside each curve.[119] By studying the waveforms compounded from the first eight harmonics, and then waveforms consisting of the odd harmonics, he consistently discovered that waveforms with a phase shift of ¼ were much stronger and more strident in tone. Waveforms based on the shift of ¾ were soft in tone, while the other patterns representing shifts of ½ and 0 were of an intermediate quality.[120]

To imitate faithfully the conditions found in nature and musical instruments, Koenig created metal waveforms from harmonics of decreasing intensity (in nature, the harmonics farther away from the prime tone generally decrease in intensity) (Fig. 7.8).[121]

He made six curves derived from the combination of the first eight harmonics with decreasing intensity, and two curves derived from the combination of the odd harmonics (1,3,5,7). He was therefore able to compare two different timbres based on the partial tones, and different timbres based on the same partials but with different waveforms. The results were similar to his first observations.[122] He also built three curves meant to imitate the vowels, "OU," "O," and "A." These waveforms derived from Auerbach's analytic studies of vowels and the relative intensities of their first eight harmonics.[123] In general, Koenig verified his earlier results showing that the phase shift of ¼ wavelength produced the greatest difference in timbre. He added, however, that these curves did not succeed in reproducing the vowels. Only the "A" curve gave something close to an "A."[124]

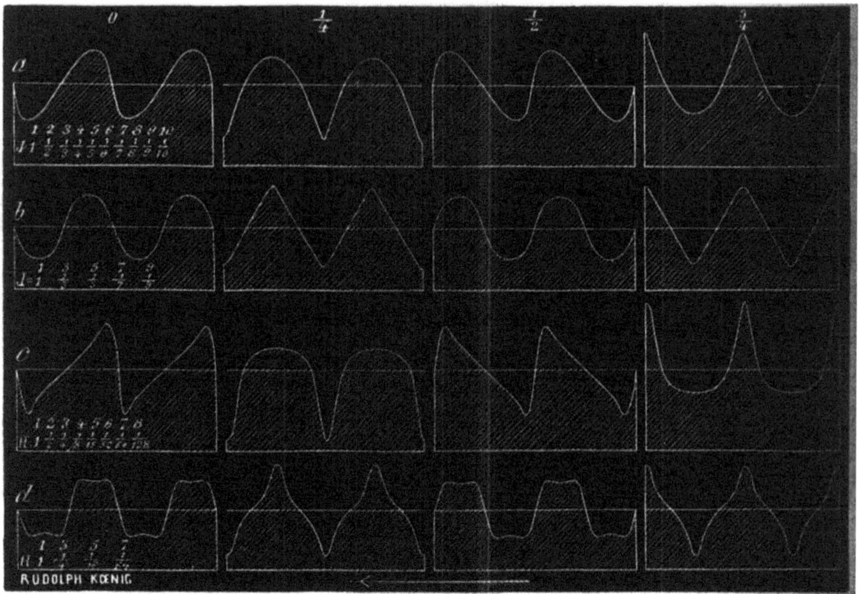

Courbes résultant de la superposition de sinusoïdes qui representent des sons harmoniques dont l'intensité décroît régulièrement.

Fig. 7.8 Compound waveforms resulting from harmonics of diminishing intensity. The harmonic series appears just under the first waveform of each row; the rows for phase shifts, 0, ¼, ½, and ¾, are above. For his commercial wave siren (see Fig. 7.9) Koenig used the first four curves of row "a" and the first two curves of row "b"
Source: Koenig (1882c, p. 228)

By 1882 Koenig had created a standard form of this apparatus for the market.[125] There were six curves – four deriving from the first twelve harmonics of decreasing intensity, and two from the odd harmonics of the same series. The standard wave siren cost 350 fr. Like other Koenig instruments its mere presence in texts, lantern slides or articles became a powerful means of illustrating controversial ideas, providing a way to conceptualise visually the contentious role of phase in timbre.[126]

Like the beat-wave siren, the phase-wave siren was open to criticisms concerning its ability to reproduce airwaves that faithfully derived from the waveform of the copper disk. Koenig believed that such problems were negligible, but he worked to build a siren with many features to compensate for any problems. His "large wave siren" (*grand sirène à ondes*) was his most elaborate and exotic instrument.[127] It was his second most expensive instrument at 6,000 fr, putting it out of the reach of most laboratories. He was particularly proud of this instrument and placed it on the cover of his book in 1882. Unfortunately, there are no surviving examples of this instrument found in present museum collections.

The *grand sirène à ondes* was 1.9 m in height. It consisted of sixteen disks cut with simple sinusoidal waveforms instead of complex curves. The first disk was a

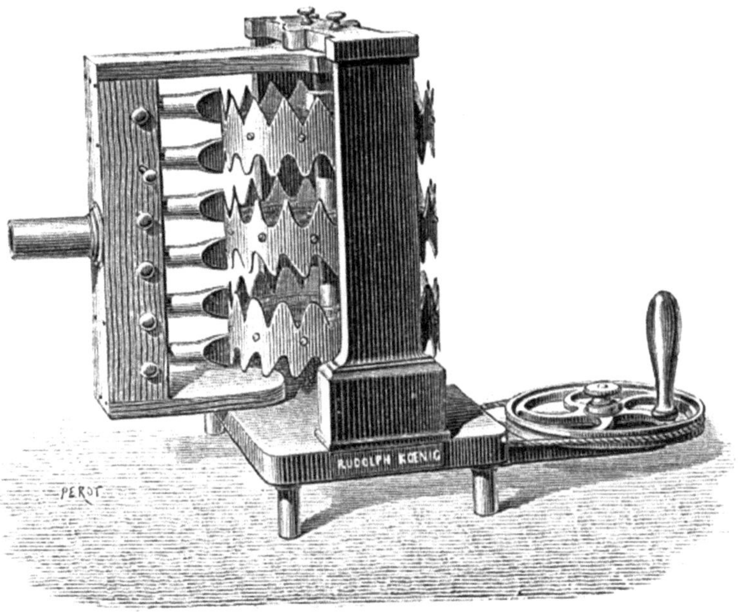

Fig. 7.9 Wave siren for studying timbre. The top two curves represent the first six odd harmonics with differences of phase of ¼ and 0 (see Fig. 7.8 row "b"). The *bottom four curves* represent the first 12 harmonics of diminishing intensity (see Fig. 7.8 row "a"). CR 60
Source: Koenig (1889, p. 28)

fundamental tone, the other fifteen were pure harmonics of that tone. Each disk had its own wind slit. A long lever connected to the slits allowed one to change the phase of each slit at will. Sixteen buttons allowed one to open or shut the flow of pressure air in the slits. One could also regulate the pressure of air to imitate varying intensity. With this instrument Koenig was able to confirm his earlier research.[128] It had the advantage of having the versatility to explore many combinations of harmonics in different situations. Koenig's main goal had been to explore the role of timbre, but he stated that some preliminary research on vowels had shown promise (Fig. 7.10).[129]

Wave sirens represented Koenig's complete departure from Helmholtz's analytic theory of sound. Koenig had done more than anyone else to transmit the analytic perspective through his tuning forks, sirens, resonators, synthesisers and analysers. But, when he designed a whole family of instruments that challenged this conception, he faced a difficult challenge. Unlike the combination-tone debate, people in the physics community did not vigorously defend, or even test, his position. Even in Britain, there was not much enthusiasm for Koenig's experiments on phase. In fact, when S.P. Thompson gave his lecture on Koenig's work in 1891 for the Physical Society of London, Koenig's position regarding the role of phase in timbre was

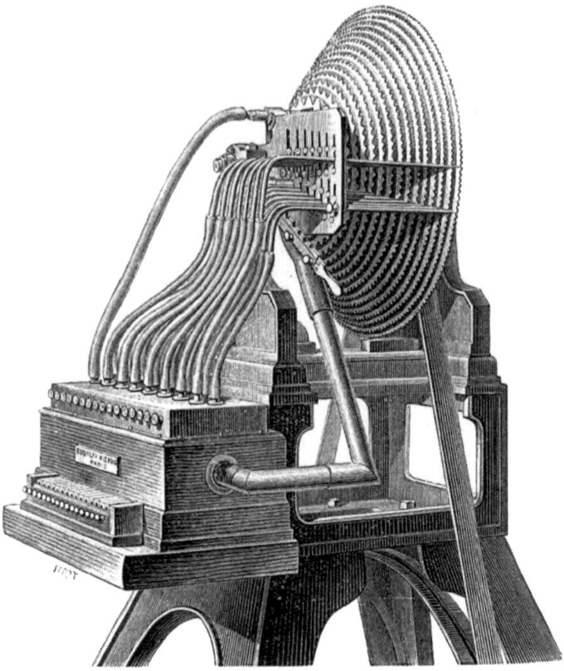

Fig. 7.10 Large wave siren for studying timbre. CR 59
Source: (Koenig 1889, p. 27)

greeted with skepticism. Thompson, for example, replied to some questions from Bosanquet:

> Please bear in mind that on Friday I spoke purely as the exponent of Koenig's views, not necessarily of my own: otherwise I should have said something in criticism of the whole method of wave sirens, and should have suppressed sundry other things that Koenig wished to be said. I wish you had been in front of the wave sirens, as they can not be heard from behind with any success.[130]

Lord Rayleigh cited Koenig's work, without taking sides, having cautioned his audience that such a view demanded a "departure from Ohm's law."[131] On the other hand, scientists in the physiological community, who were more used to the graphical approach, were more open to Koenig's holistic, pictorial perspective. At Heidelberg they invited him to speak at their own session two days after his encounter with Helmholtz, and he found them "very agreeable."[132] Physiologists such as Ludimar Hermann used graphical equipment more frequently than their colleagues in physics. Hermann in particular did a long series of graphical experiments on vowels between 1889 and 1894 which initially called for some modifications to Helmholtz's theories on the same subject, but five years later, to the dismay of

Koenig, questioned some of his findings as well.[133] He never seemed to receive the continued support he wanted from German colleagues.

Back to Vibrations

In August 1888 Koenig wrote to James Loudon that he wanted to test whether the ear could indeed distinguish differences among these complex, multiform waveforms.[134] First, he made six disks for his wave siren that resembled specific combinations of harmonious and inharmonic partials to create complex waveforms of continually changing shape. His experiments, he believed, were a success. He concluded that the "waveforms of a sound do not have to be absolutely uniform to produce musical timbre."[135]

He had also brought some of these arguments to the Heidelberg conference in 1889. On the experiment with a wave siren, Koenig reported that Helmholtz thought it was "important to observe that it [the wind slit] requires very much exactness in adjustment, and the largest difference of timbre from the two positions of the wind slit, appeared to him very weak."[136] Finally, Helmholtz commented on one of Koenig's key arguments, the role of inharmonic partials in the production of timbre. Koenig had used the example of the non-uniform, continually changing, vibrating string (Fig. 7.6). Helmholtz, not to be outdone in the domain of musical instruments, responded with his own knowledge of strings that "one can perhaps find something of this [inharmonic sounds] with the lowest cords of the piano, but they do not give in reality much musical timbre."[137]

But Koenig did not give up after these dismissals. It was now even more important for him to "show that timbre of this nature [non-uniform] are often truly produced by vibrating bodies."[138] He had already found such complex waveforms with stringed instruments, but he now wanted to find these non-uniformities in other instruments and vibrating bodies. In light of criticisms that not all seemingly periodic behaviour (such as beats) produce sounds, he set out to prove that any maximum isochronous intensity could "give birth to a sound."[139] After noticing that some of his steel cylinders (Fig. CR no. 51), when hit in different places produced two sounds, Koenig created modified steel cylinders that could emit two sounds, and if they were near unison, hear heard beats. In order to show this effect he made steel bars with the notes such as ut6 (C7) and sol6 (G7) on the respective sides thus producing a beat tone (difference tone) of ut5 (C6) (Fig. 7.11).[140] The bar at the Canada Science and Technology Museum has a point in the middle for attaching to a cast iron support (CR no. 153a). Seeing that these tones came from one source, and therefore, as Koenig thought, from one motion (just like a violin string), he interpreted this phenomenon as evidence that any vibrating body could produce these complex vibrations and that the ear could distinguish these motions as a single perceptual event. This holistic perspective was a marked contrast to Helmholtz's elemental view of timbre.

Fig. 7.11 In the summer of 1898 Koenig demonstrated a set of steel bars like this for James Loudon's graduate student, J.C. McLennan of Toronto. This one produces an ut5 difference tone. The bar would be fixed to a clamp CR 153a. Photo by author, 2008. Canada Science and Technology Museum, acc. no. 1998.0273.12.

The latter experiments on vibrations revealed Koenig's commitment to resurrecting pre-Helmholtzian conceptions of sound, where waveforms defined conceptions of sound. In a statement that recalled his lineage to Vuillaume's workshop and the early school of Parisian experimental acoustics, Koenig told Loudon that his latest round of observations on the vibrating cylinders had been partly inspired by some of Savart's previous work on the vibrations of systems. "It is a subject that interests me very much, and that I hope to pursue further."[141] He demonstrated a set of these bars in action for Loudon's graduate student, J.C. McLennan when he visited Quai d'Anjou during the summer of 1898.[142]

Ultrasonics and "Le Domaine de la Fantaisie"

The physiological community also played a major role in Koenig's last controversial project. Just before his death he took his tuning forks into the unchartered territory of ultrasonics. Previously, he had been reluctant to study tones that were in "le domaine de la fantaisie"[143] but inquiries into the psychology and physiology of higher tones by Carl Stumpf (1848–1946), Max Friedrich Meyer (1873–1967) and Franz Emil Melde (1832–1901) stimulated him to take another look at this subject.[144] These researchers had discovered that Appunn's forks, which were supposed to go up to 50,000 Hz, were in fact wrongly calibrated.[145] Preyer had used these forks in his early acoustical research, and his work had been cited uncritically by Helmholtz and Zahm.[146] In the same way he uncritically promoted Koenig's work, Zahm tried to dismiss any doubts about Appunn's forks:

> Many persons have been able to hear the note yielded by this fork [49,152 Hz]; but a question may arise whether it really gives a note of the high pitch claimed for it. Without here entering into an explanation of the manner in which the pitch of such forks is determined, I may observe that Herr Appunn, in a letter to me about this and other forks of very high pitch which he furnished me, states that he can guarantee that the frequencies of the forks correspond absolutely with the numbers stamped on them. No one can doubt the skill of Herr Appunn as a mechanician, and the delicacy of his ear for very acute sounds is, according to the testimony of all who are acquainted with him, something quite astonishing. It would probably be impossible for one with a less delicate ear to tune such a fork, even if he were familiar with the method of tuning employed in such cases. We are consequently, by the very necessities of the case, compelled to accept Herr Appunn's estimate as that of an expert and that he is an expert in his specialty no one can gainsay.[147]

For most of his career, Georg Appunn (1816–1885), and later his son Anton Appunn (1839–1900), had been Koenig's only competition in the German territories for making tuning forks. Helmholtz cited their forks several times in *Sensations*.[148] Appunn Sr. had collaborated with Preyer in the 1870s in experiments that did not fully agree with Koenig's findings on beat-tones.[149] Koenig, therefore, questioned the integrity of their tuning forks. After reading Melde's papers, he described to James Loudon how Appunn's forks were found to be "absolutely untruthful, as I had thought for a long time."[150] He also demonstrated to Mayer, who visited Quai d'Anjou in the summer of 1894, the exactness of his forks up to fa9 (21,845.3 Hz; F10), "which was for both of us the limit of our perceptibility!" He then added: "Prof. Zahm, who, like many others, was taken by the charlatanism of Appunn and Preyer, as his book shows, will be a little astonished when he realises what he must now think of their affirmations." (Fig. 7.12)[151]

In 1899 Koenig responded to suggestion that his forks were off the mark by producing a comprehensive study of the behaviour of inaudible tones up to 90,000 Hz. In his earlier work on combination tones (1874) he had produced forks up to 21,845 Hz. Those forks had been as thick as they were long (15 mm), and were almost impossible to keep vibrating. He tried making the forks thinner, but this meant they were softer and therefore could not produce a strong tone. After hearing about Preyer's experiments in 1876, Koenig had set to work to develop higher frequency forks, but having no way to verify them properly, he did not include them in his catalogue of 1882.[152] The findings of Stumpf, Meyer and Melde inspired him to develop a more objective method for verifying the highest forks. Among other tests with plates, cylindrical steel bars and whistles, Koenig used cork-dust figures that made the sound waves visible in a tube (Kundt's invention). Through this method, he was able to measure objectively the frequency of inaudible tones. He also extended (and claimed to confirm) his studies on beats and beat-tones into ultrasonic frequencies (Fig. 7.13).[153]

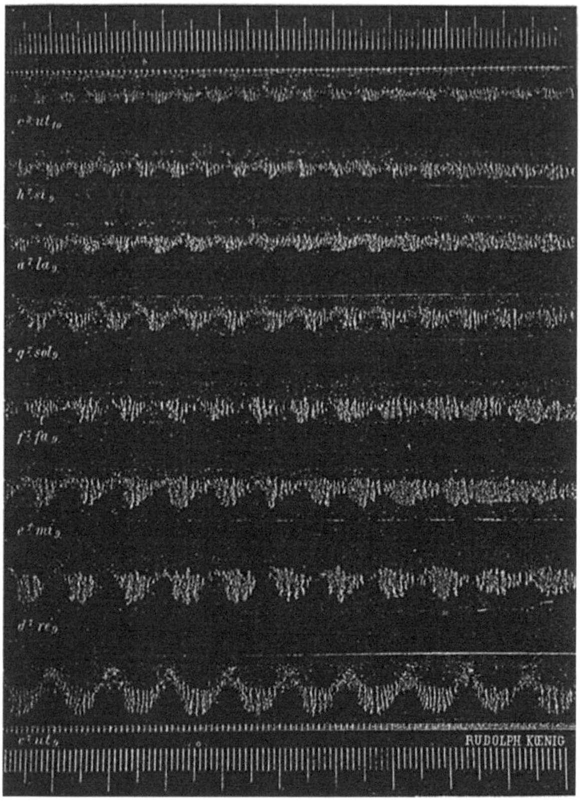

Fig. 7.12 Kundt figures for high frequencies
Source: (Koenig 1899, p. 647)

Throughout his career, Koenig made use of a mixture of thorough experiment, mechanical innovation, demonstrations and visual techniques to make his case on various subjects. His last round of experiments combined all four aspects of this approach, applied to phenomena far above the limits of human observation. On his visit to Koenig in 1898, J.C. McLennan witnessed these latest demonstrations and experiments. Koenig was quite ill by this time and living entirely on milk, but he had the energy to carry out a host of demonstrations. In one of the experiments he demonstrated a set of small forks whose "individual vibrations cannot be heard but the [lower] beat tones can."[154] The next year, 1899, McLennan visited again and Koenig showed him his completed experiments with high tones. He had now reached 90,000 Hz. As proof he gave photographs of the Kundt figures to McLennan to pass on to James Loudon in Toronto.[155] These figures, which presently rest in the archives at the University of Toronto (James Loudon Papers), are the remaining evidence that at the end of his life, Koenig became the first person to measure and record sound deep into the ultrasonic range.

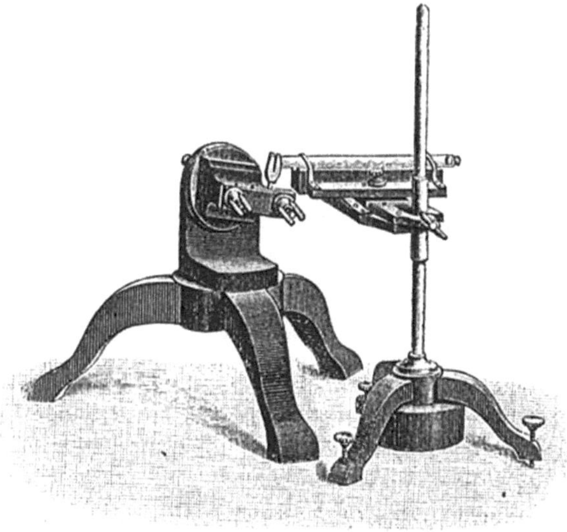

Fig. 7.13 Small tuning fork with glass Kundt tube for measuring high frequencies
Source: Koenig (1899, p. 657)

Notes

1. Thompson (1891, p. 201).
2. For a history of engineering at Toronto see White (2000).
3. Rudolph Koenig to James Loudon, Jan. 8, 1886. UTA-JLP.
4. Ibid., Jun. 22, 1883.
5. Ibid., Jun. 29, 1883.
6. Ibid., Dec. 28, 1883.
7. Ibid., Nov. 7, 1884.
8. Ibid., Feb. 20, 1885.
9. Ibid., Jun. 22, 1883.
10. Ibid., Nov. 7, 1888.
11. Ibid., Aug. 31, 1888.
12. Miller (1935, pp. 90–91).
13. In the first years of his business relationship with James Loudon, Koenig cut items from an order that were duplicated in other instruments. Loudon Papers 1878–1882.
14. Rudolph Koenig to James Loudon, Aug. 31, 1888. UTA-JLP.
15. Ibid., Aug. 31, 1888.
16. Ibid., May 23, 1889.
17. Ibid., Mar. 28, 1889.
18. Williams (1994).
19. Henry Rowland of Johns Hopkins exemplified this shift toward Germany in the 1870s. Rezneck (1962).
20. See Brenni's description of how Jules Carpentier modernized his firm in the early 1880s to keep up with competition. Brenni (1994c).
21. Paolo Brenni argues that French makers continued to make instruments with an aesthetic appeal, even though the market was demanding more functional, less decorative objects.

This stubborn refusal to adapt to the changing needs of the market was a key reason for the demise of the great French makers. Brenni (1991).
22. Rudolph Koenig to James Loudon, May 23, 1889. UTA-JLP.
23. Ibid.
24. Ibid., Aug. 31, 1888.
25. Ibid., undated letter from c. 1888.
26. Ibid., Mar. 30, 1894.
27. Ibid., Apr. 4, 1895.
28. J.C. McLennan to James Loudon, Sept. 4, 1898. UTA-JLP.
29. This quote seems to refer to the period in the 1870s and 1880s. Loudon (1901b).
30. Rudolph Koenig to James Loudon, Dec. 16, 1892. UTA-JLP.
31. Thompson (1891, p. 200).
32. For more on the English approach to science, see Merz (1976, vol. 1, pp. 227–301). For more on British acoustics during this period, see Ku (2006).
33. Cahan (1989), Staley (1994), Rocke (2001), and Smith and Wise (1989).
34. In 1898, J.C. McLennan, for example, informed James Loudon that all eh goo dinstrument makers were in Germany. J.C. McLennan to James Loudon, Sept. 4, 1898. UTA-JLP.
35. Ringer (1969).
36. Ibid., pp. 305–315.
37. Rudolph Koenig to James Loudon, Jun. 11, 1896. UTA-JLP.
38. Charles R. Cross corresponded with Levi K. Fuller of the Estey Organ Company in Brattleboro, VT. They discussed Koenig forks being used for tuning activities. Charles Cross Papers, IAMIT.
39. Rudolph Koenig to James Loudon, Jul. 26, 1894. UTA-JLP. Mayer (1894, 1896), Miller (1935, p. 89), and Zahm (1900, pp. 74–76).
40. Mayer (1896, p. 84).
41. Koenig (1896a, b).
42. Rudolph Koenig to James Loudon, Jul. 24, 1892. UTA-JLP. Mayer also returned for the summer of 1894, when the tonometer had been completed, see Ibid., Jul. 26, 1894. Mayer published portions of this research in Mayer (1894, 1896).
43. J.C. McLennan to James Loudon, Sept. 4, 1898. UTA-JLP.
44. Loudon (1901b).
45. Neumann (1932c).
46. Helene Neumann to Ernst Christian Neumann, Oct. 22, 1901, NFA. Loudon's children apparently loved the "little greens peas" from Koenig's garden so much that Koenig forwarded some seeds to Toronto with instructions for planting. Rudolph Koenig to James Loudon, May 1, 1888. UTA-JLP.
47. Rudolph Koenig to James Loudon, Dec. 17, 1896. UTA-JLP.
48. Ibid., Dec. 17, 1897.
49. Ibid., Sept. 13, 1893.
50. Ibid., Oct. 10, 1895.
51. Ibid., Apr. 4, 1895.
52. Ibid., Dec. 17, 1897.
53. J.C. McLennan to James Loudon, Sept. 4. 1898. UTA-JLP.
54. Rudolph Koenig to James Loudon, Jan. 3, 1901. UTA-JLP.
55. It appears he only made two of these large sirens.
56. Ibid., Mar. 10 and Jul. 31, 1901.
57. Ibid., Mar. 10, 1901.
58. Ibid., Mar. 26, 1901.
59. Koenig (1901).
60. Loudon (1901a). Idem. 1902.
61. His marker at the Père Lachaise is now gone. It was replaced when the family could not keep up payments during WWII.

62. Abbé Rousselot and E.B. Titchener both list Landry as Koenig's successor. Rousselot (1908, p 760). Titchener (1915, p. 13). Landry's tuning forks were almost identical to Koenig's, except they were marked "LL." Copies of them can be found in the Rousselot phonetics collection at the Biblioteque Nationale (Mitterrand Branch) in Paris.
63. Helene Neumann to Ernst Christian Neumann, Oct. 22, 1901, NFA.
64. Koenig (1876a).
65. Spottiswoode (1879).
66. Bosanquet (1881, p. 431).
67. Idem., 1881–1882, p. 19. For a detailed description of the resonator, see Idem., 1879, pp. 18–19, 21.
68. Thompson (1881, p. 352).
69. A brief summary of Preyer's research can be found in the appendix of Helmholtz (1954, pp. 531–532). Alexander Ellis described these experiments to the Musical Association following Spottiswoode's presentation in, Spottiswoode (1879, pp. 125–126).
70. Ellis in Ibid., p. 126.
71. Bosanquet (1881–1882, p. 25). For more on Blaikley see Miller (1935, p. 79).
72. Bosanquet (1881–1882, p. 25).
73. Rudolph Koenig to James Loudon, Mar. 21, 1882. UTA-JLP. Also see, Le Conte Stevens (1890, p. 548).
74. Rudolph Koenig to James Loudon, Nov. 25, 1881. UTA-JLP.
75. Mayer (1896, p. 84). For more on Mayer see Cohen (1970).
76. Zahm, pp. 322, 318.
77. Ibid.
78. Miller (1935, p. 40). Also see, Zahm (1900, pp. 8, 18, 322).
79. Le Conte Stevens (1890, p. 548).
80. Helmholtz and Ellis (1954, p. 529).
81. Thompson (1891, p. 200).
82. Ibid., p. 201.
83. Thompson (1891).
84. Rudolph Koenig to James Loudon, Jun. 26, 1890. UTA-JLP.
85. The instruments used by Thompson are now in the Science Museum at London. CR no. 194.
86. Koenig (1882c, pp. 163–166); Idem., 1882a, p. 23.
87. The ones at MIT were recently operated confirming their ability to produce strong notes. One thing that stood out during the experiment, however, was the hum and smoke of the electrical contacts, see CR no. 189.
88. Rudolph Koenig to James Loudon, May 23, 1889. UTA-JLP.
89. Ibid., Oct. 11, 1889.
90. Ibid.
91. Ibid.
92. Ibid.
93. Ibid.
94. When Hertz died in 1894, Koenig wrote to Loudon: "The year has begun very sadly through the death of Hertz at Bonn, who had not even reached his thirty-seventh year. I am profoundly sad at this, because although I only had the fortune to be in relation with him during my trip to Heidelberg in 1889, he had left me with the impression of being the most amiable and benevolent man that I had met in my long life." See Rudolph Koenig to James Loudon, Jan. 4, 1894. UTA-JLP.
95. Rayleigh (1896, pp. 461, 468).
96. See Ash (1995) for a discussion of the origins of Gestalt psychology.
97. Rücker and Edser (1895).
98. Felix Auerbach's *Akustik*in Winkelmann (1909, vol. II, pp. 623–642).
99. In the 1910s, Erich Waetzmann devised a series of experiments that attempted to tie together the results of both Koenig and Helmholtz. Waetzmann (1920a,b). For a survey of the debate

up to Waetzmann, see Ullman (1986). For developments after Waetzmann, especially those related to psychology and physiology, see Boring (1942, pp. 352–359). Even in the latter half of the twentieth century there was still some ambiguity about the nature of combination tones and aural harmonics. Plomp (1965, 1967) and Berg and Stork (1995, pp. 150–153). For more on cognitive effects and auditory scene analysis, see Bregman (1990).

100. Koenigsberger (1902, vol. 2, pp. 247–249). Idem., Engl. trans. 1965, p. 313.
101. Helmholtz (1879). For a brief review of the initial dispute, see Koenigsberger (1902, vol. 2, pp. 247–249); Idem., Eng. trans. 1965, pp. 313–314.
102. Koenig (1879, p. 175).
103. Ibid., p. 178. Koenig later used a variation of this instrument to demonstrate that a fundamental tone could excite a harmonic tone via the telephone. Koenig (1882a, p. 19). Idem., 1889, p. 59. Idem., 1882c, p. 201.
104. Kursell (2006).
105. Thompson (1878).
106. Quoted by Ellis in Helmholtz (1954, p. 535).
107. Koenig (1882c, p. 223).
108. Helmholtz (1863, p. 195).
109. Helmholtz (1863, p. 195). English translation from Helmholtz and Ellis (1954, p. 127).
110. Ibid.
111. Koenig (1882c, p. 224).
112. Koenig (1901), Deuxième partie, p. X. UTA-JLP.
113. Koenig (1882c, pp. 218–223).
114. Ibid., p. 222.
115. Kursell (2006).
116. Koenigsberger, (1902, vol. 1, p. 323). English translation from Idem., 1965, p. 180.
117. Koenig (1882c, pp. 223–225).
118. Ibid., p. 221.
119. Ibid., p. 226.
120. Ibid., p. 227.
121. Ibid., p. 228.
122. Ibid., p. 229.
123. Auerbach (1878).
124. Koenig (1882c, p. 234).
125. Koenig (1882a, p. 9). Idem., 1889, pp. 28–29.
126. Thompson (1891, p. 251) and Zahm (1900, p. 374). Other texts carried pictures of the beat-wave siren, Miller (1916). Winkelmann (1909, vol. II, p. 267).
127. Koenig (1882a, p. 9). Idem., 1889, p. 27. Idem., 1882c, p. 236.
128. Koenig (1882c, pp. 236–243).
129. Ibid., p. 243.
130. Thompson and Thompson (1920, p. 159).
131. Rayleigh (1896, p. 469).
132. Rudolph Koenig to James Loudon, Oct. 11, 1889. UTA-JLP.
133. Ibid., Jul. 26, 1894. Hermann (1889, 1890, 1893, 1894).
134. Rudolph Koenig to James Loudon, Aug. 31, 1888. UTA-JLP.
135. Ibid.
136. Rudolph Koenig to James Loudon, Oct. 11, 1889. UTA-JLP.
137. Ibid.
138. Ibid., Aug. 31, 1888.
139. Ibid.
140. The bar at CSTM, the only one known to exist, is marked 8 UT6 1096 [sic] VS; 4 = UT5 = 1024 VD; 12 SOL6 6144 VS RK. CSTM acc. no. 1998.0273
141. Rudolph Koenig to James Loudon, Aug. 31, 1888. UTA-JLP.

142. McLennan must have obtained the bar (now at the CSTM) and brought it back to Toronto. J.C. McLennan to James Loudon, Sept. 4, 1898. UTA-JLP.
143. Koenig (1889, p. 23). Idem., 1899, pp. 628–629. Koenig complained to August Zahm that working with high pitches was very unpleasant as the sounds rang in his ears for days and weeks afterwards. Zahm (1900, p. 84).
144. For overviews of auditory threshold studies, see Boring (1942, pp. 332–339), Davis and Merzbach (1975, pp. 12–14), and Feldmann (1995).
145. These forks were made by Georg Appunn (1816–1885) of Hanau.
146. Helmholtz cites these forks without question in Helmholtz (1954, pp. 18, 151). Zahm also cites these forks in Zahm (1900, pp. 83–84).
147. Zahm (1900, pp. 83–84).
148. Helmholtz (1954, pp. 18, 151, 27–28, 67).
149. This debate related to Appunn and Preyer's suggestion that summation tones were actually differential tones of the second order. See Helmholtz and Ellis (1954, p. 532) and Koenig (1882c, pp. 127–128).
150. Rudolph Koenig to James Loudon, Jul. 26, 1894. UTA-JLP.
151. Ibid.
152. Koenig (1899, pp. 627–629).
153. Ibid.
154. J.C. McLennan to James Loudon, Sept. 4. 1898. UTA-JLP.
155. Rudolph Koenig to James Loudon, Sept. 14, 1899. UTA-JLP.

Conclusion – Beyond *Sensations*

Instruments constitute a large part of what we know about Rudolph Koenig. They chronicle scientific, artisanal, social and deeply personal dimensions of their maker. And they continue to speak. I have sounded hundreds of tuning forks around Europe and North America, which resonate with a distinctively even, colourless and pure sound, one that came into being under specific conditions in the workshop culture of nineteenth-century Paris. For well over 150 years, thousands of students have been introduced to the science of sound through the influence of Koenig's atelier.

The following chapters presented a portrait of this space and its role in the establishment of a radically altered novel material foundation for the scientific study of sound between the 1860s and 1900. In the workshop, Koenig transformed acoustics into a wide-ranging line of precision instruments in the mould of other fields represented in the Parisian precision trade; in his private laboratory, he pushed the technical boundaries of the field, shaped practice, and created a visual element for studying sound; in the commercial sphere he facilitated the transmission of specific kinds of teaching and research instruments throughout Europe and North America; in the social and material realms, his atelier served as a vibrant mediating space for diverse people, skills, instruments and materials. It was "chez Koenig" that many of the world's influential scientists learned about developments in acoustics; in turn, it was "chez Koenig" that these same people influenced the products and scope of acoustical practice. Above all, Koenig's atelier served as a platform for modifying, extending, spreading and challenging Helmholtz's *Sensations of Tone*.

Koenig's workshop also recast fundamental notions of acoustical sensations. His legacy, or better, the legacy of his instruments, was complex in this regard. The rapid and popular proliferation of graphical and optical instruments altered conceptions of phenomena such as timbre and beats and reinvigorated an older conception of sound as waveforms (as opposed to discreet pulses represented in the early siren). The visual approach also inspired the study of acoustical sensations on their own terms, independent of their physical basis. On the other hand, Koenig was also the great perfector and proliferator of analytic instruments (e.g. tonometers) that, through sheer numbers and presence, reinforced, spread and stabilized features of the physical, elemental aspects of the Fourier/Helmholtz model. In this way Koenig's instruments served what Robert Silverman (1992) has shown to be two increasingly divergent points of view in acoustics – analytic and holistic. In both traditions, theory and instruments reinforced each other strengthening a point of view, and ultimately making a particular perspective seem inevitable.[1] The physicists favoured the analytic instruments, while certain physiologists and early psychologists favoured the visual methods. The confrontation between Koenig and Helmholtz was as much about bodies of instruments and practice, as about their differing social and intellectual influences.

As we saw in Chapter 7, Koenig's challenges to Helmholtz were taken seriously and revealed that the nature of the so called elements of sound (sensations of tone) were open to reinterpretation. In fact, one found a similar crisis in optics during the same period (early 1870s) that Koenig commenced his first critical studies on combination tones. As Rich Kremer and R. Steven Turner have shown, Ewald Hering famously challenged Helmholtz's mechanistic theory of optical sensations.[2] Kremer, for example, argued that Hering had been influenced by Ernst Mach in moving towards a phenomenalist approach of studying sensations. Mitchell Ash has added that this debate "mobilized alternative assumptions and theoretical models" that laid the groundwork for Gestalt psychology.[3]

Many scientists viewed the Koenig-Helmholtz debate in terms of subjective and objective sensations.[4] Everything mysterious was thrown into the subjective category, or as with Helmholtz, integrated into an equally ambiguous psychology revolving around unconscious inferences or concepts such as attention. Reacting against the latter turn in optics, Hering even accused Helmholtz of being "spiritualist."[5] In fact, this narrative remained deeply engrained in discussions about the parallel acoustical controversy. In his recounting of the debate in 1942, E.G. Boring claimed that both Koenig and Preyer "argued that combination tones are subjective, but those were the days when the dualism of mind and matter pervaded the thought of all the wise men."[6]

But much more was at stake. For Hering it was a struggle to construct a broader understanding of sensory processes where sensations were treated as a reality themselves and not defined by physics. "What transpires beyond the retina, we do not know," he wrote in 1862.[7] Ash concluded: "Whether these phenomena are "objective" or "subjective," whether they are "really" experienced directly or concluded from "unnoticed sensations," was beside the point. Accepting the psychological reality of seen objects was a methodological necessity, "an indispensible prerequisite for understanding the visual function and its laws."[8] Similarly, Koenig's adherence to the primacy of visual representations of timbre freed him from interpreting timbre on physical, physiological or psychological terms. He treated the complex waveforms as mirror reflections of his auditory observations, with effects not understood by the sum of the parts, thus opening a conceptual space for hearing that would make sense to later Gestalt followers. This view of sensations also appeared in Koenig's early vowel work where he presented numbers for the major vowels that were exactly an octave apart, governed by holistic groupings, and not, as Helmholtz advocated, physics and physiology of the larynx and mouth; Wertheimer Köhler, one of the early Gestalt pioneers found the same patterns as Koenig in his studies of 1910.[9] There are also echoes of Koenig's positions in Carl Stumpf's work on tonal fusion,[10] and Mach's view that a complete account of acoustic sensation needed to take into account relations as well as individual tones.[11]

For Hering, redefining sensations entailed heading off misguided psychophysics and what he believed to be potentially dangerous philosophical positions. In fact, Kremer has argued that a large part of the disputes derived from "different disciplinary orientations in the explanation of sensory phenomena."[12] Koenig's challenge to Helmholtz, on the other hand, seemed less about philosophical commitment, disciplinary boundaries or creating a new school of psychology, and more about defending a livelihood. He refused to venture into debates about the mind and psychology, and instead moulded acoustics to the certainties of the workshop. He worked with these phenomena daily. He watched, hour after hour, his optical and graphical instruments transcribing and displaying sound. He filed and fine-tuned thousands of tuning forks. He demonstrated his instruments to potential clients. He brought hundreds of Kilograms of instruments with him on demonstration tours. He took great offence if any one criticized one of his instruments. He almost never mentioned Mach in publications or correspondence, except for the fact he sold some of his instruments. For Koenig not going beyond the ear was more of a statement about who he was – a highly skilled artisan with deep knowledge of sound. His silence on the mechanisms underlying timbre and beat tones was a rebuke to Helmholtz who he felt boldly conjectured beyond things "as they are."

Although much attention has been paid to Hering in optics, in the equally important acoustical realm, Koenig and a handful of others exposed basic assumptions in Helmholtz's sensory physiology, which revealed deep and significant tensions in late nineteenth-century physics, psychophysics, philosophy and even ideological issues.[13] Like parallel debates in psychophysics, Koenig reacted against particular elements of Helmholtz's work (what he perceived to be overreaching theoretical and mechanistic elements), ignoring some of the subtleties and wider context.[14] As we saw in Chapter 7, he used the full influence of his workshop to win over potential converts. Rayleigh himself was cautious to take sides, faced with choice between the grand German scientist on the one hand, and the most renowned acoustical maker of his age on the other. Koenig's questions found their way into these doubts and refused to go away. One weakness for Helmholtz, as Julia Kursell has shown, was that his theory of resonance (e.g. viewing the inner ear as a piano) was basically a theory of hearing for one ear, and not capable of addressing issues such as spatial orientation.[15] Later in the twentieth century cognitive scientists developed complex theories such as auditory scene analysis that revealed a more complex, non-linear understanding of how humans process sensations.[16] Koenig's challenge to Helmholtz represented reactions and tensions that gave rise to Gestalt, and later cognitive approaches to sensory problems.[17]

Indeed, the ear and hearing remain far from understood and continue to be studied through the influence of Koenig and Helmoholtz. Present hearing technologies similar to technologies developed in the nineteenth century, will always represent the limits of the cultures of the hearing research communities and industry.[18] At a recent hearing aid trade show I observed several dimensions of the present acoustical community at play on the convention floor, including manufacturers, marketers, engineers, scientists, teachers, clients, and practitioners. Koenig's workshop in Paris represented a single space in the history of acoustics where all these influences coexisted, providing a powerful means for untangling how these forces continue to shape and create our modern soundscape.

Notes

1 This idea is modeled after Hacking (1991).
2 Kremer (1992) and Turner (1994).
3 Ash (1995, p. 52).
4 Helmholtz and Ellis (1954, p. 531). There were continuing debates about this distinction in the psychology community as well, see the *Psychological Review*, vol. 8, 1901, pp. 630–632.
5 Quoted in Kremer (1992, p. 149). Also in Ash (1995, p. 57).
6 Boring (1942, p. 357).
7 Ash (1995, p. 55). For more on Hering and Mach's views of sensations, see Kremer (1992).
8 Ash (1995, p. 55) including quotes from Hering. For more on the Hering-Helmholtz controversy, see Turner (1994).
9 Murray and Bahar (1998) and Boring (1942, p. 372).
10 Boring (1942, pp. 359–363) and Ash (1995, p. 90).
11 Murray (1988, p. 274).
12 Kremer (1992).
13 Vladimir Lenin harshly criticized Mach and his phenomenalist view of sensations: "What then is the essence of the agnostic's line? It is that he does not go beyond sensations, that he stops on this side of phenomena, refusing to see anything "certain" beyond the boundary of sensations." *Quoted in Materialism and Empirio-Criticism,* Lenin (1938, p. 171). By Lenin's standards Koenig was an agnostic.
14 The misunderstood relational aspects of Helmholtz and Wundt are stressed by Murray (1988, pp. 210, 281).
15 Kursell (2006).
16 Bregman (1990).
17 Murray (1995) has shown the historical continuities between gestalt and cognitive psychology. The grouping principles of Gestalt thinking, for example, underlie explanations of cognitive perceptual organization in "auditory scene analysis."
18 Mills (2008, 2009)
19 For cultural histories of acoustic technologies and the modern soundscape, see Thompson (2002), Wittje (2006), Idem., 2003, and (Stern 2003).

Appendix A
Key Dates in Rudolph Koenig's Life

1832 – born Nov. 26, 1832 in Königsberg, East Prussia
1840s – educated at the Kneiphöfischen Gymnasium, Königsberg. Failed abitur
1851 – moved to Paris and joined the workshop of the violinmaker Jean–Baptiste Vuillaume (1798–1875)
1858 – started his own business making acoustical instruments, Place Lycée Louis le Grand 5
1858 – began work with Édouard–Léon Scott de Martinville on the phonautograph
1859 – published first catalogue
1859 – received first commission from Hermann von Helmholtzto manufacture glass resonators
1862 – invention of the manometric flame capsule
1862 – medal of distinction at the London Exhibition
1863 – Helmholtz publishes *Die Lehre von den Tonempfindungen als Physiologische Grundlage für die Theorie der Musik*
c.1864 – moved to 30 rue Hautefeuille, next door to Gustave Courbet and near the medical faculty and the Azoux workshop for anatomical models
1865 – publication of second catalogue
1865 – Médaille d'Or from the Société d'Encouragement
1866 – collaborated in 1866 with Victor Regnault on sound experiments in the sewers of Paris
1866 – began work on *grand tonomètre*
1867 – invention of first wave siren for Terquem
1867 – gold medal at the Paris Exposition
1868 – honourary doctorate from the University of Königsberg
1870 – publication of vowel research
1870 – left Paris during Franco–Prussian war
1871 – returned to Paris after commune
1873 – published third catalogue; banking crisis in Europe and NA
1876 – publication of combination tone research
1876 – attended Philadelphia Centennial Exhibition; medal of distinction
1877 – moved from 30 rue Hautefeuille to nearby 26 rue Pontoise
1877 – began research on phase and timbre; further development of wave sirens
1879 – invention of clock fork; creation of standard tuning fork

1882 – publication of fourth catalogue; publication of book on collected research since his first year in business; return to Philadelphia; lecture series in Toronto and Montreal
1882 – moved to 27 Quai d'Anjou on the Île St. Louis
1889 – published fifth catalogue; another economic slowdown in Paris
1889 – presentation of controversial findings in front of Helmholtz at the sixty-second congress of physicians and naturalists at Heidelberg
1894 – finished the complete universal tonometer
1899 – published research on ultrasonics
1901 – death on October 2, Paris, cremated and buried in Père Lachaise cemetery (ashes and name plate removed during WWII)
1901 – L. Landry, his main collaborator for thirty years, took over the business.

Catalogue Raisonné of Koenig Instruments

Based on the Catalogue Titles in Koenig's 1889 Catalogue[1]

The following catalogue raisonné is a reference guide for the preceding chapters. I have used it primarily to document details and observations of surviving instruments. Using Koenig's catalogue 1889 as a template, I refer to specific instruments by their original catalogue numbers, preceded by *CR* which stands for *Catalgue Raisonné*. For example, *CR no.* 27 is the Helmholtz Double siren which is no. 27 in the 1889 catalogue. I have preserved the same numbers, English titles and sections, as well as prices. For each entry, I have presented where possible the history of the instrument, its function, references in primary texts and journals, and references in secondary literature. I have also described (where possible) the features, materials, markings, measurements and locations of surviving examples. The locations (listed below) at times appear with the date of purchase (if known), e.g. Toronto (1878), and at times with the date of purchase and accession number combined, e.g. Coimbra (1867: FIS.384). Organ pipe measurements include only the pipe, and not the mouthpiece. I have indentified and examined hundreds of Koenig instruments in person, others I have learned about through museum catalogues or correspondence with curators. Due to the realities of museum visits and large collections, I have examined some instruments very carefully, others under serious time constraints.

This catalogue is also meant to be a practical guide and reference for indentifying, cataloguing, researching and displaying Koenig's instruments, which are spread in collections and museums around the world. In order to create this catalogue, I have visited the major collections at the University of Toronto, Smithsonian Institution, Coimbra University, Fondazione Scienza e Tecnica, Conservatiore Nationale des Arts et Métiers, Collection of Historical Scientific Instruments, Harvard University, Musée de la Civilisation du Québec, Union College and the Univesity of Rome. I have yet to see the collection at Teylers Museum, but was able to rely on Gerard Turner's excellent catalogue of it.[2] In addition, as one can see below, I have been able to visit several medium and small Koenig collections around Europe and North America. Even with all this tracking and research, however, not all the entries are complete. In some cases I have not been able to locate an instrument or discover its function or history. Some entries, therefore, have no information at all, but still remain in the text to preserve the original order and context of the catalogue.

The database of Koenig collections is not complete enough to draw statistical conclusions. I thought originally that I could gather information on all of Koenig's surviving instruments around the world. To my delight and frustration, they keep appearing, even in well-studied collections such as Toronto. This in itself tells us something about the deceptive size of his output and operation. After persistent attempts, for example, I have been unable to list the substantial Koenig collection that reputedly survives at the Moscow State University (Chapter 7). My searches, however, have been extensive enough that the reader can see some patterns from the number of surviving instruments and their locations, as I have referred to at times in the text. Above all, the catalogue provides information about the surviving instruments as a means for enriching the story of Koenig's workshop and his clients. It offers the first comprehensive picture of the scope, practice and teaching of acoustics in the second half of the nineteenth century.

Prices appear in the original French francs. To use a simple comparison from the time, Vincent van Gogh bought a suit and six pairs of socks in Arles in 1889 for 39 fr, the same price as a standard tuning fork and resonator; a year later he sold one of his paintings for 400 fr, 50 fr more than the price of a Koenig sound analyser, which sold for 350 fr in 1889.[3] The Eiffel tower cost 7.8 million fr to build between 1887 and 1889.[4]

Note: The original 1889 catalogue is now on-line thanks to Steve Turner, Curator of Physical Sciences at the Smithsonian Institution. The reader can view the catalogue by visiting the website *Instruments for Science: Scientific Trade Catalogues in Smthsonian Collections* and calling up Rudolph Koenig's catalogues.

Locations

Amherst – Amherst College, New York, USA
Barcelona – Universitat de Barcelona, Spain
Boerhaave– Boerhaave Museum, the Netherlands
Charité– Humboldt-Universität zu Berlin, Charité, Johannes-Müller-Institut für Physiologie
Case – Case University, Ohio, USA
CNAM – Conservatiore Nationale des Arts et Métiers, Paris, France
CSTM – Canada Science and Technology Museum, Ottawa, Canada
Coimbra – Museu de Física, University of Coimbra, Portugal
Colby College, Maine, USA
Columbia – Columbia University, New York, USA
Cornell – Cornell University, New York, USA
Dartmouth – Dartmouth College, New Hampshire, USA
Dublin – University College of Dublin, Physics, Ireland
Duke – Duke University, North Carolina, USA

École Polytechnique, Paris, France
Geneve – Musée d'histoires des sciences, Geneve, Switzerland
Harvard – Collection of Historical Scientific Instruments, Harvard University, Massachusetts, USA
Henry IV, Paris, France
ISEP – Institute Superior de Engenharia do Porto, Portugal
FST – Fondazione Scienza e Tecnica, Italy
Johannes Müller Institut für Physiologie, Berlin, Germany
Kenyon – Kenyon College, Ohio, USA
Liceo Visconti, Rome, Italy, USA
Lisbon – Museum of Science, University of Lisbon, Portugal
Maynooth – National University of Ireland, Maynooth, Ireland
MCQ – Musée de la Civilisation du Québec, Québec, Canada
McGill – McGill University, Québec, Canada
Minnesota – University of Minnesota, Minneapolis, USA
MIT – Massachusetts Institute of Technology, USA
Naples – University of Naples, Italy, USA
Nebraska – University of Nebraska, Lincoln, Nebraska, USA
NMAH – National Museum of American History, Smithsonian Institution, Washington DC, USA
Oxford – Museum for the History of Science, Oxford, UK
Porto – Museum of Science, University of Porto, Portugal
QUP – Queen's University Physics, Ireland
Rennes – Faculty of Science, University of Rennes, France
Rome – La Sapienza, University of Rome, Italy
Science Museum, London, UK
St. Mary's College, University of Notre Dame, Indiana, USA
Sydney – University of Sydney, Australia
Teylers – Teylers Museum, Haarlem, the Netherlands
Toronto – Department of Physics, University of Toronto, Ontario, Canada
Tokyo – University of Tokyo, Japan
University of Mississippi at Oxford, Mississippi, USA
Union – Union College, New York, USA
Utrecht University Museum, Netherlands
Vanderbilt – Vanderbilt University, Tennessee, USA
Vermont – University of Vermont, Vermont, USA
Wesleyan – Wesleyan University in Middletown, Connecticut, USA
Western – University of Western, Ontario, Canada
Yale – Yale Peabody Museum, Yale University, Connecticut, USA

Contents

I. The Principal Means for Producing Sound

1. Eight wooden bars giving the musical scale when thrown in succession upon the floor. 6 fr

These eight wooden bars of varying thickness (each numbered) are dropped to the ground in succession emitting the musical scale or a simple melody. They were a standard illustration of the production of sound through wooden bars of varying materials, shapes and sizes. They were the first item in each of Koenig's five catalogues from 1859 and also appeared in Albert Marloye's catalogue of 1851, revealing a possible lineage to Félix Savart's lectures at the College de France. The bars, usually made of pine, are invariably the most worn in any acoustical collection, but there is still evidence, as with the examples at the Smithsonian, of each one made to specific dimensions.

D.C. Miller commented on the difference between dropping the sticks together, where one heard the effect of "noise only," and dropping them in a purposeful order, creating a musical melody. But where, he asked, does one draw the line between noise and harmony? To illustrate this difficult question, he described Wagner's controversial Tannhäuser overture from the early 1860s and how it was passionately criticized by some as "shrill noise and broken crockery effects" while being praised by others as harmonious and a "chorus of pure aspirations" "The study of noises," Miller wrote, "is essential to the understanding of the qualities of musical instruments, and especially of speech. Words are multiple tones of great complexity, blended and flowing, mixed with essential noises."[5]

Locations: Amherst (only two sticks, 1 and 8). Coimbra. MCQ (acc. no. 1993.13304). NMAH (cat. no. 87.924.5). Teylers (unsigned). Toronto.

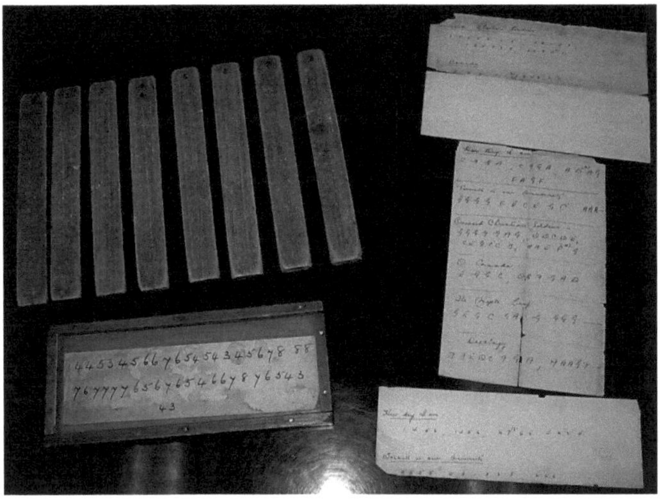

Fig. CR no. 1 Photo by Phil Scolieri, 2005. Physics Department, University of Toronto, Canada

Description: The Toronto set, all pine, carries local instructions for music by numbers (written in ink) for "How dry I am," "the Maple leaf," "Oh Canada," "Doxology" "Onward Christian soldiers" and "Toronto is our University." The Teylers set is not signed. They are different sizes made of beech, pine, oak, dense pine, and limewood.

Markings and measurements: (Toronto) Numbered 1 though 8. No. 1 is stamped "RUDOLPH KOENIG À PARIS." Each stick is 21 cm long with a slight increase in thickness as the numbers increase. No. 1 is 6 mm, and no. 8 is 12 mm thick.

References: Marloye (1851, p. 48), Miller (1916, pp. 22–24), and Turner G.L'E. (1996, p. 107).

1a. Four pieces of wood giving the major chord. 3 fr

Location: Yale (acc. no. YPM 50282).
Reference: Marloye (1851, p. 48).

2. Four tubes giving the major chord when their pistons are withdrawn in succession. 35 fr

This is a simple illustration of the production of a major chord from ut3, mi3, sol3 to ut4. In order to produce the notes, one pulls tightly fitting brass pistons from the tubes in succession. Similar to the dropping sticks (no. 1), this was a demonstration of how short duration sounds could form a musical tone. D.C. Miller, who built one of the largest collections of flutes in the world (subsequently donated to the Library of Congress), compared this demonstration to a common trick played with his favourite instrument: "A distinguishable tune can be played," he wrote in the

Science of Musical Sounds, "on a flute without blowing into it, the air in the tube being set into vibration by snapping the keys sharply against the proper holes to give the tune."[6]

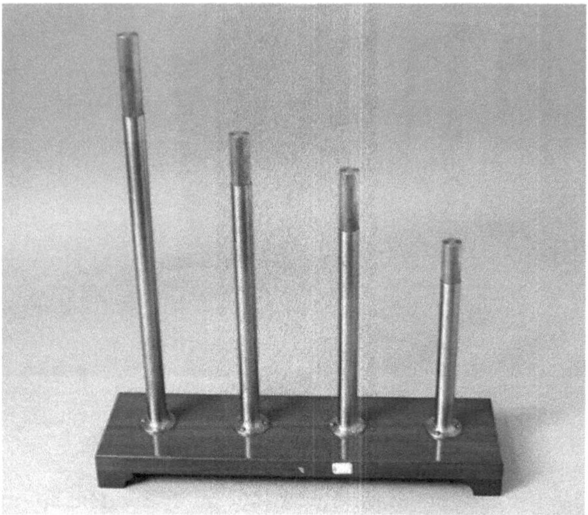

Fig. CR no. 2 Photo by author 2005, Museu de Física, University of Coimbra, Portugal

Locations: Coimbra (FIS.0369; c. 1878). Maynooth. NMAH (cat. no. 315,169). Vanderbilt (1875).

Description: The Coimbra instrument (above) consists of brass tubes and a mahogany base. Pulling the brass cylinders in succession produces the pure notes of a major chord. The Vanderbilt example has wooden organ-pipe mouthpieces as the pistons, perhaps a local adaptation.

Markings and measurements: (Coimbra) Stamped "RUDOLPH KOENIG À PARIS" in the middle of the base. Overall dimensions, 44.0 × 40.0 × 11.6 cm.

References: Daguin (1867, p. 516), Miller (1916, p. 23), and Zahm (1900, p. 36).

3. Double-Bass Bow. 6 fr

Location: NMAH (cat. no. 314,590).
Reference: Marloye (1851, p. 55)

4. Bass Bow. 7 fr

Location: NMAH (cat. no. 314,590).
Reference: Marloye (1851, p. 55)

5. Violin Bow. 7 fr

Location: NMAH (cat. no. 314,590).
Reference: Marloye (1851, p. 55)

6. Bundle of horse hairs for exciting plates at the centre. 2 fr

7. Ivory Hammer for striking forks or steel cylinders of high pitch. 6 fr

Several types of hammers were used for striking tuning forks – steel, wood, rubber and ivory. Steel produces powerful tones when striking a tuning fork, but there are sometimes unwanted higher harmonics. Rubber is too soft to produce the most powerful tones. The ivory hammer produces a pure, strong tone with few unwanted effects. It didn't appear in the catalogue until 1882, perhaps emerging as a response to controversies surrounding the purity in Koenig's forks. Koenig also sounded a tuning fork with the stroke of a violin bow.

Locations: Coimbra (FIS.0628). Harvard (acc. no. 1998-1-0713).abbe
Reference: Koenig (1882c, p. 135 (ivory) and pp. 9, 14, 22, 85 (violin bow))

8. Cagniard de Latour's whistling tube. 4 fr

9. Hélicophone. 2 fr

10. Locomotive whistle of brass. 30 fr (also see no. 204).

The locomotive whistle produced a blast of steam-powered sound that became a familiar icon of the nineteenth-century soundscape, eventually making its way into the laboratory and classroom. Koenig first made them of a beautiful, rich rosewood, but then shifted to the more functional and sturdy brass.

Fig. CR no. 10 Photo by author, 2005. Physics Department, Union College, NY, USA

Locations: Nebraska (c. 1890). QUP. Teylers (1865). Union (c. 1870).
Description: An older version at Union College (above) is made of rosewood and when tested recently produced a pure, high, fixed pitch. The example at Nebraska, a later model, is made of brass.
Markings and measurements: (Union) Stamped "RUDOLPH KOENIG À PARIS," 19.5 cm long.
References: Koenig (1882c, pp. 163–166), Marloye (1851, p. 43), Mollan (1990, p. 203), and Turner, G.L'E. (1996, p. 105)

12. Mouth-piece of organ pipe, with moveable over lip. 9 fr

A thin slab of pine slides into the opening of the lip. As it closes the gap, the sound of the whistle clearly rises in pitch.

Fig. CR no. 12 Photo by author, 2005. Physics Department, University of Toronto, Canada

Locations: NMAH (cat. no. 327,553). CSTM (acc. no. 1998.0260). Toronto.
Markings and measurements: (Toronto) Stamped "RUDOLPH KOENIG À PARIS," 3.5 × 6.4 × 28.2 cm.

13. Mouth-piece of the horn. 3 fr

References: Guillemin (1881, p. 833), Marloye (1851, p. 36), and Zahm (1900, p. 243).

14. Mouth-piece of the trumpet. 3 fr

References: Marloye (1851, p. 35) and Zahm (1900, p. 245).

15. Mouth-piece of the ophicléide. 3 fr

References: Guillemin (1881, p. 836), Marloye (1851, p. 35), and Zahm (1900, p. 245).

16. Mouth-piece of the clarionet. 4 fr

References: Guillemin (1881, p. 836), Marloye (1851, p. 35), and Zahm (1900, p. 242).

17. Mouth-piece of the hautbois. 3 fr

References: Guillemin (1881, p. 832), Marloye (1851, p. 36), and Zahm (1900, pp. 242–243).

18. Mouth-piece of the bassoon. 3 fr

References: Marloye (1851, p. 36) and Zahm (1900, pp. 242–243).

19. Cagniard Latour's mill-siren. 20 fr

This instrument works like a siren. It consists of a cylindrical tube with a fan at the open end. When one blows into the tube, the fan rotates producing a series of intermittent bursts of air which blend into a tone. Stronger blasts of air produce higher pitches.

References: Marloye (1851, p. 44) and Zahm (1900, pp. 30–31).

20. Cagniard de Latour's musical sling. 8 fr

The musical sling produces sound while being whirled around in a circle with a string. It consists of a metal plate about 15 by 7.5 cm in size attached to a string. When the plate is put in rapid circular motion, resistance to air forces it to revolve rapidly around a moveable axis. The resultant flutter produces an acute sound heard by everyone in the room. It is very similar to a bullroarer, which is found in several aboriginal cultures around the world.

References: Marloye 1851 (p. 55) and Zahm (1900, pp. 30–31).

21. Trevelyan's Rocker. 20 fr

The Trevelyan rocker, invented by Arthur Trevelyan in 1829, creates sound through a rapid rocking motion induced by heating and expansion of the metal base. The rocker (sometimes brass) has a grooved edge which rests on a lead base. When the heated rocker makes contact with the base, the point of contact on the base expands sending the rocker to its other edge. It moves back and forth at a high rate producing a sound. In the early nineteenth century when natural philosophers sought to uncover

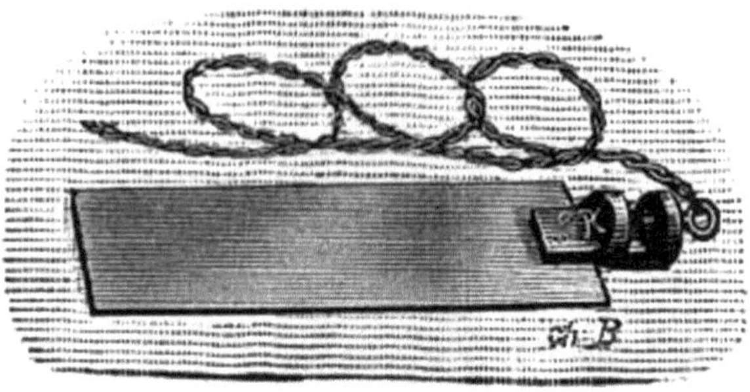

Fig. CR no. 20 Source: Koenig (1889, p. 13)

the underlying unity in physical phenomena, this was seen as a novel illustration connecting heat and sound.

Fig. CR no. 21 Source: Koenig (1889, p. 14)

References: Daguin (1867, p. 455), Deschanel (1877, p. 795), Fau (1853, pp. 404–405), Freeman (1974), Guillemin (1881, pp. 628–629), Jones (1937, pp. 238–239), Seebeck (1840), Trevelyan (1832). Idem., 1833, 1834, 1835. Tyndall (1896, pp. 81–83), Violle (1883, p. 13), and Zahm (1900, pp. 29–30).

22. Rijke's tube. 6 fr

This glass tube with an alcohol burner at the bottom produces pure sounds from movements of heated air within. The principle was discovered in 1859 by Pieter Leonhard Rijke, a professor of natural philosophy at the University of Leiden.

References: Guillemin (1881, p. 668), Jones (1937, pp. 232–234), Zahm (1900, p. 31), and (Rijke 1859).

23. Whertheim's apparatus for producing sound electrically in an iron rod. 44 fr

This apparatus represents the first attempt to connect electrical and acoustical phenomena. Guillaume [Wilhelm] Wertheim (1815–1861), its inventor, was one of a handful of specialized acoustical researchers in Paris at mid century. The apparatus works on a simple principle – changing current in the bar creates periodic constrictions which are translated into sound. The bar is made of soft iron which connects to a base through the middle. One side carries an electromagnetic coil connected to a battery. When current is sent through the coil, the bar begins to resonate producing a weak sound. The Reis telephone was based on this principle (CR no. 166).

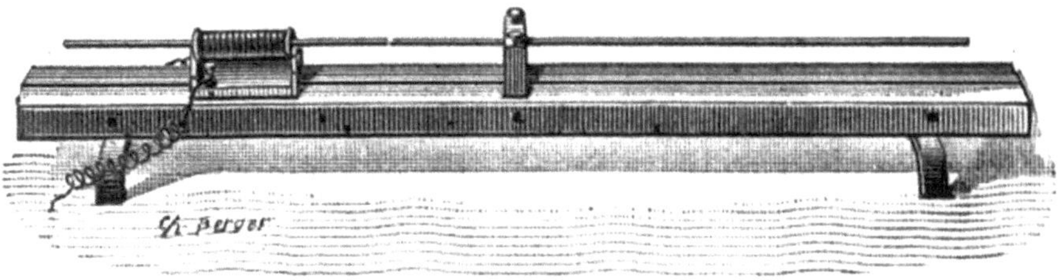

Fig. CR no. 23 Source: Koenig (1889, p. 14)

References: Wertheim (1848) and Zahm (1900, p. 32).

II. Cause and Nature of Sound

24. Cagniard de Latour's siren, with counter. 90 fr

In 1819 Charles Cagniard de Latour invented the siren, which was a revolutionary instrument for the study of sound, conceptually and practically. In contrast to traditional acoustical devices such as vibrating strings, it generated sound from discreet pulsations of air. This instrument brought about a reconceptualization of sound; it also stimulated the invention of many more types of sirens throughout the nineteenth century. Koenig, for example, created several different forms of wave siren based on Latour's original invention.

The basic Latour siren consists of a brass disk, pierced with a series of holes. Pressured air pushes against the holes, which are cut on an angle, thus moving the disk. The distinct puffs of air blend into a specific pitch depending on the number of holes and the speed of rotation. As the axle rotates a delicate counting mechanism registers each revolution of the disk. The experimenter records the duration that the siren sounds and thereby calculates the frequency (pulses/second).

Locations: QUP. Rome (1891). St. Mary's College, Notre Dame.
References: Blaserna (1876, p. 61), Daguin (1867, pp. 491–492), Desains (1857a, p. 3), Deschanel (1877, pp. 823–826), Fau (1853, p. 365), Ganot (1893, p. 225), Guillemin (1881, pp. 652–653), Ianniello (2003, p. 89), Jamin (1868, p. 503),

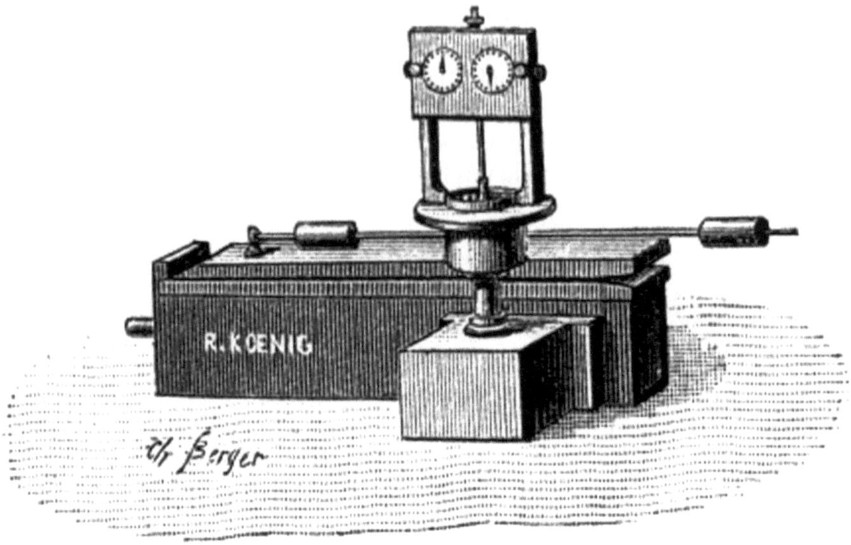

Fig. CR no. 24 Source: Koenig (1889, p. 36)

Marloye (1851, p. 54), Mollan (1990, p. 199), Tyndall (1896, pp. 91–96), Violle (1883, pp. 14–17), and Zahm (1900, pp. 62–63).

25. Siren arranged for projection. 100 fr

In order to emphasize the discreet nature of each pulse of a rotating siren disk, Koenig produced an optical demonstration of a siren in action. It consisted of a rotating, pierced disk with a leather belt (presumably on a mount similar to CR no. 30). A wind-tube with a window at one end blows pressured air on the disk. A light shines through the tube thus projecting the rapid openings and closings onto a screen.

26. Siren arranged for sounding in water. 400 fr

This siren uses pressured water instead of air. Two large reservoirs of water rest at different heights above a simple Latour siren (without a counter), connected to the end of a tube with a faucet. As the reservoirs fill, and the faucet opens, the water forces its way through the pierced holes producing a sound.

27. Helmholtz's double siren. 450 fr

The double siren was one of Koenig's more popular instruments. It consisted of two "polyphonic" or "multi-voiced" sirens with more than one series of holes, and was an invention of the German physicist and former teacher of Hermann von Helmholtz, Heinrich Wilhelm Dove (1803–1879). It produced several pure tones simultaneously, in musical chords, and under greater pressure. It was ideal for demonstrating interference effects (when sound waves combined to amplify or diminish each other) and combination tones.

Fig. CR no. 27-1 Sauerwald double siren. Courtesy of the Department of the History of Science, Collection of Historical Scientific Instruments, Harvard University, USA. acc. no. 1997-1-1799

With the help of the Berlin mechanic, Sauerwald, Helmholtz created this instrument in 1855–56.[7] The upper disk has four separate rings of holes, 9, 12, 15 and 16; the lower disk had the holes, 8, 10, 12 and 18. From these holes, one can create combination of tones differing by various intervals, some of which are musical. Each siren connects to a powerful air bellows and has four pins to activate (open) a particular circle of holes. Counting dials are in the middle of the two sirens for recording the number of turns per second and, with aid of a clock for timing the revolutions, determining the frequency of a particular row of holes. A handle at the top allows one to rotate the upper siren by graduated degrees in order to create a shift in the phase of the upper sound source compared to the lower sources (for studying interference effects). Helmholtz also added small brass resonators which covered the disks, a feature to ensure that the sounds were pure, without harmonics, and "full, strong and soft, like a fine tone on the French horn."[8] The polyphonic double siren, therefore, produced a means for investigating complex (compound) musical sounds from what was hoped to be relatively pure elements of sound.

Based on examinations of Koenig's later models (1860s and 1870s), the design was similar to the earlier models of Sauerwald, yet simpler in presentation. Sauerwald, a maker of electrical instruments, created sirens with elegant brass workmanship (such as a more refined counter face) and an extra rim of brass on the disks for aesthetics and possibly better rotation.

Fig. CR no. 27-2 Photo by author 2005. Museu de Física, University of Coimbra, Portugal. FIS.0384

There is a very large version of the double siren that has recently been discovered by Judith Pargamin, a curator at the Museum of Natural History of Lille, France. This instrument most likely came from the laboratory of Alfred Terquem, who was a professor at the University of Strasbourg in the late 1860s and then the University of Lilles following the Franco-Prussian war. In his 1882 book (p. 157) Koenig mentions collaboration with Terquem in the late 1860s. Terquem describes this instrument specifically in his article of 1870. It was either made for Terquem or made for specific research in Koenig's studio and then bought by Terquem. It measures about 40 cm in diameter, with a stand almost 3 m in height, and has a very large counterweight system for rotating the disks. In addition there is also a simpler siren of the same diameter that replaces the traditional siren chambers. It has holes of various shapes (e.g. diamond and triangles).

Fig. CR. no. 27-3 Photo by Judith Pargamin, Museum of Natural History of Lille, France

Locations: CNAM (inv. 12602). Columbia. Coimbra (FIS.0384). Dublin. ISEP. Lisbon. McGill. Maynooth. NMAH (cat. no. 80.98.2). Oxford (acc. no. 17376). QUP. Rennes. Rome. Toronto (1878). Vanderbilt (1875). Vermont. Wesleyan. Original Sauerwald double sirens as described by Helmholtz can be found at the Boerhaave, Harvard (by Sauerwald, acc. no. 1997-1-1799), Müller Institut, and Teylers.

Markings and measurements: (NMAH, originally at Smith College) This instrument was sold by "N.H. EDGERTON PHILA, PA." (Toronto). Stamped "RUDOLPH KOENIG À PARIS" on the wooden base and on the top elbow of the brass frame. 46 × 25 × 26 cm.

References: Auerbach in Winkelmann (1909, pp. 590–591), Blaserna (1876, pp. 96–100), Helmholtz (1863, pp. 241–242), Jamin (1868, pp. 591–592), Loudon and McLennan (1895, pp. 102–103), Mollan (1990, pp. 199, 328), Pisko (1865, pp. 48–52), Terquem (1870, p. 280), Turner, G.L'E. (1996, p. 110), Tyndall (1896, pp. 103–106, 385–392), Violle (1883, pp. 101–103), Vogel (1993, p. 267), and Zahm (1900, pp. 402–404)

28. Large siren for Seebeck's experiments with key-board and counter. 1,400 fr

The Seebeck siren came from the research of August Seebeck, director of the technical school at Dresden, who had introduced modifications to Cagniard de la Tour's

Fig. CR no. 27-4 Photo by Judith Pargamin, Museum of Natural History of Lille, France

basic siren for his studies on the nature of pitch. In the early 1840s he designed a siren to investigate the relations between the spacing of the holes and variations of sound produced. Koenig was the first maker to offer this siren to the Parisian market in the late 1850s. It was a research apparatus, but also served as a dramatic demonstration of isochronism (regular spacing of pulses) and interference effects (the combination of waves to produce beats, silences and augmentations). Even in the absence of sound, the beautiful patterns of concentric circles and changing positions of holes on the disks evoke an underlying mechanical structure of sound.

Koenig's models came with nine disks (1865 catalogue) and seven (1889 catalogue). Four disks tested various distortions from isochronous settings (using slightly unequal time spacings) to see if the listener could distinguish these differences; one tested interference effects (when one wave pattern imposed on another diminished or augmented vibrations); one reproduced the musical scale with eight series of holes; one reproduced eight harmonics of a fundamental note; and one produced beats. In earlier models the disks were firm cardboard (CR 28c), and brass in later versions.

The complete apparatus went through a few transformations during Koenig's career. The earlier instruments came with a simple rotation device on a cast iron stand. Later examples from the early to mid 1870s (Porto, NMAH and Rome) have built-in wooden wind chests with counters. (The counter, supposed to be mounted below the disk, is missing on the Porto model pictured below). The one pictured in the 1889 catalogue is more refined with the parts and chambers concealed resembling a tambour style clock. The latter changes especially show that the instrument was still an important part of the acoustical cabinet and worthy of functional and aesthetic refinement. In the late 1860s, it was a vital piece of research equipment. French experimenter, Alfred Terquem, used the Seebeck siren along with other Koenig sirens to study timbre and test the controversial claims of Ohm, Seebeck, and Helmholtz.

Fig. CR no. 28-1 Photo Courtesy of the Museum of Science, University of Porto, Portugal

Location: Maynooth. NMAH (cat. no. 314584). MIT. Porto (c. 1875). Rome (c. 1873). Vanderbilt (c. 1875).

Markings and measurements: (Porto) 62 × 57 × 47 cm. (MIT) One brass disk marked, "RK" with a series of eight holes marked, 24, 36, 12, 12, 12, 12, 12, 12. Diameter = 29.6 cm. (NMAH) overall dimensions, 55.9 × 45.7 × 50.9 cm.

Fig. CR no. 28-2 Source: Koenig (1889, p. 16)

References: Blaserna (1876, pp. 125–126), Daguin (1867, p. 509), Guillemin (1881, pp. 653–654), Jamin (1868, pp. 505–506), Koenig (1865, p. 7), Seebeck (1841, 1843), Terquem (1870, pp. 279–280), Turner (1977), and Zahm (1900, pp. 61–62)

28a. The same apparatus with simpler wind-chest. 1,200 fr

28b. The same apparatus without clock-work and counter. 800 fr

28c. The same apparatus in simpler form. 400 fr

Locations: Coimbra (FIS.0734). Harvard (acc. no. 1997-1-1009). MCQ (acc. no. 1993.13295). Utrecht.

Markings and measurements: (Coimbra) cardboard disks stenciled in block letters, "RUDOLPH KŒNIG À PARIS" on one side. Opposite sides have brief instructions written by hand with underlined titles which read: "Série de sons harmoniques;" "Les chocs peuvent partir de différents centres pour concourir à la formation d'un même son, pourvu qu'ils soient suffisamment isochrones et produits dans la même direction;" "Gamme;" "Effets d'interférence;" "Effets produits si l'isochronisme des chocs n'est pas parfait I;" "Effets produits si l'isochronisme des chocs nest pas parfait II;" "Effets produits si l'isochronimie des chocs n'est pas parfait III." Diameter of disks = 31.5 cm.

28d. Siren disk giving the scale. 50 fr

This disk carries eight series of holes that produce the physicist's scale.

Location: Coimbra (FIS.0734).

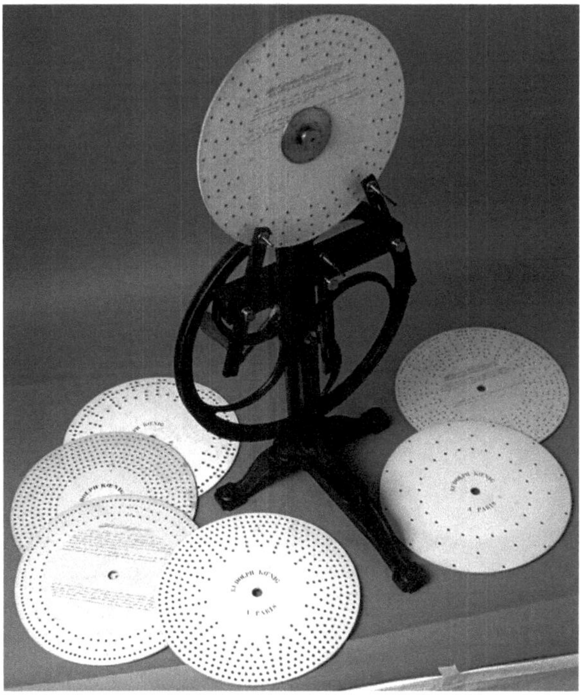

Fig. CR no. 28c Photo by author 2005, Museu de Física, University of Coimbra, Portugal. FIS.0734

29. Oppelt's Siren. 90 fr

Oppelt's siren, first described by Friedrich Oppelt in 1852, has a series of 24 holes that demonstrate a number of acoustical effects. The first fifteen holes produce simple tones, the next five give different intervals from the scale, and the remaining holes provide various musical chords and harmonies. One could attach it to a Savart rotation machine (CR no. 30).

Locations: CSTM (acc. no. 1998.0245). Harvard (acc no. 1997-1-1018). FST. McGill (only brass disk). Naples.

Description: Harvard has a cardboard version from Koenig's earlier workshop. The disk at the FST in Florence (50 cm dia.) is made of zinc and the first series of holes is 6, 9, 12, 15, 18, 24, 30, 36, 48, 60, 72, 96, 120, 144, and 192 holes. The next series of holes give the intervals, 5/4 (third) 24 and 30 holes, 4/3 (forth) 24 and 32 holes, 3/2 (fifth) 24 and 36 holes, 5/3 (sixth) 24 and 40 holes, and 23/16 (diminished seventh) 32 and 46 holes. Each ring of holes consists of two sets of holes combined into one that create a combined sound. The next two series of holes (24, 32, 40, 48 and 24, 30, 36 and 48) produce musical chords, ut1, mi1, sol1, ut2 and ut1, fa1, la2 and ut2.

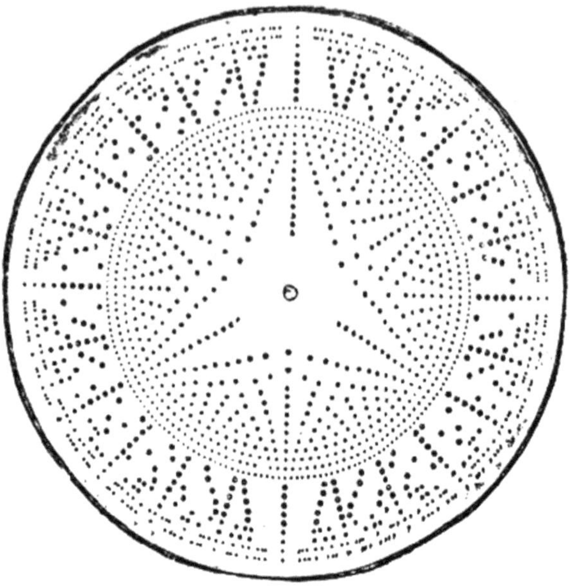

(d = 0^m,50) (N° 29).

Fig. CR no. 29 Source: Koenig (1889, p. 17)

Markings and measurements: (CSTM) 50 cm dia. (Harvard) Stamped in ink "RUDOLPH KOENIG À PARIS." Inscribed on back, "Rapports des nombres du trous, 2, 3, 4, 5, 6, 8, 10, 12, 16, 20, 24, 32, 40, 45, 54, 5/4, 4/3, 3/2, 5/3, 7/4, 6543, 8654, 6543, 8654."

References: Giatti (2001, p. 89), Opelt (1852, 1855), and Zahm (1900, pp. 400–401).

30. Savart's toothed wheel, with bar and counter. 1,200 fr

30a. The same apparatus mounted on wood (old model). 800 fr

30b. The same apparatus smaller, without bar and counter. 250 fr

The French scientist, Félix Savart, designed a rotating toothed wheel to produce sounds by discreet, periodic grating against a tongue of metal or wood. As the wheel increases in speed, the grating turns into a continuous sound, rising in pitch with faster revolutions. Savart wanted to test the limits of human sound perception. Marloye, who had collaborated with Savart, made an improved version with a counter. Koenig made one with a counter that could be adapted for use with other sirens, such as the wave and Opelt sirens.

Fig. CR no. 30b Photo by author, 2005. Physics Department, Union College, USA

Locations: NMAH (cat. no. 328742). Union College (only the four-part brass wheel remains). Lisbon.

Description: The one at the NMAH (a smaller model without counter) has been adapted for use with a wave siren (no. 62). It comes with an oak frame, bar, tube and slit for producing pressured air on the rotating disk (no. 30bb). The one at Lisbon (identical frame) has four brass toothed wheels that produced a major chord when played simultaneously.

Markings and measurements: (NMAH) air tube, wooden frame, and handle are stamped "RUDOLPH KOENIG À PARIS." Overall dimensions: 48 × 44 × 102 cm.

References: Daguin (1867, pp. 493–494), Fau (1853, p. 355), Ganot (1893, pp. 224–225), Guillemin (1881, p. 651), Jamin (1868, p. 506), Marloye (1851, p. 53), Savart (1830), Idem., 1831, Violle (1883, pp. 11–13), and Zahm (1900, p. 29).

30bb. Bar with slit to be fixed upon the preceding apparatus. 30 fr

30c. Rotatory apparatus of preceding without the toothed wheels. 180 fr

III. Pitch of Sounds

31. Chart giving the vibration-frequency of sounds. 2 fr

This is a reference table for a variety of musical sounds in the tempered scale. It is based on the forks ut3 = 512 v.s. (physicist's tuning fork); la3 = 870 v.s. (the official French standard tuning fork); la3 = 880 v.s. (German standard tuning fork); la3 = 888 v.s. (English standard tuning fork). It also provides the length of waves for the notes of the physicist's scale, based on the tuning fork ut3 = 512 v.s., and the range of the principle musical instruments and the human voice.

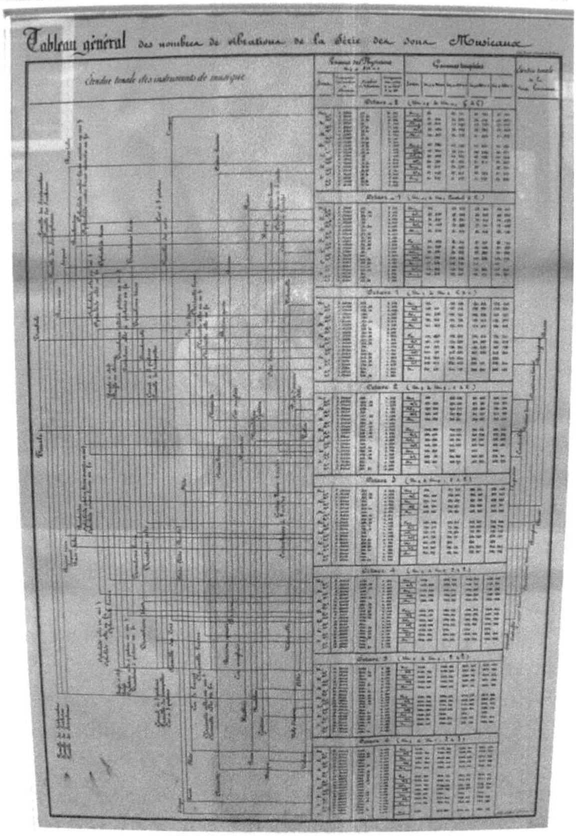

Fig. CR no. 31 Photo by author, 2005. Physics Department, University of Toronto, Canada

Location: Toronto.

Markings and measurements: In frame, 50 × 73 cm. Hand signed, "Rudolph Koenig."

32. Clock Fork of 128 single vibrations. 2,000 fr

In the summer of 1879, in the wake of his disagreement with Alexander Ellis regarding the precision of his standard forks, Koenig started experimenting with a new instrument for determining pitch. He borrowed the idea for this clock-like instrument, in which the seconds are produced by vibrations of a tuning fork, from Niaudet, who had presented his invention to the Academy of Sciences in 1866, and subsequently displayed it at the Paris and Vienna Exhibitions (1867 and 1873 respectively). Koenig, however, was not interested in making a precision clock. He wanted to use it as a comparison tool, along with an actual chronometer, for measuring the frequency of a tuning fork.

Testing the true vibrations of an unknown fork involved a comparison between a real chronometer and the tuning-fork chronometer. For the latter, Koenig attached the tuning fork of unknown frequency to the escapement of a clock that moved 1/60th of a division (second) on the clock dial for every 128 vibrations. The number of hours, minutes and seconds would then be translated into vibrations by multiplying the total (in seconds) by 128. One hour on the dial of the clock fork would be the equivalent of 460,800 vibrations (3,600 by 128). In other words, a tuning fork of 128 v.s. would produce one hour on the clock fork. A reading of 1 h and 28 s compared with 1 h on the actual chronometer would mean that a faster fork had been employed, producing more vibrations. In such a situation there would have been 464,384 total vibrations (3,628 times 128) during a period of one hour on the chronometer, which would mean the unknown fork was vibrating at 129 v.s. (464,384 v/3,600 s). Therefore, with the simple comparison of clock fork time to real time, the exact pitch could be determined.

Koenig attached special micrometer screws to the prongs in order to adjust the frequency to the exact number needed for calibrations. The fork could be adjusted until it was at the exact pitch of 128 v.s. After setting this standard by using comparisons with the chronometer, he employed the Lissajous optical method (with Lissajous microscopes and mirrors) to compare and tune unknown forks. He then used the Lissajous optical method to tune other forks. He boasted that this apparatus was not only remarkable for its "extraordinary precision" but it also operated with "little complication or difficult manipulation."[9] He claimed that it was almost completely automatic and thus free of human error.

Location: The only known clock fork, by any maker, is one made by Max Kohl at Case University in Ohio.

References: Auerbach in Winkelmann (1909, p. 190), Ellis (1877a,b), Koenig (1877, p. 162), Idem., 1882, pp. 173–175. Kohl (1909), Loudon and McLennan (1895, pp. 118–120), Miller (1916, pp. 38–42), Niaudet–Breguet (1866), Rayleigh (1877), Zahm (1900, pp. 419–420)

33. Clock fork of 145 single vibrations. 2,000 fr

34. Standard Fork, ut3 = 512 s.v., with compensation for temperature between 5° and 35°C. 200 fr

Following his studies on the relations between temperature and pitch, and using the clock fork for determining pitch, Koenig built a standard tuning fork with an adjustment for temperature. He found that for temperatures under 50°C a change in temperature of one degree created a change of 0.0143 v.s. in the fork ut1 = 128 v.s. (64 Hz; C2), and 0.0572 per one degree for the ut3 fork = 512 v.s. (256 Hz; C4). In total he conducted over 300 observations between Jul. and Dec. 1879.

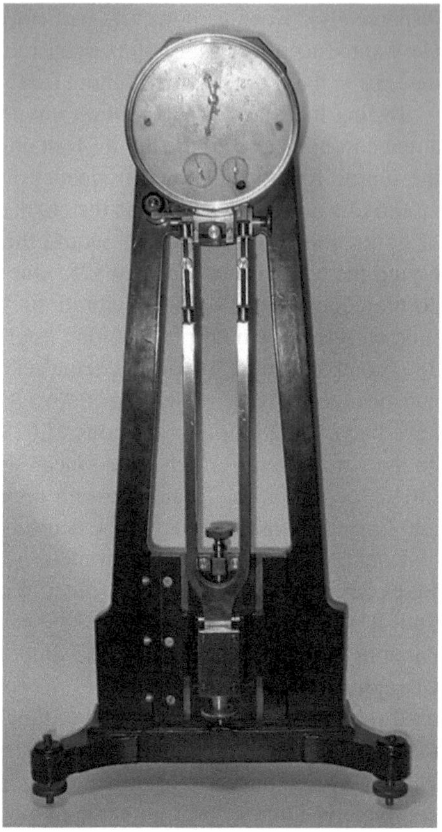

Fig. CR no. 32 Clock fork by Max Kohl. Photo by Bill Fickinger, Case University, Ohio, USA

The fork at the University of Rome has graduated aluminum dials attached to the side of each prong. They have a small brass weight on the side and are marked from 5 to 35°C. Using the dial, therefore, one can set the fork to the proper temperature to ensure a frequency of 512 v.s. The fork is mounted in a cast iron stand in front of a brass, cylindrical resonator. There are two other forks of the same construction, with different standards: la3 = 870 v.s. (435 Hz; A4) and si3 = 921.7 v.s. (460.9 Hz; B4).

Location: Rome.
Markings: Forks are marked with "UT3 512 vs 5°–35°C RK" "LA3 870 vs 5°–35°C RK" "SI3 921.7 vs 5°–35°C RK."
References: Ianiello (2003, p. 93), Koenig (1882c, pp. 182–189), Marloye (1851, p. 48)

35. Standard Fork, ut3 = 512 s.v. at 20° cent. 100 fr

Mounted on a cast iron stand with a brass cylindrical resonator. Pitch set at 20°C.

Fig. CR no. 35 Source: Koenig (1889, p. 20)

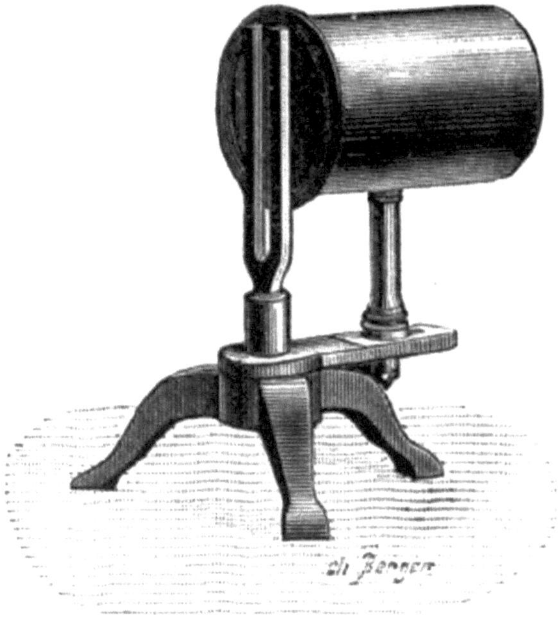

Locations: Case. Teylers (1889).
Markings: (Teylers) Resonator is stamped "RUDOLPH KOENIG À PARIS." Fork: "UT3 512 vs RK."
References: Marloye (1851, pp. 47–48) and Turner, G.L'E. (1996, p. 113).

36. Complete universal tonometer, proceeding from 32 to 43690, 6 s.v. [Not complete and without price as of 1889]

This was Koenig's masterpiece consisting of 154 tuning forks that ranged from 32 to 43,690 v.s. covering the full range of human hearing. This range was extended beyond 65,000 v.s. at its completion in 1894. The forks came with stands and sliding weights to adjust the frequency, some of them having resonators. In total it produced 1618 notes. It was originally priced at 50,000 fr, but Abbé Rousselot, a phonetics researcher at the Collège de France, bought the instrument from Koenig's family for 25,000 fr shortly after his death.[10]

Koenig had been working on this tonometer since 1877. He announced in his catalogue 1889 that he had nearly finished the job, but realized that he had to overcome technical difficulties for making and fine tuning forks in higher frequencies. He renewed his efforts in 1891 and finally completed it in 1894. In 1889 it comprised of the following forks:

(1) Four forks with steel mirrors and sliding weights gave the notes ut-2 (32 v.s.) to ut1 (128 v.s.) or (64 Hz; C2). Notes on the first two forks were separated by one half of a *vibration simple*, and for the second pair by one "v.s." No resonators.
(2) 132 forks of decreasing size with sliding weights. 127 forks of which were the first harmonics of ut-1 (64) or (32 Hz; C) starting at ut1 (128 v.s.), five of which were doubles at ut2, ut3, ut4, ut5, ut6. Each fork differed from the following fork by 64 v.s., the total range being 128–8,192 v.s. The positions were marked on the fork prongs. Between ut1 and ut3, the forks advanced by 2 v.s., from ut3 to ut5 by 4 v.s., from ut5 to ut7 by 8 v.s.
(3) 40 cylindrical resonators with adjustable pistons for reinforcing the tuning forks in group 2. Cast iron supports and tripods for the resonators and forks.
(4) 18 forks for the notes ut7 (8,192 v.s.) to fa9 (43,690.6 v.s.).

The forks at the Biblioteque Nationale go up to ut10 (65,536 v.s.). The ut10 fork is the highest surviving Koenig fork.

Location: Part of this tonometer (from group 4) is in the Rousselot collection of instruments at the Mitterrand Branch of the Biblioteque Nationale in Paris, "département de l'audiovisuel." The location of the remaining forks is unknown.
Markings and measurements: (Biblioteque Nationale) "UT10 65,536 vs RK" (1.8 cm long; 1.5 cm depth of prong).
References: Miller (1935, p. 89) and Zahm (1900, pp. 74–76).

Grand Tonomètre (1867–1876), Smithsonian Institution

At the London exhibition of 1862 Koenig displayed a 65-fork tonometer. By 1867 he had expanded this range to an apparatus with 330 forks. The first four octaves (32–512 v.s.) comprised four forks with graduated limbs and sliding weights. They had intervals of one half, one, two and four *vibrations simple* respectively. Next was a series of 65 forks up to 1,024 v.s., separated by eight v.s. There were then 86 forks from 1,024 to 2,048 v.s., and 172 forks from 2,048 to 4,096 v.s., separated by 12 v.s. (or six beats). Because of the difficulty of making tuning forks for higher notes, 86 steel rods raised the frequency to 8,192 v.s. The rods were separated by 48 v.s., giving 24 beats per second. These rods were excited by rubbing (friction) and sounded by longitudinal vibrations. Ten more rods gave the notes, 8,192, 10,240, 12,288, 16,384, 20,480, 24,576, 32, 768, 40,960, 49,152, 65,536 v.s., which represented the notes of the common chords in three octaves. The highest notes were above the audible range. These rods vibrated laterally.

This entire tonometer does not seem to have been sold and was incorporated into another, even larger tonometer displayed at Philadelphia in 1876. Between 1867 and 1876, Koenig added another 350 forks in the upper range from 4,096 to 8,192 v.s., each separated by 6 v.s. The report of the 1876 jury stated that the tonometer contained 670 tuning forks. Professor Barker, who took care of the collection in Philadelphia from 1876 to 1882, reported that there were 692 forks in total.[11] The

Fig. CR. No. 36 Rack is 36 inches high. Photo courtesy of the National Museum of American History, Smithsonian Institution, Washington DC, cat. no. 315716, neg. 70524

22 fork difference probably came from the 22 steel cylinders that Koenig sold from ut7 to ut10 (see #51).

Location: NMAH (cat. no. 315716).
Measurement: Rack is 24 inches at base, 36 inches high.
Description: 661 forks remain at the NMAH. The first 61 forks (rows 1–3) appear to be from a 65-fork tonometer, with the classic elongated yoke face. The forks between 512 and 4,096 v.s. (rows 4–11) are "U" shape, and appear to have been made from bending a straight rod. They make up the middle range of the large tonometer of 1876. The higher forks (rows 12–18) again have the elongated yoke face, but are much smaller. The whole array of forks on the rack range in height (excluding stem) from approximately 14 cm to 2 cm. There is also a series of lower forks that are stored separately.

The overall patterns of intervals between the forks tell us much about the practical limitations of tuning with beats in Koenig's workshop. The lower notes could be tuned with more ease and certainty, as they only differ by one quarter of a vibration. The lowest forks (about 55 cm in length) vibrate longer (over 60 s) and subsequently produce beats for a longer period of time making them easier to count (using a clock). The highest forks (about 2 cm in height) vibrate for approximately 3 s offering little time for counting beats.

The walnut rack itself has spaces for 677 forks.[12] It is divided into four distinct groups: rows 1–3 have 22 spaces per row; rows 4–6 have 29 spaces per row; rows 7–11 have 35 spaces per row; while rows 12–18 have 50 spaces per row. It appears that the forks in each group come from the same-sized blank and were each fine tuned according to their neighbours (using beats). Extensive measurements were made of several individual forks (length of U, width of U, width of space inside U) and indeed the general proportions are very similar within each group, with significant differences between each group. The one variable that consistently changes in one direction within each group, by very small amounts, is the length of the prongs. As one would expect, as the prongs become minutely smaller, the pitch becomes higher (shorter prongs result in more cycles per second). Rough file markings across the plane of the top of the prongs reveal how Koenig or his workers shortened the prongs to obtain a rough estimate of the desired pitch. But not all successive forks were shorter, showing that it was not the only variable involved in the tuning of the final pitch (e.g. other key variables included: the thickness of prongs, the equal mass of prongs, the distance prongs were apart, and the kind of steel). Furthermore, the forks within each group seem to have been sufficiently different in overall shape making tuning by shortening more complicated. The inside base of the prongs, inside the "U", often reveals small amounts of filing activity which would have lengthened the prongs slightly lowering the pitch. Furthermore, the width of the prongs at these points is sometimes found to be thinner than the middle or top. This filing would have weakened the stem/yoke creating longer vibrations, and lowering the pitch.

References: Barnard (1870a, pp. 504–506), Lissajous (1868, p. 481), Kielhauser (1907, pp. 17–19), Pantalony (2003), and Richardson (1927, pp. 113). Also see F.A.P. Barnard in Walker (1880, pp. 488–489)

37. Scheiblers Tonometer. 3,000 fr

The basic tonometer was invented in 1834 by the silk manufacturer Johann Heinrich Scheibler (1777–1837), who developed a series of tuning forks, separated by a consistent number of vibrations, as a more reliable means for tuning and setting a standard for pitch. He started with a fork representing the average "a" of three concert pianos in Vienna which was approximately 440 Hz.[13] He then tuned a second fork to be one octave lower (a/2) than the "a" fork, using his ear and some signature combination tones that often appeared with an octave.[14] In order to determine the absolute number for these forks he built a series of 56 tuning forks, the first one being four vibrations (or beats) sharper (higher) than the lower "a" and the last one being a few (beats) flatter (lower) than the higher "a." The sum of the fifty-five sets of beats was the difference between the lowest and highest forks. Because lower "a" multiplied by two was the higher "a" or the next octave, the sum of beats or difference was the actual frequency of the lower "a." In terms of numbers, he found the 55 sets of beats added to 220, which meant that a/2 was also 220 and therefore concert "a" was 440.

One could perform similar experiments with any other known ratio of the musical scale and, in fact, Scheibler developed sophisticated schemes for tuning instruments using beats. He presented these findings to the congress of physicists at Stuttgart in 1834, and "a-440" came to be known as the "Stuttgart pitch."[15]

Koenig was the first Parisian maker to commercially produce the tuning-fork tonometer. He displayed it at the London International Exhibition in 1862 and it would soon sell for 2,000 fr, which was 20 times the price of the average instrument in his 1865 catalogue, revealing the amount of work he put into it and the high cost of the steel. It consisted of 65 tuned forks, covering one octave, separated from each other by only four complete vibrations, each mounted on a beautifully finished pine resonator box. The jury awarded Koenig a medal of distinction (*médaille unique*) commenting: "By aid of this instrument, and a *practised*ear, very delicate gradations of pitch may be obtained."[16] They also held out the hope that "an authoritative establishment of international uniformity would confer an inestimable public advantage."[17] Shortly following the exhibition, Rodolphe Radau, a Königsberg physicist living in Paris, introduced Koenig's tonometer to readers of the weekly scientific journal *Cosmos*, claiming that his instrument would now make it possible to popularize Scheibler's invaluable method of tuning.[18]

Locations: École Polytechnique. MIT. NMAH (cat. no. 315725). Rome. Toronto (1878). Vanderbilt (1875).

Description: The forks at the École Polytechnique in Paris make up one of the earliest tonometers by Koenig, probably the original one shown at the 1862 exhibition in London. The U-shaped forks rest on the hour-glass, turned wooden stems (the characteristic shape of his forks in the early 1860s). All 65 forks rest in an ornamental wooden container with glass sides. The design of forks resemble other forks (no. 44) found at CNAM that were displayed at the 1862 exhibition.

The tonometer at the University of Toronto consists of 66 forks and resonant boxes. As the forks rise in pitch (by 8 v.s.) the prongs diminish in length by less than a millimeter in length, each resonant box gradually diminishing in size as well. There is some variation in these changes due to fine tuning at the top and inside base of the prongs. Each box has two rubber hoses at the base for cushion.

One of the forks (512 vs) has recently been examined in a material science laboratory at MIT. Analysis of the steel has revealed that Koenig chose the steel to balance the efficiency of vibrations (fairly hard steel; cooled but not quenched) with the ease of filing (soft enough to shape). It appears that he selected a bar stock, forged or cold worked it into rough shape and then applied a heat treatment with annealing and slow cooling over a long period of time. Micro-hardness tests on two phases revealed an average 120.58 w/25 g at the ferrite area; an average 144.48 w/25 g for the pearlite area. The fork was approximately 0.55% annealed carbon steel (hypoeutectoid).[19]

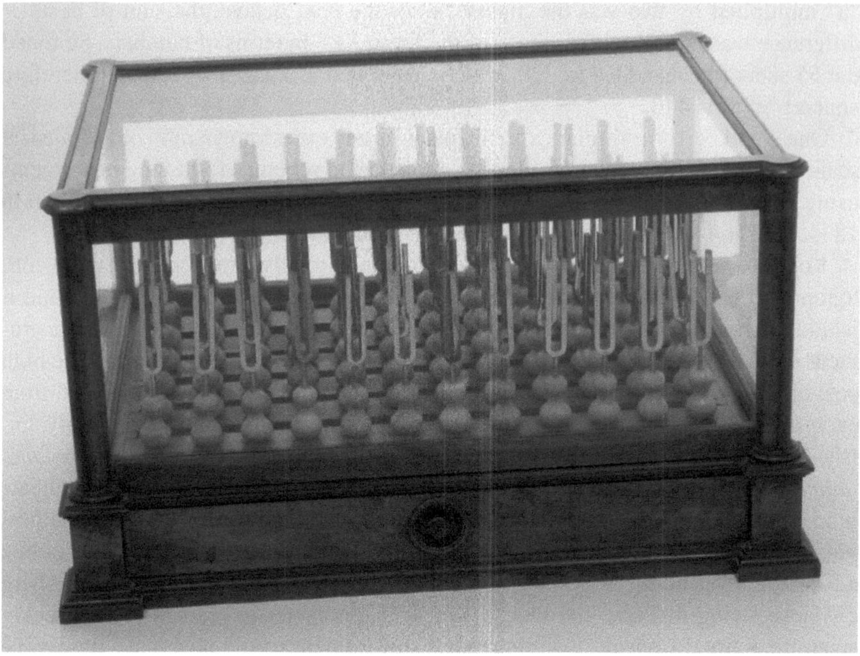

Fig. CR no. 37-1 The tonometer probably displayed at 1862 Exhibition. Ministère de la Culture, Inventaire générale. © Collections Ecole polytechnique under photography

Fig. CR no. 37-2 Photo by Louisa Yick. Courtesy of the Physics Department, University of Toronto, Canada

Markings and measurements: (Toronto) Fork No. 2 is stamped, "2 520 vs RK," (15.8 cm long); the box is stamped "2 520 RUDOLPH KOENIG À PARIS" (6.8 × 11.7 × 30 cm). Fork No. 65 is stamped "65 UT4 1024 vs RK" (11.7 cm long); the box is stamped "65 UT4 1024 RUDOLPH KOENIG À PARIS" (5.3 × 9 × 15.8 cm). La3 is set at 853.3 v.s. To provide an idea of the scale that Koenig worked on for his tuning forks, the following are the measurements for one octave of forks and boxes, respectively (all boxes have "RUDOLPH KOENIG À PARIS" stamped below the numbers): Fork: "33 SOL3 768 vs RK" (13.4 cm long) Box: "33 SOL3 768" (5.7 × 9.7 × 21.2 cm); "34 776 vs RK" (13.3 cm) "34 776" (5.7 × 9.7 × 20.5 cm); "35 764 vs RK" (13.3 cm long) "35 784" (5.7 × 9.8 × 20.4 cm); "36 792 vs RK" (13.2 cm) "36 792"; "37 800 vs RK" (13.1 cm) "37 800"; "38 808 vs RK" (13.1 cm) "38 808" (5.5 × 9.5 × 19.9 cm); "39 816 vs RK" (13.0 cm) "39 816" (5.5 × 9.5 × 19.7 cm); "40 824 vs RK" (13.0 cm) "40 824" (5.5 × 9.5 × 19.4 cm); "41 832 vs RK" (13.0 cm) "41 832" (5.5 × 9.5 × 19.3 cm); "42 840 vs RK" (13.0 cm) "42 840" (5.5 × 9.5 × 19.2 cm); "43 848 vs RK" (12.9 cm) "43 848" (5.5 × 9.3 × 18.9 cm); "43–44 LA3 853,3 vs RK" (12.8 cm long) "43–44 LA3 853,3" (5.5 × 9.3 × 18.7 cm). 40, 41, and 42 each have the same length of prongs on the outside, but differ inside due to filing, 11.1, 11.0, and 10.9 cm.

(École Polytechnique) Glass case has a display sign (not in picture above) that reads, "TONOMÈTRE D'APRÉS SCHEIBLER."

References: Ellis in Helmholtz (1954, pp. 443–446), Ellis (1968, pp. 17–18), Miller (1935, pp. 55–56), Radau (1862a, p. 112), Scheibler (1834), Jackson (2006, pp. 151–181), and Zahm (1900, pp. 74–76).

37a. The same apparatus with smaller forks and without the resonators. 1,500 fr

Locations: CNAM (inv. 12603). Rome.

Description: The tonometer at the University of Rome consists of four rows of tuning forks on a wooden rack.

References: Ianniello (2003, p. 102).

38. Twelve forks with resonance boxes giving ut2, ut3, mi3, sol3, ut4, mi4, sol4, 7th harmonic of ut2, ut5, re5, mi5. 485 fr

This series of forks is based on the harmonics of the fundamental ut2. They demonstrated that one can sympathetically excite a harmonic series with the base note, ut2.

Locations: Case. CSTM (acc. no. 1998.0247; ut3, mi3, mi4, sol4). Dartmouth has 3 forks and resonators (acc. nos. 2002.1.34159; 2002.1.34160; 2002.1.34161) Toronto (4 forks and resonator boxes).

Markings and measurements: (Toronto) Pine boxes stamped "RUDOLPH KOENIG À PARIS." "7," (3.7 × 6.8 × 15.8 cm); "UT4," (5.0 × 8.8 × 16.0); "MI4," (4.5 7.8 × 21.5); "SOL4" (4.2 × 7.1 × 8.5). Forks: "7 1792 vs RK" 9.4 cm long;

"UT4 1024 vs RK" 11.8; "MI4 1280 vs RK" 10.7; "SOL4 1536 vs RK" 10.0. (Dartmouth) "SOL4 1536 vs"; "MI4 1280 vs"; "UT5 2048 vs"

References: Fau (1853, p. 354) and Koenig (1889, p. 56). Idem., 1882c, pp. 194–195. Miller (1916, p. 212) and Zahm (1900, p. 24).

38a. Fork ut2 = 256 s.v. with resonance box. 110 fr

Fig. CR no. 38a Photo by author 2005. Museu de Física, University of Coimbra, Portugal. FIS.0387

Locations: Coimbra (FIS.0387). CSTM (acc. no. 1998.0247). Kenyon. Nebraska. NMAH (cat. no. 315725.44). Rome. Tokyo. Union. Vanderbilt (1875).

Markings and measurements: (Coimbra) Box: "UT2 RUDOLPH KOENIG À PARIS." 16.7 × 28.7 × 62.2 cm. Fork: "UT2 256 vs RK." Height of fork from where stem meets yoke to top of prongs, 31.5 cm, 5.5 cm wide.

38b. Four forks ut3, mi3, sol3, ut4, with resonance boxes. 145 fr

Tuning forks representing the major chord.

Locations: Coimbra (FIS.0865; FIS.0864; FIS.0385; FIS.0386; date, 1867). NMAH.

Reference: Marloye (1851, p. 48).

38c. Fork, ut3 = 512 s.v. with resonance box. 40 fr

Location: Coimbra (FIS.0866). Dartmouth (acc. no. 2002.1.34034).
Reference: Marloye (1851, p. 48).

38d. Four forks mi4, sol4, 7th harmonic of ut2, ut5, with resonance boxes. 130 fr

38e. Fork ut4 = 1024 s.v. with resonance box. 35 fr

Location: Coimbra (FIS.0867). Dartmouth (acc. no. 2002.1.34035).

39. Four forks re3, fa3, la3, si3, with resonance boxes. 140 fr

These forks combine with 38b to complete the physicist's scale.

Location: NMAH (cat. no. 314952).

40. Two forks giving tempered mi3 and sol3 with resonance boxes. 70 fr

41. Thirteen forks giving tempered scale, ut3 to ut4, ut3 = 512 s.v. with box. 180 fr

Location: Vanderbilt (1875).
Description: Smaller forks with "U" shape on top of stem. Coated, polished steel

42. Fork la3 = 870 s.v. at 20°C with resonance box. 35 fr

43. Fork la3 = 870 s.v. at 15°C with resonance box. 35 fr

Following his experiments with the clock fork in the late 1870s, Koenig developed standard forks for commercial use. Forks 42 and 43 were set to the French standard pitch of 435 Hz (A4). Other forks found at the *Museo di Fisica*in Rome, not made by Koenig, show that their Koenig forks served as a standard for making and testing other forks from around the country.

Location: Rome.
Description: (Rome) This fork is gilded to preserve pitch, set at 15°C. A second gilded fork has a small electromagnetic coil between the top of the prongs, presumably for prolonged vibrations. There are also two forks set at 20°C with the "RK" monogram, LA3, 870, 20°C. They each have the Italian royal crest stamped on the yokes. These forks are usually mounted on a pine resonator box, but in this case they are mounted on a cast iron stand.
References: Koenig (1882c, p. 190) and Miller (1916, pp. 50–51).

43a. Fork la3 tuned to the official French standard, with resonance box. 35 fr

In his experiments with the clock fork in 1879, Koenig discovered that the official French standard fork housed at the Conservatoire de Musique in Paris, which was said to be 870 v.s. at 15°C, was in fact 870.9 v.s. at 15°C, or 870 v.s. at 24.3°C. The earlier French standard derived from the work of Jules Lissajous. In 1858 the French government established a commission to create a standard French pitch. In response to this challenge, Lissajous of the Lycée Saint-Louis developed his visual method for making precision tuning forks and created a standard tuning fork, la3 = 870 v.s (435 Hz; A4). Lissajous collaborated with the instrument maker Marc François Louis Secretan (1804–1867) and the resultant fork came to rest in the Conservatoire National de la Musique. Lissajous's method of making forks became the technical catalyst for a revolution in precision tuning under Koenig.

Location: FST
References: Brenni (1994a), Giatti (2001, p. 98), Koenig (1882c, pp. 190–191), Miller (1916, pp. 50–51), and Turner, S. (1996).

44. Thirteen forks giving the tempered scale ut3 to ut4, la3 = 870 v.s. with case. 180 fr

This set was used for tuning musical instruments. Numbers 44–47 established and tested the chromatic scale of equal temperament by using the method of beats. The first set of thirteen is tuned to the chromatic scale from ut3 to ut4. Each auxiliary fork is tuned exactly four beats higher and used for comparison with an organ or piano.

Locations: CNAM (inv. 07052; date, 1862). Porto. Toronto (1878). Vanderbilt.
Description: The forks at Toronto have a highly polished chrome coating with a brass ball on the stem. There is filing on the inside of the yoke, revealing fine tuning after it had been coated. The ones at CNAM are set on a rack first displayed at the 1862 exhibition. The rack holds twenty-six forks, thirteen on each level. The stem of the forks have a wooden collar with an hour-glass shape, characteristic of Koenig's earliest work.
Markings and measurements: (Toronto) Each are stamped "RK". "UT4" 9.8 cm long; "SI3" 10.0; "LA#3" 10.2; "LA3 8760 vs" 10.6; "SOL#3" 10.9; "SOL3" 11.2; "FA#3 11.5; "FA3" 11.9; "MI3" 12.3; "RÉ#3" 12.7; "RÉ3" 13; "UT#3" 13.3; "UT3" 13.7. (CNAM) The sign above the rack reads, "GAMME TEMPÉRÉ ET DIAPASON AUXILIAIRES LA-870 CONSTRUITE PAR RUDOLPH KOENIG À PARIS."
References: Miller (1916, pp. 34–37) and Zahm (1900, p. 308).

Fig. CR no. 44 Photo by author, 2005. Physics Department, University of Toronto, Canada

45. Thirteen auxiliary forks, each one being tuned to give exactly 4 beats per second with the corresponding ones of the preceding. 180 fr

Location: CNAM (inv. 7053, goes with 7052; date, 1862).
References: Miller (1916, pp. 34–37). Zahm (1900, p. 308).

46. Thirteen forks giving the tempered scale ut3 to ut4 on the basis of any assigned la3. 200 fr

Reference: Miller (1916, pp. 34–37)

47. Thirteen auxiliary forks giving four beats with preceding. 200 fr

Forks that deviated from standard pitch by a set number of beats were convenient for tuning. A musical note tuned to the standard la3 (435 Hz; A4), therefore, would beat four times when placed next to the "LA3 + 4 VD" fork.

Location: Case
Description: The forks at Case resemble no. 44 at the University of Toronto, but have the extended yoke with a steel cylindrical stem and brass ball at the end.

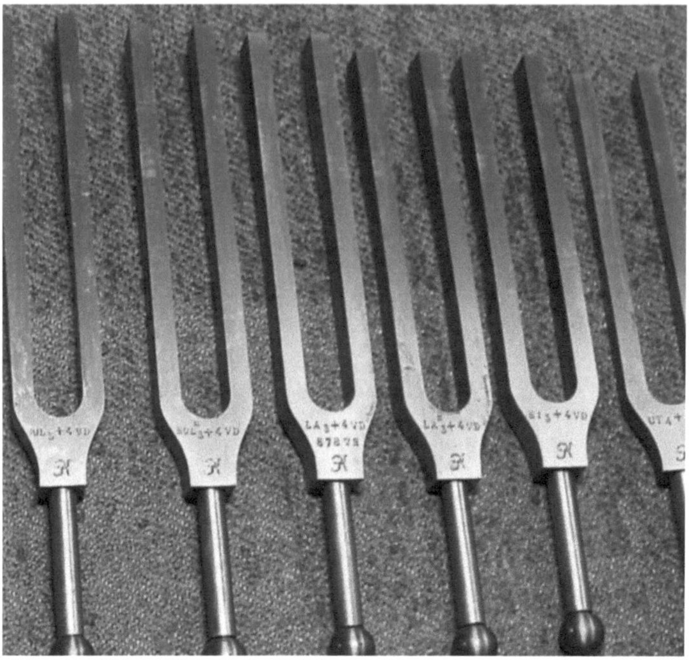

Fig. CR no. 47 Photo by Bill Fickinger. Physics Department, Case University, USA

Description: All stamped "RK" "UT3" "UT3 + 4 VD" "RÉ3 + 4 VD" "RÉ#3 + 4 VD" "MI3 + 4 VD" "FA3 + 4 VD" "FA#3 + 4 VD" "SOL3 + 4 VD" "SOL#3 + 4 VD" "LA3 + 4 VD 878 vs" "LA#3 + 4 VD" "SI3 + 4 VD"
Reference: Miller (1916, pp. 34–37).

48. Large fork from 32 to 48 s.v. to determine the lowest limit of sound. 300 fr

Threshold studies and demonstrations of the highest and lowest limits of hearing were an important part of nineteenth-century acoustics. These forks would be set at the front of a large amphitheatre and used to demonstrate successively lower and lower notes, until one reached the limit of hearing at 16 Hz. Most listeners can not hear the lowest notes.

Locations: Cornell. Harvard (acc. no. WJ0003). Rennes.
Description: (Harvard) The stem runs through a hole at the base of the U of the fork, showing that the U is a separately made piece. The massive sliders are brass with a wedge system for keeping it in place. Two large, roughly shaped cast iron disks are attached to the weights.

Fig. CR no. 48 Courtesy of the Department of the History of Science, Collection of Historical Scientific Instruments, Harvard University, USA. acc. no. WJ0003

Markings and measurements: (Harvard) A massive steel tuning fork, 1.3 m in height, with a cast iron stand. It is graduated from 32 to 48 v.s., marked every two v.s., for testing the lowest limits of sound. The branches of the forks are about 0.73 m in length. The fork at Harvard has the following markings: The top of the left prong is marked "1", below it on the face "vs" and then the graduated numbers (top to bottom) 32, 34, 36, 38, 40, 42, 42.6, 44, 46, 48, and 50. The last three graduations have longer spaces between numbers (more length is needed on the lower part due to the increasing firmness of the prongs). The top of the right prong is marked "2", below it on the face is "vs" followed by UT-2, 34, RE-2, 38, MI-2, 42, FA-2, 44, 46, SOL-2, 50. The inside of the left prong is stamped: "BEST WARRAN[T]E[D] CAST STEEL SHEFFIELD."

References: Miller (1916, p. 43). Zahm (1900, p. 85).

49. The eight high forks of no. 206, to determine the lowest limit of sound by the sound of beats. 340 fr

50. Series of 18 forks from ut7 = 8192 s.v. to fa9 = 43690.6 s.v. with case and iron stand, to determine the limit of high sounds. 900 fr

Koenig designed this series of thick-pronged forks (**see Fig. CR 201**) for determining the limit of high frequencies. They came with a cast-iron stand for putting two side by side (no. 50c). One compared the forks using beats, because as Koenig wrote, it was difficult to distinguish intervals at that pitch. He did not go beyond fa9 saying that ordinary people do not hear the last three forks beyond ut9. Anything above that, he wrote in the 1889 catalogue, would be in "the realm of fantasy."[20] The thick prongs, designed to reduce unwanted harmonics, derived from his studies in beats where he wanted to ensure a pure tone.

Locations: CNAM (inv. 12612; date, 1894). Nebraska.
Description: Nebraska has the last four forks, "UT9 32768," "RE9 36864," "MI9 40960," and "FA9 43690,6." Each stamped "RK"
Reference: Koenig (1889, p. 23)

50a. Series of 15 forks from ut7 to ut9. 670 fr

50b. Series of 7 forks giving ut7, mi7, sol7, ut8, mi8, sol8, ut9. 310 fr

This smaller set was also used for testing the highest audible frequency. They are thick-pronged forks with a brass collar for fitting to the iron stand (no. 50c). They are pictured in D.C. Miller's book in a stand (similar to **CR no. 201**) attached to a small Kundt tube for confirming their wavelength.

Location: Case. Harvard (acc. no. 1998-1-0134).
Description: The forks in both collections are stamped "UT7 8192," "MI7 10240," "SOL7 12288," "UT8 16384," "MI8 20480," "SOL8 24576," "UT9 32768." All stamped "RK"
Reference: Miller (1916, p. 47).

50c. Iron stand for fixing two forks beside one another. 50 fr

Location: Case. Harvard (acc. no. 1997-1-1076).
Description: Cast-iron tripod which was used as base with several Koenig apparatus, nos. 35, 48, 78, 123, 126, 157, 161, 162a, 189, 201, 245, and 241.

Fig. CR no. 50b Courtesy of the Department of the History of Science, Collection of Historical Scientific Instruments, Harvard University, USA. acc. no. 1998-1-0134

51. Series of 22 steel cylinders giving notes from ut7 to ut10, with steel hammer. 150 fr

These cylinders demonstrated the highest threshold of hearing and beyond. They were the first means by which Koenig developed high-frequencies, before he developed ways to make high-frequency tuning forks. In fact, he had to use these cylinders to expand the range of his early tuning-fork tonometer that was displayed at the 1867 Paris Exposition (see CR no. 36). The last cylinder and the shortest, "UT10 65536 v.s." (32,768 Hz) is well above human hearing (roughly 18,000 Hz). The main cylinders are suspended by a fine thread, attached at both nodes. The frequency is proportional to the inverse square of the length (if diameter remains constant). The other seven cylinders are suspended by a fine thread to be held close to the ear. The slight differences between the cylinders reveals the time-consuming, precision workmanship that went into these instruments. Harvard has a steel cylinder that was used for making and calibrating these types of steel cylinders (Harvard acc. no. WJ0059).[21]

Locations: CNAM (inv. 12613). Toronto.
Markings and measurements: (Toronto) Each frequency in Hz (half v.s.) is written in ink on a label on the wooden mount. The length between the strings appears to be based on the same ratio, 0.55 of the total length. Each cylinder is 2.0 cm in diameter (revealing that they came from the same stock of steel rod) and is

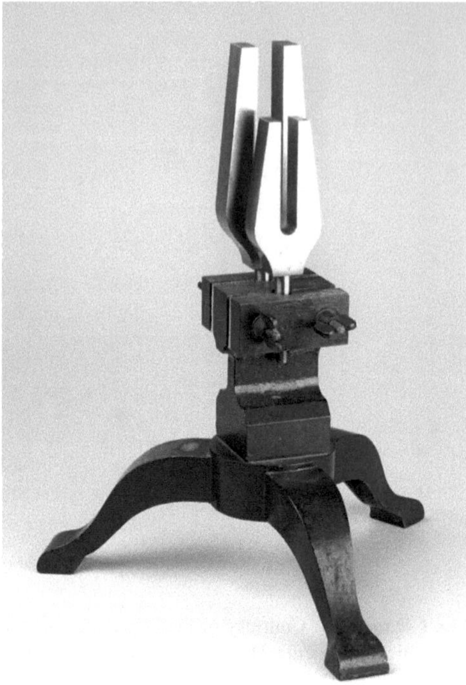

Fig. CR no. 50c Courtesy of the Department of the History of Science, Collection of Historical Scientific Instruments, Harvard University, USA. acc. no. 1997-1-1076

stamped "RK." "UT10 615536 vs" (5.1 long, 2.8 cm between the strings); "SI9 61040 vs" (5.25, 2.8); "LA9 54613,3 vs" (5.5, 3.0); "SOL9 49152 vs" (5.85, 3.2); "FA9 43690,6 vs" (6.3, 3.25); MI9 40960 vs" (6.5, 3.5); "RÉ9 36864 vs" (6.9, 3.8); "UT9 32768 vs" (7.3, 4.1); "SI8 30520 vs" (7.5, 4.2); "LA8 27306,6 vs" (7.9, 4.4); "SOL8 24576 vs" (8.4, 4.7); "FA8 21845,3 vs" (8.9, 5.0); "MI8 20480 vs" (9.1, 5.0); "RÉ8 18432 vs" (9.75, 5.3); "UT8 16384 vs" (10.3, 5.6); "SI7 15260 vs" (10.6, 6.0); "LA7 13653,3 vs" (11.2, 6.3); "SOL7 12288 vs" (11.8, 6.9); "FA7 10922,6 vs" (12.6, 7.0); "RÉ7 9216 vs" (13.8, 7.5); "UT7 8192 vs" (14.7, 8). The steel hammer, made from the same steel rod as the cylinders above, measures 2.0 cm diameter.

Reference: Barnard 1870a, pp. 504–506. Miller (1916, p. 46). Zahm (1900, pp. 86–87).

51a. Series of ten steel cylinders, without hammer. 80 fr

Location: Case. Coimbra (FIS.0628). Dartmouth (acc. no. 2002.1.34153). NMAH (cat. no. 87.924.6).

Fig. CR no. 51 Photo by Louisa Yick. Physics Department, University of Toronto, Canada

Markings and measurements: The cylinders at the NMAH (originally from the Worcester Polytechnic Institute) are ut7, ut8, ut9, ut10, sol7, sol8, sol9, mi7, mi8, and mi9

51b. Steel hammer. 6 fr

Location: Dartmouth (acc. no. 2002.1.34153).

52. Large siren disk to determine the highest limit of sounds. 300 fr

Used with the Savart rotation machine (CR no. 30). It carries ten circles of holes, whose number vary from 8 to 1024.

52a. The same but smaller. 200 fr

53. Galton's whistle with divisions. 20 fr

Sir Francis Galton invented this whistle in 1876 for testing the upper limits of sound in animals and humans. He demonstrated it at the South Kensington Exhibition in May of 1876. Tisley and Co. were the first to commercially produce it, but several companies made it soon thereafter, including Koenig.

Locations: MIT. Nebraska

Description: (MIT) Bulb missing. All remaining parts are made of brass. In order to lower the pitch, one rotates the graduated dial outward thus opening the aperture

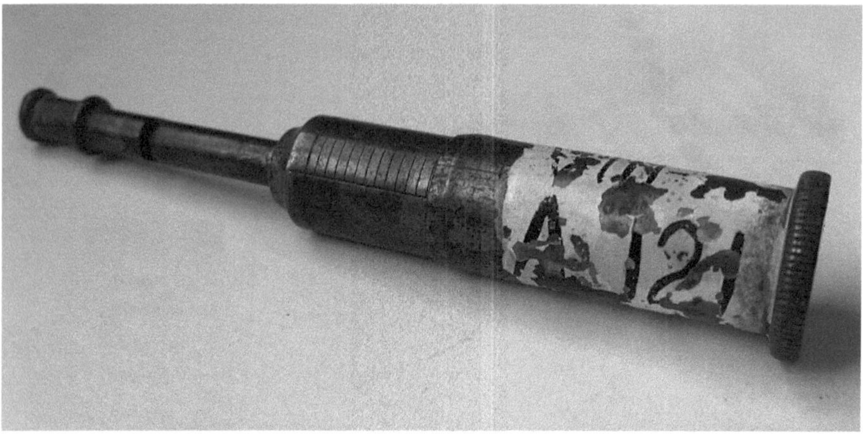

Fig. CR no. 53 Photo by author, 2005. Physics Department, MIT, USA

at the lip. When the dial is moved inwards, closing the size of the opening at the lip, higher notes are heard.

Markings and measurements: The MIT whistle is 72 mm long when set at "0." The vertical scale on the main tube numbers from 1 to 12 (12 mm in length). The scale that wraps around the outer tube has 10 divisions. The flute stem is punched "17." The body is marked in paint, "VIII A 121" presumably referring to its location in the original MIT physical cabinets.

References: Auerbach in Winkelmann (1909, pp. 208–210), South Kensington (1876, p. 61), and Zahm (1900, p. 87).

53a. Galton's whistle without divisions. 12 fr

IV. Timbre of Sound

54. Series of 19 Helmholtz resonators. 170 fr

Just as Newton used the prism to break light into the spectrum, Helmholtz invented a the resonator for filtering specific frequencies of sound. Resonators with a certain volume, size of neck and opening amplify vibrating columns of air of a specific wavelength. In the same way that a musical string of certain length, tension, mass and diameter has a natural vibrating frequency, the aerial resonator, with specific dimensions, has a natural frequency at which it vibrates most efficiently. The listener places one end (the nipple) in the ear, seals it with wax to keep unwanted sounds out, and listens for the specific resonating tone. When listening to music or singing, the resonance appears in the form of a sudden amplification or popping sound.

This instrument derived from Helmholtz's theory of timbre that stated that all compound sounds (vowel or musical sounds) could be broken down into simple, pure notes. Initially, he used any available spherical glass chambers that were an appropriate size, such as the receivers of retorts, but in 1859 he commissioned Koenig to make spherical resonators with specific dimensions. Koenig first made glass ones and switched to brass in 1865. The subsequent series of nineteen resonators consisted of all the harmonics of the ut1 = 128 v.s. (64 Hz; C2).

Fig. CR no. 54 Photo by author, 2005. Psychology Department, University of Toronto, Canada

Locations: Charité. Henri IV (Paris). NMAH (cat. no. 314957). Rome. Teylers (1865). Toronto (c. 1892, Psychology). Toronto (1878, Physics). Vanderbilt (1875). St. Mary's.

Description: The brass resonators appear to have been made in two halves and spun into a mould (on a lathe) and then joined. The nipple for inserting the resonator into the ear is very similar in each model, but the other opening or neck (which is one of the variables in the formula for resonance) varies in size by measurable amounts. These latter openings have evidence of hand filing, revealing that this was possibly a focal point for fine tuning. Koenig stamped the monogram "RK" and the pitch number on the base of this neck. The University of Toronto has two complete sets with only slight differences in measurements (millimeter range), revealing the consistency of the construction process.

Markings and measurements: (Toronto, Psychology) "UT2 RK 2" stamped near base of neck, (3.85 cm diameter of opening, 0.20 height of neck, [no measurement] diameter of sphere); "SOL2 RK 3" (3.62, 0.20, -), "UT3 RK 4" (3.00, 0.45, -), "MI3 RK 5" (2.45, 0.40, -), "SOL3 RK 6" (2.01, 0.30, -), "7 RK 7" (1.72, 0.10, 7.85), "UT4 RK 8" (1.40, 0.08, 6.75), "RÉ4 RK 9" (1.50, 0.20, 6.20), "MI4 RK 10" (1.40, 0.22, 5.71), "11 RK 11" (1.45, 0.25, 5.25), "SOL4 RK 12" (1.50, 0.25, 5.25), "13 RK 13" (1.45, 0.10, 5.0), "14 RK 14" (1.45, 0.20, 4.600), "SI4 RK 15" (1.40, 0.05, 4.45), "UT5 RK 16" (1.30, 0.03, 4.20), "17 RK 17" (1.30,

0.15, 3.85), "RÉ5 RK 18" (1.22, 0.10, 3.70), "19 RK 19" (1.23, 0.20, 3.45), "MI5 RK 20" (1.20, 0.20, 3.40).

References: Blaserna (1876, p. 55), Deschanel (1877, pp. 855–856), Ganot (1893, p. 237), Guillemin (1881, p. 735), and Helmholtz (1863, pp. 74, 561–562). Idem., 1954, pp. 43, 372–374. Jones (1937, pp. 139–145), Miller (1916, pp. 68–69), Turner G.L'E. (1996, pp. 114–115), Tyndall (1896, pp. 204–206), Violle (1883, pp. 285–286), and Zahm (1900, pp. 274–276).

54a. Series of 10 Helmholtz resonators. 110 fr

This set contains the harmonics of ut2 = 256 v.s. (128 Hz; C3)

Locations: Harvard (acc. no. WJ0011). QUP. Sydney. French Lycées: Ampère (Lyon), Buffon (Paris), Decour, (Paris), Molière (Paris), Paré (Laval), Voltaire (Paris).[22]

Description: The resonators at Sydney are all marked with the "RK" monogram on the base of the neck. They are stamped as follows: UT2, UT3, SOL3, UT4, [missing], unmarked, UT5, RE5, and MI5. They rest on a mahogany baseboard with the plaque: "W. Ladd & Co/11 & 12 Beak St/Regent St W." Ladd was a dealer of Koenig's instruments in England.

Reference: Mollan (1990, p. 199).

55. Series of 14 universal resonators, graduated, from sol1 to mi5. 380 fr

Koenig invented a cylindrical resonator that could be adjusted to cover a small range of notes, approximately half an octave. These resonators consist of two brass tubes that slide into each other and thus change the volume and frequency. Each one has a range of four to six notes, with the sides of the inner tube graduated and stamped with the frequencies. The series of 14 tubes has an overall range from sol1 to mi5. The Koenig sound analyser also uses these 14 universal resonators (see CR no. 242).

Locations: McGill. Nebraska. Vermont (only seven).

Description: (from the Toronto analyser) The first resonator has "RK SOL1-SI1" marked on the outside. Inside, it reads "SI1, LA#1, SOL#1, SOL1" (# represents a sharp musical note). Each subsequent resonator has both the RK monogram and the range of notes stamped on the outside: 2) SI1 – RE2; 3) RE#2 – F#2; 4) FA#2 – LA2; 5) LA2 – UT3; 6) UT3 – MI3; 7) MI3 – LA3; 8) LA#3 – RE4; 9) UT4 – MI4; 10) RE4 – FA4; 11) MI4 – SOL#4; 12) FA4 – LA4; 13) SOL#4 – UT5; 14) UT5 – MI5.

References: Ganot (1893, p. 237) and Helmholtz (1863, pp. 74, 561–562). Idem., 1954, pp. 43, 372–374. Loudon and McLennan (1895, p. 115) and Zahm (1900, pp. 274–276).

Fig. CR no. 55 Courtesy of the McPherson Collection, Physics Department, McGill University, Canada

56. Helmholtz large apparatus for compounding timbres of 10 harmonics. 1,500 fr

The sound synthesiser was Helmholtz's clearest instrumental expression of his theory of timbre, or sound quality. Whereas his spherical resonators dissected compound sounds (vowels or musical sounds) into elemental frequencies, the synthesiser did this by building up complex sounds from simple frequencies. In 1857 he went to the instrument maker Friedrich Fessel of Cologne to turn this idea into reality. The initial instruments used a combination of electrically driven tuning forks, resonators and piano keys to synthesise compound sounds. When the system was on, all of the forks would vibrate in series. To activate a sound, however, one needed to press an ivory piano which moved a circular lid away from the opening of the resonators thus activating the sound.

Based on Helmholtz's published descriptions and correspondence with the German scientist, Koenig produced this instrument commercially as early as 1860. Whereas Helmholtz had used the eight notes of B (120 Hz) and its harmonics, Koenig used a different standard based on ut3, 512 v.s. (256 Hz; C4).[23] He claimed that his starting note of ut2 (128 Hz; C3) was only 8 Hz different from Helmholtz's starting note of 120 Hz.

The instrument at the University of Toronto is in excellent condition and has been operated recently.[24] Following the directions from Helmholtz's paper on vowels (1859), where he provided specific combinations to play and their relative intensities (strong or weak), we were able to create combined sounds that had distinctive qualities, but not necessarily closely resembling vowels (at least to the ears of contemporary English speakers). There were a few challenges that made recreation quite difficult – the loud buzz of the electrical forks, the rattle of the interrupter (Loudon and McLennan solved this problem by putting it in a separate room),[25] and the emission of mercury vapour from the interrupter. One can see why Koenig

Fig. CR. No. 56-1 Interrupter. Photo by author, 2005. Physics Department, University of Toronto, Canada

diplomatically wrote to Helmholtz in 1861 that even though it was difficult to reproduce vowels, the apparatus was still useful for illustrating the basic ideas of his theory. In 1867, Alexander Graham Bell marvelled that these "tuning forks speak vowel sounds" when he first witnessed the apparatus being operated in London by Alexander Ellis.[26]

Locations: Harvard (acc. no. 1997-1-0893). Toronto. Science Museum (acc. no. 1885-1). Vanderbilt.

Description: In the later model Koenig used ten tuning forks and resonators. The one at the University of Toronto includes ten cylindrical brass resonators and ten electrically driven tuning forks resting on a mahogany base. The forks are accompanied by coils with finely-wound green-silk insulation. The resonators are activated with a metal stopper that covers the aperture and is moved by a string which is connected to the keyboard. Each resonator slides towards or away from the fork apparatus for adjusting intensity. A mercury interrupter with a tuning fork set at 128 Hz connects all the forks in series. A mirror rests on top of the fork for adjusting its frequency using Lissajous calibrations. An optical comparator or vibration microscope would be used to tune the interrupter fork.

The example at the Science Museum (Wroughton Location) also has the same arrangement as the picture in Koenig's 1889 catalogue. "R=55 ohms" is

Fig. CR no. 56-2 Photo by author, 2005. Physics Department, University of Toronto, Canada

scratched on wood base of the interrupter. The largest resonator is stamped "SKMUS" with a royal crown (South Kensington Museum). This instrument was used for quantitative study, revealed by home-made scales of graph paper pasted on the moveable wooden base of each resonator. They mark the distance the resonator moved away from the tuning fork, and thus the changes in intensity. A few of the forks have small, locally-made tin clips attached to the prongs as weight adjustment mechanisms.

In addition to the ten tuning forks and resonators, Harvard and Vanderbilt have two electromagnetic devices connected to wooden resonators. The wooden resonators appear to be activated by a telegraph-like mechanism.

Markings and measurements: (Toronto) Overall dimensions: 41.5 × 58.7 × 106.5 cm; resonators marked with "RK", "1 UT2" (128 Hz), "2 UT3," "3 SOL3," "4 UT4," "5 MI4," "6 SOL4," "7," "8 UT5," "9RE5" and "10 MI5" (1,280 Hz). Mercury Interrupter (13 × 24.3 × 33.0 cm).

References: Bell, A. G. *Article, Feb. 6, 1879*. Bell Papers, Library of Congress. Ganot (1893, pp. 239–240), Helmholtz (1863, pp. 184–186, 566–567). Idem., 1954, pp. 121, 377. Loudon and McLennan (1895, pp. 122–123), Miller (1916, pp. 245–246), Pisko (1865, pp. 20–30), Turner, G.L'E. (1996, p. 116), and Zahm (1900, pp. 365–366).

56a. The same apparatus for compounding timbres of 8 harmonics

Locations: ISEP. Teylers.

Markings: The eight resonators and forks at the Teylers Museum are marked: 1/UT2, 2/UT3, 3/SOL3, 4/UT4, 5MI4, 6/SOL4, 7[nil], 9/UT5. The mercury interrupter uses a fork marked, "RK" and "UT2 256 vs". This apparatus rests on a mahogany baseboard.

References: Turner, G.L'E. (1996, p. 116).

57. Five forks with resonators tuned to the characteristic notes of the vowels u, o, a, e, i. 175 fr

These five forks and resonators imitated the vowels OU (U in English and German), O, A, E, I. They derive from Helmholtz's theory of vowel sounds that stated that there was one frequency, a "fixed pitch," (among other weaker frequencies) which determined the distinctive character or timbre of the vowel. In the 1850s, Helmholtz employed his resonators to listen for these sounds along with tuning forks to activate the resonance of the mouth cavity. He held a series of tuning forks to the mouth when it was in the shape of an "O" and discovered, by trial and error, the "characteristic pitch" of the resonant cavity and thus produced a list of characteristic vowel frequencies.

Koenig transformed these findings into a commercial instrument. The first apparatus, which appeared in the 1865 catalogue, was modeled after Helmholtz's figures for the vowels ("OU" fa2, 175 Hz, "O" si3 flat, 466 Hz, "A" si4 flat, 932 Hz, "E" si5, 1976 Hz, and "I" re6, 2349 Hz). As a result of his own research in the late 1860s, Koenig changed these figures to 224, 448, 896, 1,792, and 3,584 Hz. (The figures in 1870 were 225 (OU), 450 (O), 900 (A), 1800 (E), and 3600 (I), but he modified his instruments to fit his preferred physicist's scale based on 256 Hz).

Locations: Coimbra (FIS.0388). CNAM (inv. 12635). Harvard (acc. no. 2000-1-0010). MCQ (acc. no. 1993.13811; c. 1865). Rome. Teylers (1865). Toronto.

Description: The set at the MCQ and Teylers have an hour-glass-shaped wooded stem and bent U-shape fork, which dates them to Koenig's early workshop.

Markings and measurements: (Toronto) Five forks: "I" 6.0 cm long; "E" 7.0; "A" 9.3; "O" 12.5; "OU" 170. Five Resonators: "I" 2.5 cm diameter; "E" 3.0; "A" 4.5; "E" 3.0; "OU" 7.6.

References: Boring (1942, pp. 367–375), Helmholtz (1863, pp. 167–173). Idem., 1954, pp. 105–110. Koenig (1870) and Turner, G.L'E. (1996, p. 115).

58. Free reed surmounted by a resonator to produce the vowel sounds u, o, a. 30 fr

This instrument is an elegant mechanical model of human vocal production as it was understood in the mid nineteenth century. The resonator, when set into a free-reed pipe and windchest, produced a rich, compound tone. By closing the opening

Fig. CR no. 57 Photo by Louisa Yick. Courtesy of the Physics Department, University of Toronto, Canada

of the resonator with varying amounts one could imitate the sound of the vowels U (French OU), O, and A. The reed pipe is a free reed, whereby the reed oscillates freely and rapidly through an opening sending continuous vibrations of air into the pipe. The resonator, like the cavity of the mouth, reinforced certain regions of the harmonic spectrum, imitating the desired sound or vowel.

Reference: Zahm (1900, p. 241).

59. Large apparatus based on the principle of the wave-siren for the synthetical study of the timbre of sounds. 6,000 fr

Koenig's *grand siréne à ondes* (large wave siren) for reproducing timbre was his most elaborate and exotic instrument. It combines up to 16 notes and derives from his visually based apparatus for reproducing sound from actual waves of brass. This model from the early 1880s was his second most expensive instrument, putting it out of the reach of most laboratories. He sold another version of it for 10,000 fr in 1897.[27] He placed an engraving of it on the cover of his 1882 book.

The large wave siren was 1.9 m in height. It consisted of sixteen disks cut with simple sinusoidal waveforms. The first disk produced a fundamental tone, the other fifteen produced harmonics of that tone. Each disk had its own wind slit that blew pressured air against the rotating wave. Sixteen buttons allowed one to open or shut the flow of pressure air in the slits. A long lever connected to the slits allowed one to change the phase of each slit at will. The pressure of air could also be regulated to imitate varying intensity. Koenig's main goal had been to explore the role of timbre, but he stated that some preliminary research on vowels had shown promise.

Koenig based the large wave siren on an earlier model from 1867, with aluminum waves in cylindrical form, which was displayed at London in 1872 and brought to the 1876 Exhibition in Philadelphia (Fig. 6.4).

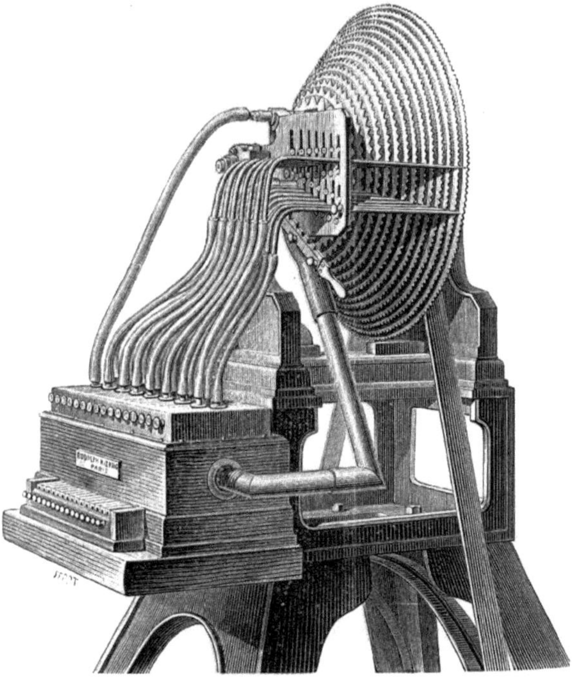

Fig. CR no. 59 Source: Koenig (1889, p. 27)

References: Auerbach in Winkelmann (1909, pp. 183–184) and Koenig (1882a, p. 9). Idem., 1882c, pp. 157, 236–243. Miller (1916, pp. 244–245) and Zahm (1900, pp. 375–376).

60. Wave-siren for studying the different timbres produced by varying the phases of the same harmonics. 350 fr

This instrument uses brass wave patterns to reproduce timbre. Like a siren, pressured air pushes against the rotating, brass wave thus creating a distinctive sound. The waves represent the mathematical combination of several harmonics with different phase relations (displacement of the waves along the *x*-axis). Because of the phase differences, two complex waves can have the exact same number and intensity of harmonics, but look quite different in waveform. But could the human ear detect these differences? In contrast to Helmholtz, Koenig believed that phase changes did cause noticeable differences in quality of sound. Helmholtz, on the other hand, stated that only the basic ingredients mattered to the ear (number and intensity of harmonics), not how they were arranged in time. Koenig tested this idea by creating brass wave forms which included harmonics of equal intensity, but with shifted phases.

In order to make these waveforms, he produced graphical inscriptions (with the help of photography) and put together compound waveforms under various phase conditions. He then traced and cut these figures on the circumference of a cylindrical band of thin brass. This model, first advertised in 1882, was a commercial version of earlier prototypes discussed in his research papers. The top two curves represent the first six odd harmonics with differences of phase of ¼ and 0. The higher harmonics diminished in intensity to imitate nature. The bottom four curves represent the first 12 harmonics of diminishing intensity. They have differences of phase of ¾, ½, ¼, and 0. One thing that stands out about the surviving instruments is the extremely smooth, quiet and rapid rotation of the wheels.

Locations: CNAM (inv. 12610). Harvard (acc. no. 1997-1-0993). Oxford (acc. no. 61236). Rome. Science Museum (acc. no. 1890-14).

Markings and measurements: (Oxford) "RUDOLPH KOENIG À PARIS" on the top knobs of the frame. Also marked "10" by local department. The numerical markings from top to bottom are as follows ("D DE PH" stands for "difference de phase"): "1,3,5… D DE PH ¼, ¾." "1,3,5…D DE PH 0, ½" "1,2,3…D DE PH ¾" "1,2,3…D DE PH ½" "1,2,3…D DE PH ¼" "1,2,3…D DE PH 0." The frame is 40 cm in height, as listed in the 1889 catalogue.

References: Auerbach in Winkelmann (1909, pp. 267–269), Miller (1916, p. 245), Thompson (1891, p. 251), and Zahm (1900, pp. 373–374).

61. Iron pulley, mounted, for the movement of preceding. 50 fr

62. Wave-siren disk with sinuous contour. 70 fr

This was a simpler demonstration of the relations between timbre and phase differences. This disk was placed on the Savart rotation machine (no. 30b) along with a wind tube and slit (no. 63) for providing pressured air. If the slit was placed perpendicular to the wave form, one obtained a simple tone. If it were inclined in the direction of the rotation (thus, according to Koenig, imitating a change in phase) the simple tone transformed into a timbre of a fundamental accompanied by a series of harmonics of decreasing intensity with the phase difference of ½. If one inclined the slit in the other direction, one returned to a difference of phase of 0.

In 1999 a group of researchers at the Smithsonian Institution operated this siren. There was a slight change of the siren tone when the wind slit was rotated, e.g. it seemed a little less "clean" or more raucous, but it was very difficult to characterize the change.[28]

Fig. CR no. 60 Courtesy of the Department of the History of Science, Collection of Historical Scientific Instruments, Harvard University, USA. acc. no. 1997-1-0993

Fig. CR no. 62 Photo courtesy of the National Museum of American History, Smithsonian Institution, Washington, DC, cat. no. 328742, neg. 70277

Locations: Harvard (acc. no. 1997-1-1010). NMAH (cat. no. 328742).
Markings and measurements: (NMAH) "RK" 40 cm diameter.
Reference: Koenig (1882c, pp. 161–162, 241–243) and Zahm (1900, p. 377–378).

63. Wind-tube with slit opening for preceding. 50 fr

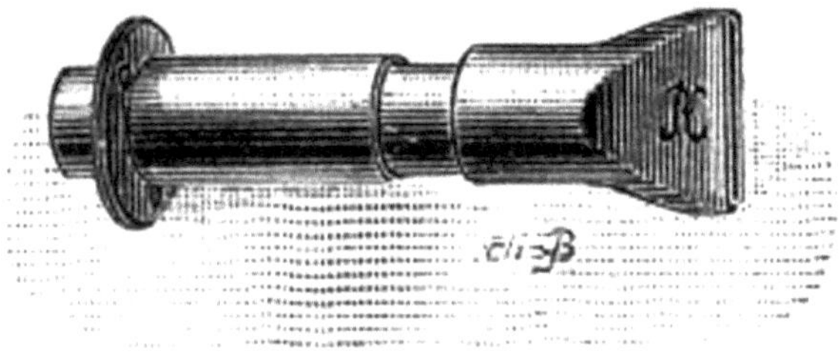

Fig. CR no. 63 Source: Koenig (1889, p. 29)

Location: NMAH (cat. no. 328742).
Marking: Stamped "RUDOLPH KOENIG À PARIS".

V. Propagation of Sound

64. Bell suspended in a glass balloon, to show the enfeeblement of sound in a vacuum. 22 fr

This was a classic demonstration dating back to the scientific revolution. A bell sounds in a vacuum but can not be heard. There is no medium to carry the sound.

Location: Rome (c. 1874).
References: Blaserna (1876, p. 30), Daguin (1867, p. 450), Deschanel (1877, pp. 797–800), Fau (1853, pp. 351–353), Jamin (1868, p. 501), Marloye (1851, p. 45), Violle (1883, pp. 4–6), and Zahm (1900, pp. 40–41)

65. Bell with clock movement for the same purpose. 35 fr

This apparatus demonstrated the same effect as CR no. 64 but with a bell and clockwork.

Reference: Daguin (1867, p. 450), Ganot (1893, p. 205), Tyndall (1896, pp. 36–39), Violle (1883, pp. 4–6), and Zahm (1900, pp. 39–40).

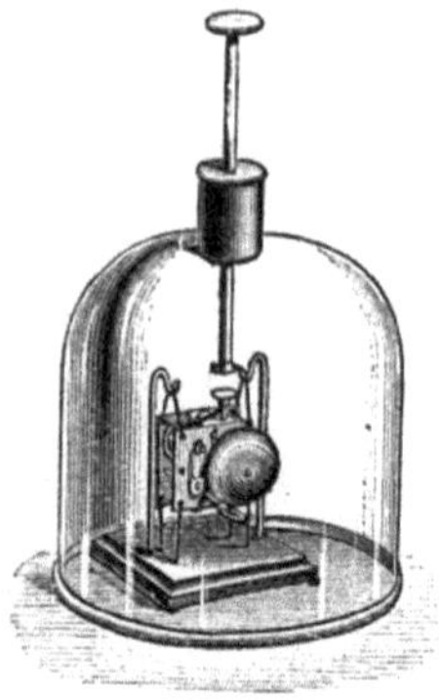

Fig. CR no. 65 Source: Koenig (1889, p. 30)

66. Tyndall's apparatus for showing the acoustic opacity of a mass composed of air at different temperatures, or of gases of different densities. 185 fr

Fog was a major concern for ships in the nineteenth century. It seriously impeded or blocked signals from light houses. In an effort to develop alternative signals, Joseph Henry in America and John Tyndall in England investigated powerful fog horns. In the 1870s Tyndall performed studies on "acoustic clouds" or inhomogeneities in the atmosphere that thwarted or redirected sound waves. He did outdoor experiments on the coast with extremely powerful sirens.

With the help of his assistant, John Cottrell, he then designed and built this table-top version of the apparatus for more controlled experiments and demonstration. A bell sounds at one end and the waves travel through a sealed, centre chamber. Carbonic gas flows in from the upper tubes, and coal gas flows up from the bottom tubes. These tubes, each flowing with gas at different densities, cause fluctuations and inhomogeneities in the central pipe thus changing the transmission of sound. The detector consists of a funnel and a sensitive flame. Indeed, Tyndall confirmed his earlier outdoor studies that the different densities impeded and blocked sound transmission.

References: Beyer (1998, pp. 77–78) and Tyndall (1896, pp. 312–320).

67. Apparatus to measure the velocity of sound at short distances. 350 fr

References: Bosscha (1854), Koenig (1882c, pp. 30–31), and Pisko (1865, pp. 207–208).

68. Large tube mounted upon an iron stand, with receiving capsule for the study of propagation of sound. 750 fr

According to the catalogue, this zinc coil was 30 m long, 2.10 m in height, with 12 "elbows." It was 0.7 m in diameter. Combined with Koenig's graphical recorders, it was used to repeat the experiments of Regnault, Violle, Tyndall and LeRoux on the propagation and reflection of sound. Only one survives at the muse des arts et métiers in Paris.

Fig. CR no. 68 Source: Koenig (1889, p. 32)

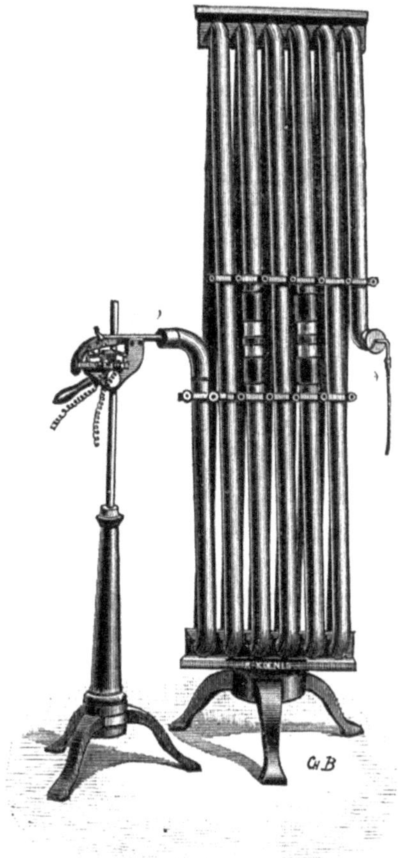

Location: CNAM (inv. 12611-001; date, 1890).
Measurements: (CNAM) h = 2.5 m.
Reference: Loudon and McLennan (1895, pp. 134–35).

68a. Same apparatus smaller. 500 fr

A smaller version of no. 68, with 23 m in length and tubes of 0.5 m in diameter.

Location: CNAM (two instruments, acc. nos. 12611-002 and 12611-003; c. 1894).
Measurements: (CNAM) h = 170 cm.

69. Electrically mounted pistol. 120 fr

70. Two membranes arranged according to Regnault's method for measuring the velocity of sound. 50 fr

These membranes accompanied the Regnault chronograph (CR no. 216) for measuring the speed of sound. One marked the beginning of the sound signal and the other the end. The whole set-up derived from the experiments carried out by Regnault and Koenig in the sewers of Paris in the 1860s.

Reference: Regnault (1866).

71. Chladni's apparatus for measuring the relative velocities of sound in different gases. 35 fr

At the end of the eighteenth century, E.F.F. Chladni devised a way to measure the velocity of sound using an organ pipe filled with various kinds of gas. The pitch of the pipe changed according to the different compositions of the gases used.

Reference: Chladni (1809, pp. 273–276).

72. Ten rods of the same length of different kinds of wood. 25 fr

The ten rods illustrated that sound propagates at different speeds in different mediums. These experiments were based on the work of E.F.F. Chladni.

Reference: Chladni (1809, pp. 106–108).

73. Cottrell's apparatus to show the law of reflection of sound. 75 fr

Sound, like light, reflects off surfaces. In this instrument, which resembled a spectroscope, sound produced by a reed travels though a tube and reflects off a mirror into another tube where it is detected by a sensitive flame. The angle of reflection can be measured from the graduated support base. It confirmed the law of "sonorous rays" that the angle of incidence and reflection are equal. John Tyndall's assistant, John Cottrell, created this apparatus.

Reference: Tyndall (1896, pp. 439–440). Zahm (1900, pp. 116–117).

74. Savart's large bell-jar and resonator. 440 fr

Savart's bell-jar and resonator (*grand appareil de timbre*) was a simple, but powerful demonstration of the rich, acoustic qualities of a bell.

Fig. CR no. 74 Courtesy of the Department of the History of Science, Collection of Historical Scientific Instruments, Harvard University, USA. acc. no. 1997-1-0302

Location: Harvard (acc. no. 1997-1-0302).
Measurements: 112 cm = H; 32 cm = diameter of bowl; 21 cm = diameter of resonator.
Description: The one at Harvard is mounted on a wooden tripod. There is a brass cylindrical resonator with a piston for adjusting the volume. In contrast to the brass resonator, the brass bowl is a light yellow/pinkish colour. It sounds easily and powerfully, with many harmonics.
References: Blaserna (1876, p. 53), Daguin (1867, pp. 544–543), Desains (1857a, pp. 119–120), Fau (1853, p. 400), Ganot (1893, pp. 208–209), Jamin (1868, pp. 535–536), Marloye (1851, p. 44), Tyndall (1896, pp. 203–204), Violle (1883, pp. 279–280), and Zahm (1900, pp. 269–270).

74a. The same apparatus with bell-jar of 0.22 m diameter. 160 fr

74b. The same apparatus for placing on a table. 100 fr

74c. The same apparatus with bell-jar of 0.16 m diameter. 65 fr

Location: NMAH (cat. no. 328479). Teylers (1865).
Description: The one at the NMAH came from Weston College, Massachusetts. It has a sturdy octagonal base and turned bell support. The wood is walnut. The Teylers instrument is similar with a brass bell mounted on a octagonal wooden base. A rectangular wooden resonator attached to a wooden slider moves toward and away from the bell.
Markings and measurements: (NMAH) Base stamped, "RUDOLPH KOENIG À PARIS." The diameter of the bell is 18.5 cm. The resonator is 11.4 × 11.4 × 22 cm.
References: Koenig (1865, p. 15), Turner, G.L'E. (1996, p. 114), and Zahm (1900, pp. 269–270).

75. Acoustical turbine of Drovàk and A. Mayer. 60 fr

The acoustical turbine was something like the radiometer in optics. An ut4 tuning fork with resonator was placed in front of four aluminum resonators on a wheel. The activated resonators, all set at ut4, propel the turbine around the axle. It was simultaneously discovered by Alfred Mayer of the Stevens Institute and V. Drovàk of Austria.

Fig. CR no. 75 Source: Koenig (1889, p. 33)

Location: Toronto (missing).
References: Auerbach in Winklemann (1909, p. 489), Drovàk (1876, p. 42), Ganot (1893, p. 274), Mayer (1878, p. 328), Miller (1935, p. 73), Violle (1883, p. 288), and Zahm (1900, p. 281).

76. Electrical fork ut4 with resonance box. 100 fr

Alfred Mayer used this instrument as a continuous sound source for his sound reaction wheel (CR no. 75). The example at Rome is on a cast-iron tripod and must have been paired with a brass, cylindrical resonator.

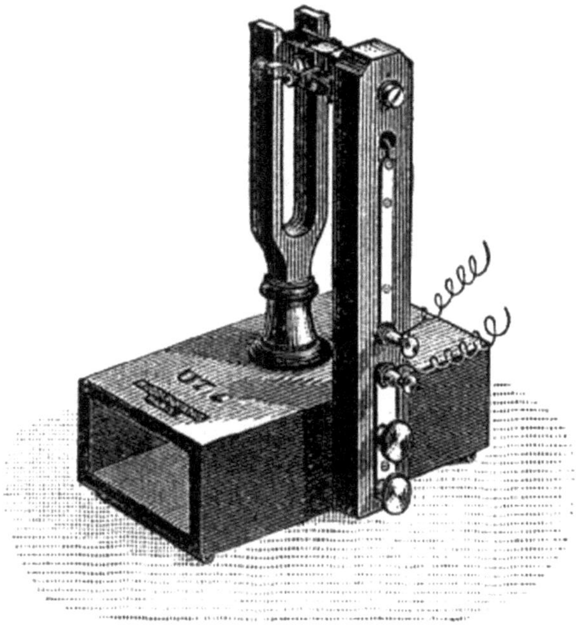

Fig. CR no. 76 Source: Koenig (1889, p. 33)

Location: St. Mary's College, Notre Dame. Rome.
References: Mayer (1878, p. 328) and Zahm (1900, p. 283).

77. Two forks ut4 and ut4 + 4 d.v. on resonance boxes, to show the influence of the movement of a vibrating body on the pitch. 70 fr

When a sound source moves toward the ear, the sound waves compress making a higher pitch. If the source moves in the opposite direction, the pitch lowers. This effect came into prominence in the 1840s with the work of Christian Doppler. The Dutch scientist, Christoph Buys Ballot, tested the idea with sound waves by using trumpets on moving trains. The phenomena came to be known as the Doppler effect.

In 1865, Koenig advertised two tuning forks ut4 and ut4 + 4vd with resonance boxes as a small demonstration of this effect (he did not use the term Doppler effect). When the two forks were sounded next to each other, they produced 4 beats per second (they were separated by 4 Hz). When the lower note was moved toward the listener, its pitch increased and therefore lowered the number of beats heard; when the high-pitched fork moved toward the listener, its pitch increased and therefore raised the number of beats.

References: Koenig (1882c, p. 41) and Doppler (1843). Idem., 1846. Ballot (1845).

77a. Fork ut4 + 4 d.v. on resonance box. 35 fr

The ut4 fork is the same as that in CR no. 38e.

78. Mach's apparatus for the same purpose as no. 77. 100 fr

This was another demonstration of the Doppler effect. The long, hollow brass tubes rotate around an axis fixed to a heavy stand. The tubes connect to a wind bellows that supplies pressured air. Reed pipes attached to the ends emit a sound and the pitch changes as the branches approach and move away from the listener. Franz Joseph Pisko pictured an older version with a wooden frame in his 1865 book. The Coimbra instrument shows Koenig's connection to the musical instrument trade through J. Jaulin, who was a musical instrument maker in Paris. He exhibited a reed instrument called the "panorgue" at the 1851 Exhibition in London (entry no. 1274 in the official catalogue). He was listed as "Julian Jaulain 11 rue d'Albony, Faubourg St. Martin, Paris."

Location: Coimbra (FIS.1283).
Description: Brass pipes with steel reed pipes on the end.
Markings and measurements: (Coimbra) "RK" on the end of one tube. Each tube is one meter in length. A steel reed pipe is screwed to the end of each tube. The reed pipes are signed, "J. Jaulin Bte. S.G.D.G." [brevète sans garantie du gouvernement].
References: Ellis (1851, p. 1238), Loudon and McLennan (1895, pp. 135–136), Mach (1861, pp. 66–68). Idem., 1862, pp. 335–336. Pisko (1865, pp. 222–225) and Zahm (1900, p. 113).

79. Axis and handle for preceding, which is to be mounted on one of the stands no. 194a or b. 25 fr

VI. Simple Vibrations of the Different Bodies

∗At the end of this section Koenig notes that all of his wooden pipes come without varnish. Varnish on the pipes from UT2 to UT3 would be an extra 5 fr and varnish above the notes UT3 (smaller lengths) would be 3 fr.

Fig. CR no. 78-1 Source: Koenig (1889, p. 34)

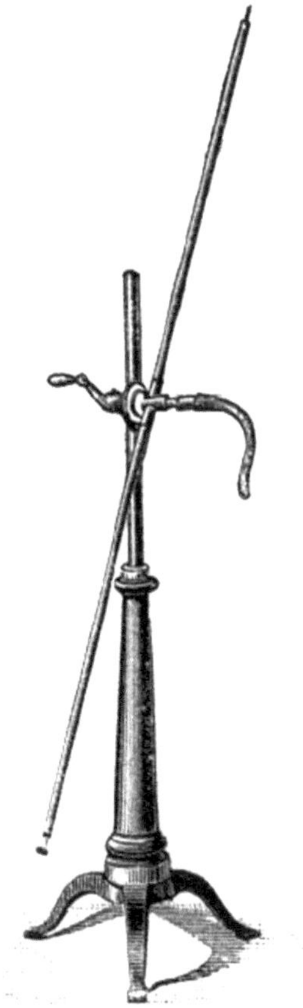

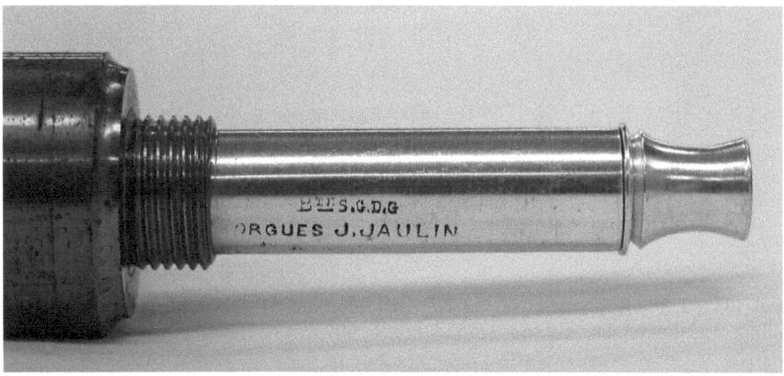

Fig. CR no. 78-2 Photo by Gilberto Pereira, Museu de Física, University of Coimbra, Portugal. FIS.1283

Vibrations of Air

80. Bellows with regulator and wind chest, large model. 650 fr

Bellows could be used for any experiment that needed a continuous, powerful source of air. The regulator, which derived from organs, controlled the variations of pressure. Organ pipes, reed pipes, manometers and sirens were placed in a series of holes on top. There are also two outlets for connecting the pressured air to other instruments via rubber tubes.

Fig. CR no. 80 Photo courtesy of the National Museum of American History, Smithsonian Institution, Washington, DC, cat. no. 327553, neg. 60507

Locations: CNAM (inv. 40159). NMAH (cat. no. 327553). Rome.

Description: The NMAH instrument came from Union College, New York. Although unsigned, it is identical to Koenig's pictured in the 1889 catalogue.

It has 12 outlets and 12 keys. The table is pine, the outlets and the regulator are oak. The hinges are vellum.

Markings and measurements: (NMAH) Unsigned. 100 × 114.5 × 58.5 cm.

References: Daguin (1867, p. 531), Fau (1853, pp. 360–365), Ganot (1893, pp. 226–227), and Marloye (1851, p. 35).

80a. The same apparatus of smaller size. 400 fr

This model has 8 holes.

80b. The same apparatus as no. 80a, without regulator. 300 fr

Location: University of Mississippi, Oxford.

81. Large bellows, 1 m in length by 0.75 m in width, without regulator and windchest. 500 fr

82. Manometer to measure the pressure of air. 10 fr

Oak pipes with glass tubes.

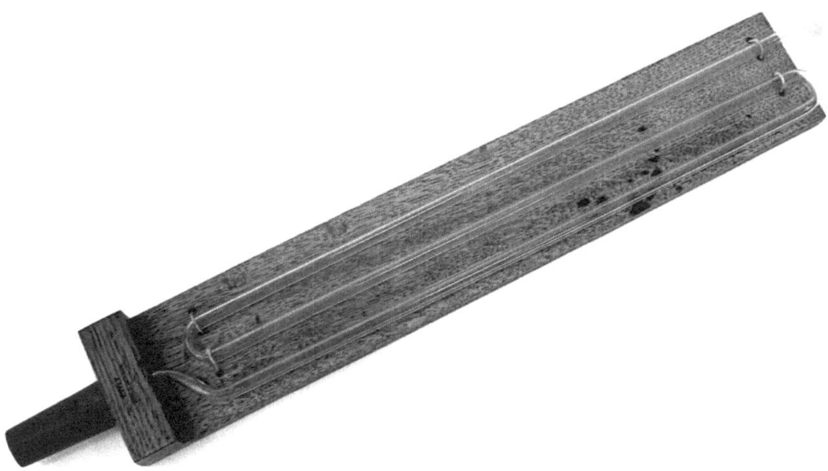

Fig. CR no. 82 Photo by author, 2005: Physics Department, University of Toronto, Canada

Location: Teylers. Toronto (1878).

Markings and measurements: (Toronto) marked "77" in ink referring to the 1873 catalogue. Stamped "RUDOLPH KOENIG À PARIS." 2.5 × 5.9 × 41.5 cm

Reference: Turner, G.L'E. (1996, p. 125).

83. Cavaillé-Coll's small air regulator. 8 fr

A small regulator invented by the Parisian organ maker, Aristide Cavaillé-Coll. This instrument could control the input from a large bellows for use with a siren, manometer or organ pipe. As air pressure builds, the hinged container inflates and is controlled by the sliding brass weight.

In September 1862, Cavaillé-Coll (1831–1899) collaborated with the scientist, Léon Foucault, the instrument maker, Gustave Froment, and the astronomer, Urbain-Jean-Joseph Le Verrier, on an experiment to measure the speed of light. Cavaillé-Coll, who had just completed his masterpiece organ of over 7,000 pipes at Saint Sulpice, joined these sessions in order to help operate a wind-driven siren to calibrate the rotation of a small mirror. The regulator controlled the rotation of the mirror.

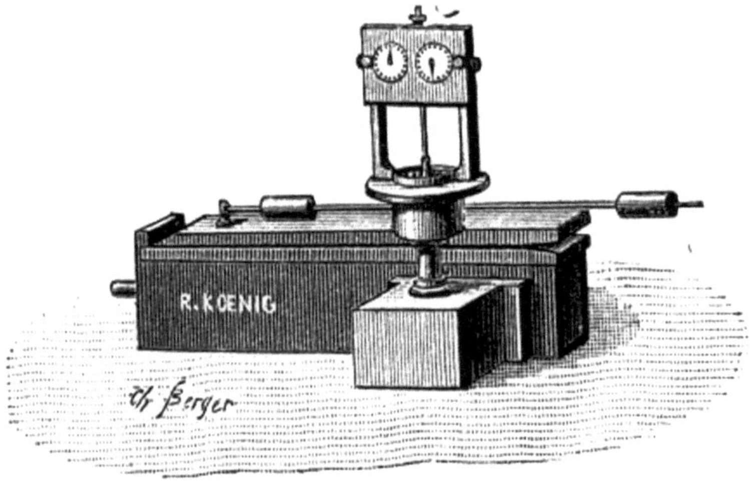

Fig. CR no. 83 Source: Koenig (1889, p. 36)

Location: NMAH (cat. no. 328423.2).
Description: (NMAH) The regulator is made of oak with a brass intake and fixtures. The bellows are made of vellum or thin parchment.
Markings and measurements: Stamped "RUDOLPH KOENIG À PARIS." 10.3 × 38.7 × 16.5 cm.
References: Turner, G.L'E. (1996, p. 184).

83a. Cock to regulate the wind. 8 fr

84. Large organ pipe in water trough, for experiments on the vibrations of air columns. 400 fr

Koenig invented this instrument to study the vibrations of air in a large organ pipe. Using the long pipe and open windows for displaying nodes and ventral segments, he mapped the internal vibration patterns with great precision and thus demonstrated a timeless problem in acoustics, that the theoretical values for the position of nodes and vibrating segments do not match experimental findings.

The key part of the instrument is a small, brass tube that runs into the middle of the pipe. It displays the internal vibrations through a direct tube to the ear or a manometric flame or manometer. An observer can also see the interior through a glass window. In smaller pipes, one cannot introduce an indicator or membrane without disturbing the flow of air. Koenig felt he successfully avoided this problem by creating the large pipe (ut1 when open) about 2.5 m long, by 0.12 width and 0.12 m depth with the small tube that runs under the pipe, through an open slit in the bottom of the pipe and into the middle of the interior. The pipe rests in a trough of water acting as a seal for the exposed slit. The small tube can be moved along the length of the pipe to observe and measure the vibrations of air inside. In addition, membranes and drum devices can be placed inside the large space without worrying about disruptions of the air columns.

It was a versatile experimental chamber permitting a few different experiments. Like CR no. 237, one could detect nodes (places of pressure change) with vibrating flames. One could also use the ear tube to locate nodes and antinodes. However, Koenig found that it was difficult to find the position of nodes with precision. The antinodes (ventral segments where there was longitudinal movement, but no pressure changes) could be positioned with much more precision, because as one moved back and forth through the antinode, there was a sudden increase of tone on the edges that was as "clear as the strokes of a bell."[29] In another set of experiments, he studied the phase relations of vibrations inside the pipe.

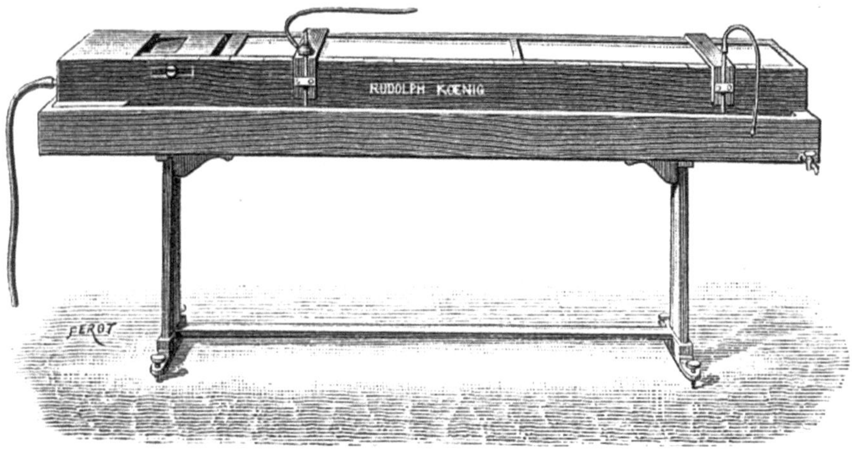

Fig. CR no. 84 Source: Koenig (1889, p. 37)

References: Auerbach in Winkelmann (1909, p. 429), Jones (1937, pp. 158–161), Koenig (1882c, pp. 208–217), Violle (1883, p. 129), and Zahm (1900, p. 233–234).

84a. The same apparatus without the turning mirror. 350 fr

84b. The same apparatus without the means for observing the direction of the vibrations. 300 fr

85. Organ pipe with glass window and small membrane. 20 fr

This simple visual demonstration of vibrating air inside an organ pipe derived from the work of Félix Savart. Albert Marloye first sold this apparatus in Paris. As the membrane covered with sand is lowered into the sounding pipe, one sees the sand dance or agitate as it approaches the nodal point. A node of vibration corresponds to a place where there is changing density or pressure, yet no longitudinal vibration. For example, at the centre of the pipe two longitudinal segments push into each other creating a dead zone in the middle. The continuous squeezing and pulling back create the pressure changes, and cause the membrane to vibrate. The ventral segments were quieter, with little pressure change.

Locations: CSTM (acc. no. 1998.0261). Teylers. Rome.
References: Blaserna (1876, p. 21), Daguin (1867, p. 449), Deschanel (1877, p. 794), Fau (1853, p. 377), Ganot (1853, p. 253), Guillemin (1881, p. 631), Marloye (1851, p. 41), Turner, G.L'E. (1996, p. 120), Tyndall (1896, p. 214), Violle (1883, p. 125), and Zahm (1892, p. 226).

86. Long pipe giving a harmonic, with one very thin side. 16 fr

The thin side is sprinkled with sand to reveal the nodal points in the pipe.

87. Kundt's stopped pipe with three manometers. 80 fr

The three manometers demonstrate dilations and compressions of vibrating air in the organ pipe. Water inside the manometer tubes move in accordance with changes in air pressure. The water level stays the same in the manometer connected to the pipe during both dilations and compressions; the water lowers in one under the influence of dilations; and the water rises in the other under the influence of compression.

Location: Toronto.
Markings and measurements: Stamped "RUDOLPH KOENIG À PARIS" on the pine. 10 × 10 × 48 cm.

88. Pipe which can be closed at the node. 10 fr

If an open organ pipe is fully closed at the middle node it still plays the same note because the node remains in the same position. This pipe has a wooden slider that bisects the pipe thus creating a closed pipe that is half the length of the open pipe. The note is the same.

Fig. CR no. 85 Source: Koenig (1889, p. 39)

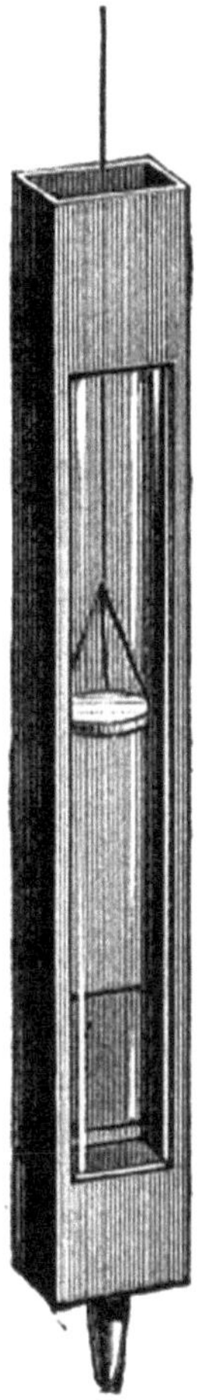

Locations: Sydney. Rome. Teylers. Toronto (1878). Union. Wesleyan.

Description: (Toronto) Pine pipe with mahogany lip at the base. Wooden slider.

Markings and measurements: (Toronto) Marked "83" in ink referring to the 1873 catalogue, stamped "RUDOLPH KOENIG À PARIS" on the pine. 5.0 × 4.2 × 43.1 cm.

References: Daguin (1867, p. 533), Desains (1857a, pp. 55–56), Fau (1853, p. 375), Jamin (1868, p. 538), Marloye (1851, p. 41), Turner, G.L'E. (1996, p. 120), Violle (1883, p. 127), and Zahm (1892, pp. 226–227).

Fig. CR no. 87 Source: Koenig (1889, p. 39)

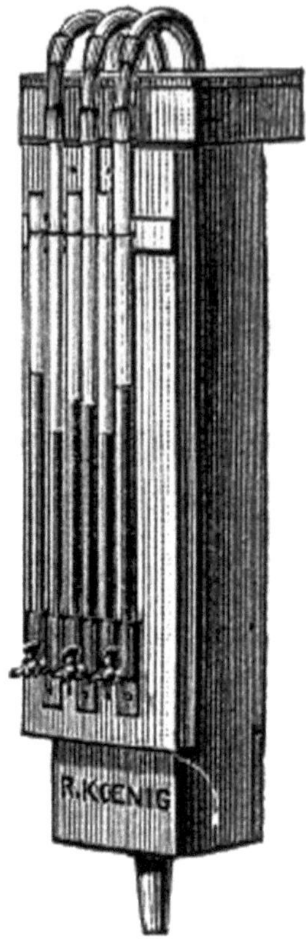

89. Pipe arranged to give the second harmonic, with opening at a loop. 8 fr

The natural note of this pipe jumps an octave when the wooden lever at the midpoint is opened. The pressure falls to zero at the node creating a ventral section, thus doubling the frequency.

Locations: MIT. Rome. Toronto (1878). Union.
Markings and measurements: (Toronto) marked "84" in ink referring to the 1873 catalogue. Stamped "RUDOLPH KOENIG À PARIS." 4.3 × 5.0 × 43.0 cm.
Reference: Zahm (1892, p. 227).

90. Pipe with different openings at the node. 20 fr

There were many demonstrations that manipulated pressure changes at the nodal points. Changes in pressure alter pitch. Different sized holes at the node, therefore, produce different musical notes. Larger holes produce higher pitch.

Fig. CR no. 89 Photo by author, 2005. Physics Department, University of Toronto, Canada

Fig. CR no. 90 Photo by author, 2005. Physics Department, University of Toronto, Canada

Location: Harvard (acc. no. 1997-1-0945). Toronto (1878).
Description: (Toronto) Pine with a mahogany lip and slider.
Markings and measurements: (Toronto) Marked "85" in ink referring to the 1873 catalogue. Stamped "RUDOLPH KOENIG À PARIS" on the pine. The holes

from largest to smallest are SOL3, FA3, MI3, and RE3. There is no hole at the last position, UT3. The whole pipe measures 6.5 × 5.6 × 60 cm.

91. Tube with different openings at end. 18 fr

Changes in pressure at the end of the pipe create different nodal and ventral relations and thus different notes. A sliding wooden strip, with circular holes, moves across the end of the pipe. As the holes increase in diameter, the note increases in pitch. The pipe plays five notes, ut2 (when fully closed), sol2, la2, si2 and ut3.

Location: Toronto (1878).
Description: Pine with mahogany lip and sliding strip. Leather seal on underside of slider.
Markings and measurements: Marked "87" in ink referring to the 1873 catalogue. Stamped "RUDOLPH KOENIG À PARIS" on the pine. 5.5 × 5.5 × 63.0 cm.

92. Cube arranged as preceding. 18 fr

93. Three equal pipes with mouth-pieces of different lengths. 20 fr

As the width of the mouth piece increases, the pitch rises.

Fig. CR no. 93 Photo by author, 2005. Physics Department, University of Toronto, Canada

Location: Toronto.
Markings and measurements: (Toronto) Marked "86" referring to the 1873 catalogue. Stamped "RUDOLPH KOENIG À PARIS." First pipe: 5.0 × 4.3 ×

43.0 cm. Width of mouthpiece, 1.9 cm. Second pipe: 5.0 × 4.3 × 43.0 cm. Width of mouthpiece, 2.7 cm.

94. Pipe with a moveable lip. 20 fr

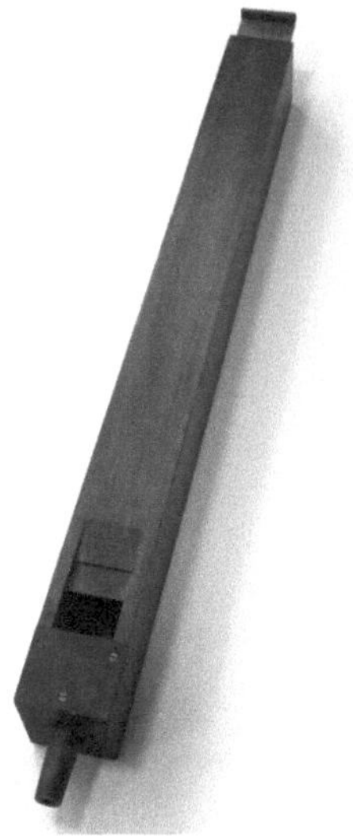

Fig. CR no. 94 Photo by author, 2005. Physics Department, University of Toronto, Canada

Location: Toronto (1878).
Description: Pine pipe with mahogany lip.
Markings and measurements: (Toronto) Marked "91" in ink referring to the 1873 catalogue. Stamped "RUDOLPH KOENIG À PARIS" on the pine. 6.6 × 5.7 × 62 cm.

94a. Same apparatus of smaller size. 12 fr

95. Four equal pipes, three in wood of different thickness, and one lined with cloth. 35 fr

These pipes demonstrated the changes in pitch due to changing pipe thickness. The 1889 catalogue stated that "the two pipes with sides of medium and strong thickness give the same sound, the other two give lower, less clear sounds."

95a. Three pipes in wood of different thickness. 24 fr

Differences in thickness change the quality of tone, or timbre. A thin wall will vibrate more freely thus producing more harmonics.

Location: Toronto (1878).

Fig. CR no. 95a Photo by author, 2005. Physics Department, University of Toronto, Canada

Description: (Toronto) When played recently, the pipe with thin walls produced a more reedy timbre.
Markings and measurements: (Toronto) Two pipes survive, both are stamped "RUDOLPH KOENIG À PARIS" and marked "94" in ink referring to the 1873 catalogue. 4.0 × 3.5 × 43 (0.4 cm thick); 5.5 × 5.0 × 43 cm (1.2 cm thick).

95b. Two pipes, one being lined with cloth. 16 fr

Lining the inside of a pine pipe changes the tone.

Location: Toronto (1878), Union.
Description: (Toronto) The lining is a soft, cream-coloured cloth. It produces a lower note when played.
Markings and measurements: (Toronto) Two pipes are stamped "RUDOLPH KOENIG À PARIS" and marked "94" in ink referring to the 1873 catalogue. 5.0 × 4.2 × 43.0 cm; 5.0 × 4.2 × 43.0 cm. (Union) Marked "no. 94" in pencil.

96. Three equal pipes in brass, wood and card-board. 30 fr

According to the 1889 catalogue, these pipes give "*sensiblement*" the same sound. The tubes are made of brass, mahogany and cardboard respectively. The mouth-piece, lip and foot of all the tubes are mahogany.

Locations: MIT. NMAH (cat. no. 87.924.7). CSTM (acc. no. 1998.0253; only wood pipe). Toronto (1878).

Fig. CR no. 95b Photo by author, 2005. Physics Department, University of Toronto, Canada

Markings and measurements: The tubes at the NMAH measure 4.2 cm (inside diameter), l=29.5. The pipes at Toronto are stamped "RUDOLPH KOENIG À PARIS" and marked "95" in ink referring to the 1873 catalogue. Tubes measure 31 × 3 cm.
References: Fau (1853, p. 372) and Marloye (1851, p. 40).

97. Nine pipes, five of the same depth but of different lengths, giving ut3, ré3, mi3, fa3, sol3, and four of the same length but of different depths, giving ré3, mi3, fa3, sol3. 30 fr

These pipes, according to the 1889 catalogue, demonstrate an "empirical law" established by the organ maker Cavaillé-Coll, that the length of the pipe is equal to the theoretical length of the wave of the fundamental, minus two times the depth. The ones at the CNAM were displayed at the 1862 Exhibition in London.

Location: CNAM (inv. 07056; date, 1862). Toronto (1878). Union.
Markings and measurements: (Toronto). They are all marked "96" in ink referring to the 1873 catalogue. All are stamped "RUDOLPH KOENIG À PARIS." The first five pipes with different lengths are as follows, "UT3" (5.6 × 5.6 × 62 cm); "RÉ3" (5.6 × 5.6 × 55 cm); "MI3" (5.6 × 5.5 × 55 cm); "FA3" (5.6 × 5.5 × 46.0 cm); "SOL3" (5.6 × 5.5 × 40.6 cm). The next four have the same length:

Fig. CR no. 96 Photo by author, 2005. Physics Department, University of Toronto, Canada

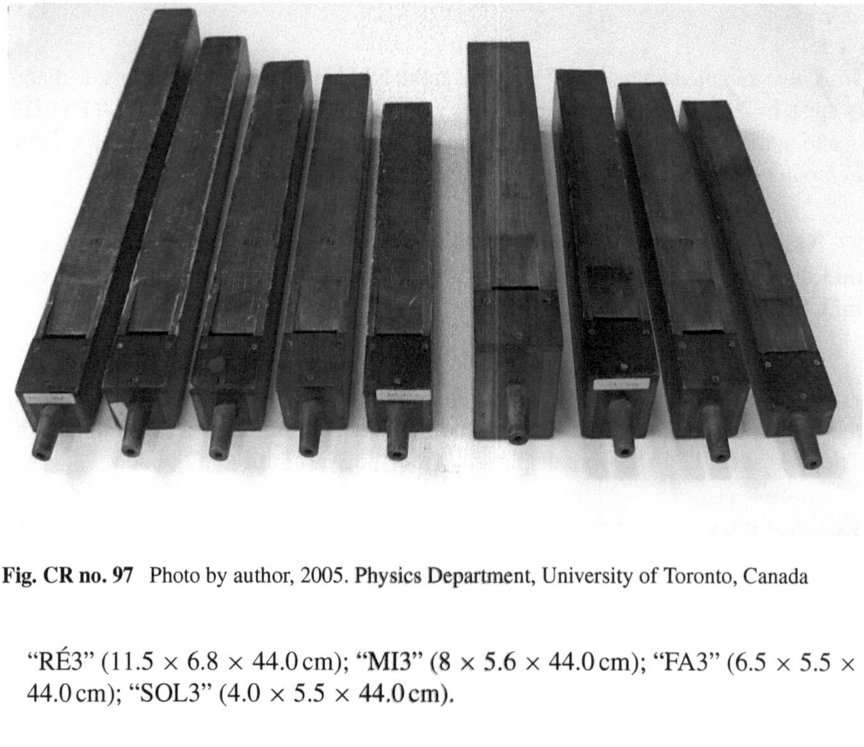

Fig. CR no. 97 Photo by author, 2005. Physics Department, University of Toronto, Canada

"RÉ3" (11.5 × 6.8 × 44.0 cm); "MI3" (8 × 5.6 × 44.0 cm); "FA3" (6.5 × 5.5 × 44.0 cm); "SOL3" (4.0 × 5.5 × 44.0 cm).

98. Eight rectangular stopped pipes, one of which is cubical. 80 fr

In these pipes the product of the length by the depth is constant. Koenig wrote that according to Savart these pipes gave the same notes under most conditions. They apply the same principle as the preceding pipes.

Fig. CR no. 98 Photo by author, 2005. Physics Department, University of Toronto, Canada

Location: Toronto.
Markings and measurements: The pipes at the University of Toronto are marked "99" referring to the 1873 catalogue. They are each stamped, "RUDOLPH KOENIG À PARIS." 5.7 × 5.7 × 16.2 cm; 7.7 × 7.5 × 13 cm; 4.6 × 4.6 × 19 cm; 8.7 × 8.6 × 12 cm; 9.5 × 9.4 × 11.5 cm; 10.6 × 10.4 × 11.2 cm; 6.7 × 6.4 × 14.3 cm; 3.8 × 3.6 × 25.1 cm.

98a. Four rectangular stopped pipes, one of which is cubical. 40 fr

Reference: Marloye (1851, p. 40).

99. Six rectangular stopped pipes to show the influence of the three dimensions. 60 fr

The 1889 catalogue states: "Two of these pipes have the same width and depth as the cubic pipes but different length, giving the third and fifth. Two others have the same width and length but different depth, giving the same notes as the preceding; the last two have the same length and depth, but different widths. The same diminution of length or depth produces the same changes of sound, which can be as much as an octave, while fro dimunition of the width, along with the size of the lip, the sound only rises a semi-tone."

100. Two equal pipes with mouth-pieces in different positions. 18 fr

These pipes demonstrate that the position of the mouth-piece has no effect on sound.

Fig. CR no. 100 Photo by author, 2005. Physics Department, University of Toronto, Canada

Location: Toronto.
Markings and measurements: Marked "92" in ink referring to the 1873 catalogue. Stamped "RUDOLPH KOENIG À PARIS" on the pine. Mahogany lip at base. Bent pipe = 10.0 × 5.6 × 39.0 cm; straight pipe = 5.6 × 5.5 × 43.3.
References: Daguin (1867, p. 539) and Violle (1883, p. 144).

101. Two stopped cubical pipes. 20 fr

These two pipes are different sizes for studying relations between volume and pitch.

Locations: NMAH (cat. no. 315727). Toronto. Union.
Markings and measurements: (Toronto) Marked "100" referring to the 1873 catalogue. Stamped "RUDOLPH KOENIG À PARIS" on the pine. Mahogany lip at base. 6.5 × 6.5 × 12 cm; 11.5 × 11.3 × 16.8 cm.
References: Deschanel (1877, p. 842), Desains (1857a, p. 73), Guillemin (1881, p. 688), and Marloye (1851, p. 40).

102. Two stopped triangular prismatic pipes. 22 fr

These pipes demonstrate the relations between volume and pitch.

Location: Toronto. Union.
Markings and measurements: The ones at Toronto are marked "101" referring to the 1873 catalogue. Stamped "RUDOLPH KOENIG À PARIS" on the pine. Mahogany lip at base. 6.3 × 6.3 × 12.8 cm; 10.7 × 10.7 × 17.9 cm.
References: Deschanel (1877, p. 842), Desains (1857a, p. 73), Guillemin (1881, p. 688), and Marloye (1851, p. 40).

Fig. CR no. 101 Photo by author, 2005. Physics Department, University of Toronto, Canada

Fig. CR no. 102 Photo by author, 2005. Physics Department, University of Toronto, Canada

103. Two long pipes of brass, one open, the other stopped, to give the succession of harmonics. 12 fr

A number of harmonics sound when one blows strongly into these pipes. They were originally made of glass.

Locations: CSTM (acc. no. 1998.0262). Teylers (c. 1865).

Markings and measurements: The CSTM has two long, brass pipes, l=68.5, diameter=1.6 cm. One pipe is closed. Two tubes at the Teylers Museum are made of glass. One is open, the other closed.

References: Deschanel (1877, pp. 839–840), Marloye (1851, p. 36), and Turner, G.L'E. (1996, p. 120).

104. A long open pipe, giving the sounds 1,2,3,4. 21 fr

This pipe has five wooden levers at the nodal points for creating selected harmonics. By opening the lever, the node turns into an anti-node of the ventral, thus changing the frequency. A long piston extends through the pipe.

Fig. CR nos. 104 and 107 Physics Department, University of Toronto, Canada

Locations: Dartmouth (acc. no. 2002.1.34047). NMAH (cat. no. 87.924.9). Toronto.

Description: There is soft cloth (for sealing the opening) on end of the piston and underneath the levers. The pipe is pine with a mahogany a lip.

Markings and measurements: (Toronto) Marked "104" in ink, referring to the 1873 catalogue. Stamped "RUDOLPH KOENIG À PARIS." Rectangular pipe is 67.4 cm long (71.5 with mouth-piece). Nodal and ventral points marked on both sides of pipe, from mouthpiece to end, "N/4, N/3, N/2, N/4, N/2, N/4, N/2, N/3, N/4. V/[illegible], V/4, V/2, V/3, [illegible]." The one at the NMAH measures 37 mm h, 48 mm w, 716 mm d, wt. 385 gr.

References: Daguin (1867, p. 533), Desains (1857a, pp. 56–57), Jamin (1868, p. 542), Violle (1883, p. 127), and Zahm (1900, p. 228).

105. A long stopped pipe, giving the sounds 1, 3, 5, 7. 21 fr

This closed pipe has six wooden levers and a piston for producing the odd harmonics.

Location: NMAH (cat. no. 315727). Dartmouth (acc. no. 2002.1.34052). Toronto (1878).

Markings and measurements: (Toronto). Stamped "RUDOLPH KOENIG À PARIS," marked "105" in ink referring to the 1873 catalogue, and is stamped on the sides: "N/7, N/3, N/3, N/7, N/5, N/7, N"; and, "V, V/7, V/[?], V/3, V/5, V/7." 3.3 × 3.5 × 72.9 cm.

106. A long pipe, stopped at both ends, giving the sounds 1, 3, 5 when the mouth-piece is fixed, and the sounds 1,2,3,4 when moveable. 52 fr

The mouth-piece and positions can be adjusted to make different sized pipes. There are eleven wooden levers for opening holes at nodal points.

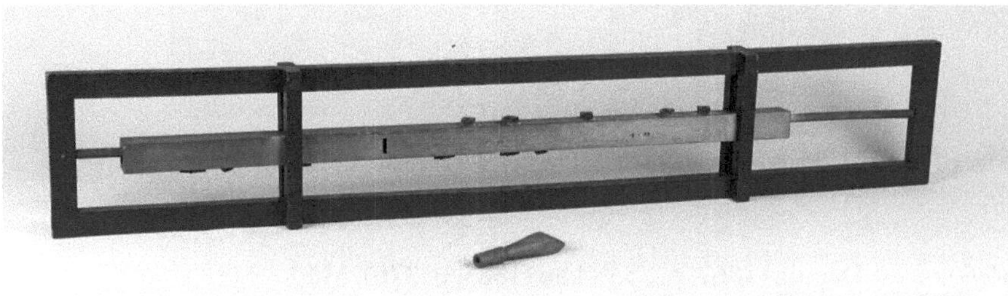

Fig. CR no. 106 Courtesy of the Department of the History of Science, Collection of Historical Scientific Instruments, Harvard University, USA. acc. no. 1997-1-0923

Location: Harvard (acc. no. 1997-1-0923). Toronto.

107. Circular pipe giving the sounds 1, 3, 5. 28 fr

This pipe is essentially a closed pipe, giving only odd harmonics. There are six wooden levers to produce these sounds. A sliding door divides the pipe in half creating a node at that point.

Location: Harvard (acc. no. 1998-1-0226). NMAH (cat. no. 327553). Toronto.

Description: There are a few markings that remain from the construction process. Faint pencil lines drawn through the center of the holes mark the precise nodal placements. The pine circle is warped slightly.

Markings and measurements: (Toronto) Signed "RUDOLPH KOENIG À PARIS." Diameter of circle from mouthpiece to divider, 21.0 cm. Markings clockwise from divider, "N, V/5, V/3, V/3, V, V/5, V/3, V/5."

108. Flute in four parts. 12 fr

This flute consists of a tube with a mouthpiece, two open tubes (each the theoretical wave-length of sound), and a closed tube the length of a half-wave.

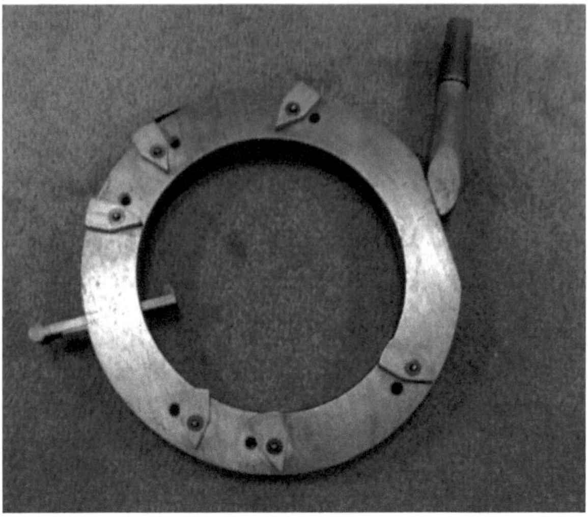

Fig. CR no. 107 Photo by author, 2005. Physics Department, University of Toronto, Canada

References: Daguin (1867, p. 533), Fau (1853, p. 376), Marloye (1851, p. 37), Violle (1883, p. 130), and Zahm (1900, p. 229).

109. Apparatus with water-stopped pipes. 120 fr

This is another way to demonstrate the relations between volume and pitch. In this case there are two pipes of different diameters filled with water. A graduated metal rod rests between the pipes for measuring the height of the water in millimeters. The beauty of this apparatus is that one can control the volume with great precision by employing the stop-cocks. In one experiment, the water is lowered in both tubes until the fundamental tone is produced (e.g. 256 Hz). The resultant column of air will be shorter in the wide tube, and longer in the thin tube. Lower the water in both tubes until the octave sounds (128 Hz). Measure the lengths on the graduated rod. The length of each column will be exactly double the first, providing another measurement of the wave-length.

Location: NMAH (cat. no. 315175).
Measurements: The example at the NMAH measures 91.1 cm in height and rests on a cast iron stand.
References: Jones (1937, p. 235) and Zahm (1900, p. 28).

109a. The same apparatus mounted in wood

Location: Harvard (acc. no. 1997-1-1805).

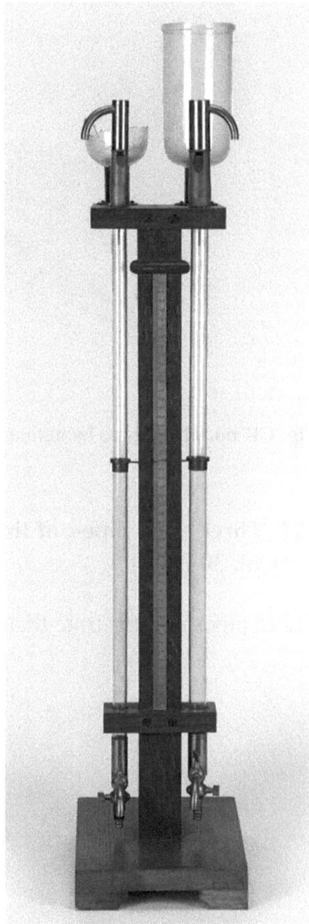

Fig. CR no. 109a Courtesy of the Department of the History of Science, Collection of Historical Scientific Instruments, Harvard University, USA. acc. no. 1997-1-1805

110. Four stopped pipes, tetrahedral, cubical, cylindrical and spherical, having equal volumes. 50 fr

These pipes demonstrate that equal volumes produce similar sounds.

Location: Dartmouth (acc. nos. 2002.1.34092 to 94). Harvard (acc. no. 1997-1-1939). NMAH (cat. no. 327553). Toronto.

Markings and measurements: (Toronto) Prism, 18.5 × 12.5 × 14.5 cm; sphere, 10 × 20 cm; cube, 15 × 7.5 × 7.5 cm; cylinder, 14.3 × 14.3 (diameter) × 8.5 cm.

References: Daguin (1867, p. 540), Fau (1853, p. 373), and Zahm (1900, pp. 236–237).

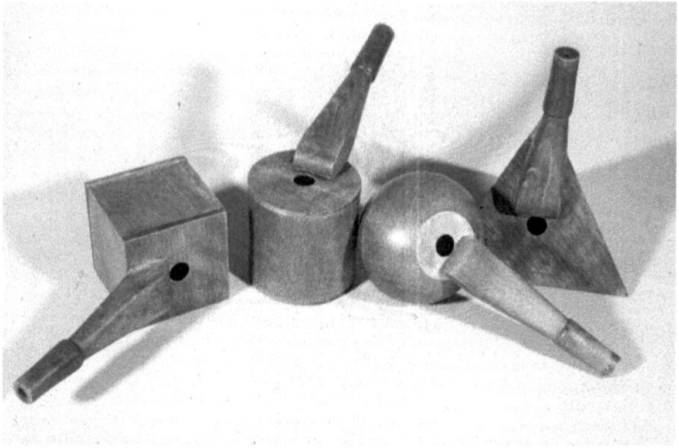

Fig. CR no. 110 Photo by author, 2005. Physics Department, University of Toronto, Canada

111. Three open pipes, of the same length and volume, one prismatic the others conical. 30 fr

These pipes demonstrate that equal volumes produce similar sounds.

Fig. CR no. 111 Photo by author, 2005. Physics Department, University of Toronto, Canada

Location: Toronto (1878).

Markings and measurements: (Toronto) Marked "111" in ink referring to the 1873 catalogue. Each pipe stamped "RUDOLPH KOENIG À PARIS." 4.0 × 3.9 × 41.8 cm (7.0 × 7.0 cm at open end); 5.5 × 5.5 × 41.8 cm; 7.8 × 7.7 × 41.8 (4.5 × 4.3 cm at open end).

112. Nine open pipes giving the scale, ut2–ut3, the fundamental being duplicated. 150 fr

These were most likely the largest pipes made by Koenig. They produce powerful low notes. Each of the pipes carries a sliding wooden door for altering the pitch by a semi-tone (*demi-ton*) or 1/12th of an octave (which is the same as the interval between two piano keys).

Fig. CR no. 112 Photo by author, 2005. Physics Department, University of Toronto, Canada

Location: Toronto.

Description: Oak pipes, each with varying grain patterns. These pipes have a wooden, sliding trap door on the end for adjusting pitch, which rises as the door is opened. [One of the ut2 pipes did not work for several years until we discovered a mouse nest in the chamber below the lip].

Markings and measurements: (Toronto) Marked "96a" in ink. Each one stamped "RUDOLPH KOENIG À PARIS." "UT3" 7.9 × 6.4 × 59.4 cm; "SI2" 7.9 × 6.6 × 64.5; "LA2" 8.6 × 7.1 × 71.5; "SOL2" 9.1 × 7.9 × 79.0; "FA2" 9.8 × 8.2 × 87.8; "MI2" 10.2 × 8.3 × 93.0; "RÉ2" 10.1 × 8.5 × 108.6; "UT2" 10.2 × 8.5 × 123.2; "UT2" 10.2 × 8.5 × 123.2.

112a. Five open pipes giving ut2, ut2, mi2, sol2, ut3. 85 fr

This is a smaller set of no. 112.

Location: MIT. Union.
Description: The pipes at Union College (sol2, mi2, and ut2) have mahogany sliders at the open end. The pitch rises as the slider is opened. The lowest note occurs when the slider is fully closed.
Markings and measurements: Stamped "RUDOLPH KOENIG À PARIS". "SOL2" 10.0 × 8.0 × 92.7 cm; "MI2" 10.7 × 8.9 × 109 cm; "UT2" 12.6 × 10.6 × 134.4 cm. There is a long Koenig pipe at MIT that measures 134.0 cm in length. It is marked "UT1" but it is the same length as the ut2 pipe at Union College, so perhaps it was once a closed pipe (a closed pipe of the same length would produce a note one octave lower than an open ut2).

112b. Two open pipes giving ut2. 42 fr

Reference: Marloye (1851, p. 41).

113. Eight open pipes giving the scale ut3 to ut4. 60 fr

The openings at the end of these pipes can be altered with a moveable lead cover. By making slight adjustments, one could change the pitch and bring the various pipes into harmony or out of harmony. The slight differences would presumably be detected using beats. The larger the opening, the higher the pitch (with a change being no larger than a semitone or 1/12th of an octave).

Location: Toronto.
Description: Pine pipes with mahogany lips. The pipes at Toronto have an opening on one side at the top of the pipe. On all pipes the holes were covered by a lead sheet. Presently, only fa3 has a full sheet remaining.
Markings and measurements: Each are stamped "RUDOLPH KOENIG À PARIS" and marked "113" in ink referring to the 1873 catalogue. Ut4 is missing. "UT3" (6.4 × 5.5 × 62.2 cm); "RÉ3" (6.0 × 5.0 × 56.2 cm); "MI3" (5.5 × 5.0 × 56.2 cm); "FA3" (5.2 × 4.5 × 48 cm); "SOL3" (4.9 × 4.3 × 43 cm); "LA3" (4.8 × 4.0 × 38.2 cm); "Si3" (5.2 × 4.5 × 48 cm).

113a. Four open pipes giving ut3, mi3, sol3, ut4. 30 fr

114. Eight stopped pipes giving the scale ut3 to ut4. 60 fr

The stopper could be adjusted in order to change the pitch of the pipe and bring the pipes into or out of harmony with each other.

Location: Harvard (acc. no. (WJ0021-28)). Toronto (1878).

Fig. CR no. 113 Photo by author, 2005. Physics Department, University of Toronto, Canada

Description: (Toronto) Six closed pine pipes from a set of eight giving the scale ut3–ut4. They each have mahogany lips. The wooden stoppers have knobs for removal/adjustment at the opening; a soft cloth material around the edge of the stopper ensures a tight fit.

Markings and measurements: (Toronto) Stamped "RUDOLPH KOENIG À PARIS" and marked "113a" in ink referring to the 1873 catalogue. "UT3" 6.3 × 5.6 × 29.5 cm; "RÉ3" 5.8 × 5.0 × 27.2 cm; "MI3" 5.5 4.6 × 25.2 cm; "FA3" 5.0 × 4.5 × 24.2 cm; "SOL3" 5.0 × 4.3 × 22 cm; "LA3" 4.8 × 4 × 19.7 cm. Si3 and Ut4 missing.

114a. Four stopped pipes giving ut3, mi3, sol3, ut4. 30 fr

115. Free reed pipe with two conical resonators. 30 fr

A free reed rapidly oscillates back and forth through a similarly shaped opening as pressured air is blown against it. The resultant pulses of air move into the pipe thus resonating and producing a sound. The free reed was first tried in European organs in the latter part of the eighteenth century. Some players found them more expressive than the popular beating reeds (CR no. 116) which produced more powerful, but harsher sounds. The Paris organ maker, Aristide Cavaillé-Coll used free reeds;

Hermann von Helmholtz also declared that they were superior to beating reeds. But even later in the nineteenth century, there was still debate about the advantages of free over beating reeds. The use of differently shaped resonators demonstrated their influence on timbre.

Location: Coimbra (FIS.0401). Teylers. QUP.
Description: The oak pipe at Teylers Museum comes with three pyramid-shaped oak resonators.
Markings and measurements: 52 × 52 × 250 mm
References: Auerbach in Winkelmann (1909, pp. 464–465), Daguin (1867, pp. 547–548), Deschanel (1877, pp. 846–847), Fau (1853, pp. 403–404), Ganot (1893, pp. 250–251), and Guillemin (1881, pp. 830–831). *Grove Dictionary of Music*, “Organ.” Helmholtz (1863, pp. 154–155), Jackson (2006), Jamin (1868, p. 533), Marloye (1851, p. 43), Mollan (1990, p. 195), Pantalony (2005b, pp. 140–142), Turner, G.L’E. (1996, p. 121), Tyndall (1896, pp. 220–223), and Violle (1883, pp. 147–148).

116. Striking [beating] reed pipe with resonators. 30 fr

A beating reed completely covers the aperture which leads into the pipe. Pressured air blows against it causing it to rapidly oscillate and thus “beat” against the aperture. Pulses of air then move into the pipe, or resonator, thereby creating a tone. Beating reeds are found in instruments such as the clarinet or saxophone. In the nineteenth century there were heated debates about the advantages and disadvantages of beating versus free reeds (see CR no. 115). The beating reeds were reputed to produce more powerful tones. The resonators that came with this reed instrument demonstrated the production of different timbres. The glass sides allow one to view the reed mechanism.

Location: Toronto (1878). QUP.
Description: Oak pipe and oak resonators. Original black cloth tape on glass sides.
Markings and measurements: Marked “115” in ink referring to 1873 catalogue. Stamped “RUDOLPH KOENIG À PARIS.” Oak cone, 41.5 × 9 × 9 cm; pipe, 5.6 × 5.6 × 27.1 cm.
References: Auerbach in Winkelmann (1909, pp. 464–465), Daguin (1867, pp. 547–548), Deschanel (1877, pp. 846–847), Fau (1853, pp. 403–404), Ganot (1893, pp. 250–251), and Guillemin (1881, pp. 830–31). *Grove Dictionary of Music*, “Organ.” Helmholtz (1863, pp. 154–155), Jackson (2006), Jamin (1868, p. 533), Marloye (1851, p. 43), Molan (1990, p. 195), Pantalony (2005a, pp. 140–142), Turner, G.L’E. (1996, p. 121), Tyndall (1896, pp. 220–223), and Violle (1883, pp. 147–148).

Fig. CR no. 116 Photo by Louisa Yick. Physics Department, University of Toronto, Canada

Vibrations of Membranes

117. Circular rubber membrane, which can be stretched at will. 11 fr

This instrument demonstrated basic Chladni-like vibration patterns of a membrane at different tensions. Helmholtz claimed to have used "tuned" membranes to test for the objective existence of a specific tone. The wooden screws tighten the ring to stretch the membrane.

Location: Coimbra (FIS.0394; date, 1867). Teylers (1865)
Description: The catalogue advertises a rubber membrane, however the membrane at the Teylers is pig's bladder and paper. The frame at Teylers is walnut, whereas the one at Coimbra appears to be a light mahogany. The one at Coimbra is 180 mm in diameter.
References: Helmholtz (1863, p. 234). Idem., 1954, p. 157. Turner, G.L'E (1996, p. 107) and Zahm (1900, p. 271).

Fig. CR no. 117 Photo by author 2005. Museu de Física, University of Coimbra, Portugal. FIS.0394

118. Circular paper membrane, 30 cm in diameter. 7 fr

Location: Amherst.
Reference: Marloye (1851, p. 45).

119. Square paper membrane. 6 fr

Fig. 61 (Nos 119, 122, 123).

Fig. CR nos. 119, 122, and 123 Source: Koenig (1889, p. 48)

120. Triangular paper membrane. 6 fr

121. Three small paper membranes, circular, square, and triangular. 10 fr

122. Stand for membranes. 45 fr

There are three cast-iron stands with adjustable tips for changing the height.

123. Windtube mounted on stand. 10 fr

Membranes, unlike vibrating rigid plates, required different means for eliciting vibrations. This windtube, mounted on a cast-iron tripod, stimulated the above membranes into vibration. Various patterns emerged depending on where one placed the windtube (nodal or anti-nodal points).

124. Ellipsoidal bell mounted on a handle. 12 fr

This bell produced "strident" timbre-rich sounds that were very good at stimulating intricate vibrational patterns on a membrane. August Zahm stated that the "harsh, creaking sound" emitted by the bell produced "the most complicated patterns" on a membrane (CR no. 117).

Fig. CR no. 124 Photo by author, 2005. Physics Department, Union College, USA

Location: Union.
Description: Wooden handle, wrought iron bell.
Markings and measurements: Stamped "RUDOLPH KOENIG À PARIS" on the wooden handle, which has a height of 17.5 cm. The bell is an ellipsoidal shape (10 × 4 × 8 cm) with a height of 12 cm.
Reference: Zahm (1900, p. 271).

125. Open whistle with different holes. 5 fr

126. Sedley Taylor's apparatus to show the vibrations of liquid films. 25 fr

Some of the best physics demonstrations (and research) derive from careful observation of everyday phenomena. Sedley Taylor, the inventor of this apparatus, was a popular science lecturer in Victorian England. He used this apparatus to show the effect of sound on one of the thinnest possible membranes – the film of liquid soap. Sound pulses sent through the air chamber come into contact with the opening that is covered with the thin film. If the opening is then projected onto a screen, noted August Zahm, "we obtain, by speaking or singing into the resonant cavity of the apparatus, the most gorgeous kaleidoscopic effects conceivable. Every note, and every vowel sounded on the same note, instantly evokes the most marvelous figures, tinted with all the hues of the rainbow. There is nothing in the whole range of physics more beautiful then the phenomena here exhibited." The apparatus came with three different metal coverings – circular, square, and triangular.

Fig. CR no. 126 Source: Koenig (1889, p. 49)

Reference: Zahm (1900, pp. 271–272).

Vibrations of Strings

127. Differential sonometer of Marloye with weights. 110 fr

Charles Barnes of Oxford called this "one of the most useful and valuable instruments in acoustics."[30] Albert Marloye invented this form of two-string sonometer for demonstrating the laws of vibrating strings. It was used to determine the relations between frequency and certain characteristics of the string-length, tension, diameter, and density. Two strings run the length of the resonance box and pass over wooden bridges which are a meter apart. They are attached to a tuning key and weights and pulleys. By altering the above variables and comparing the frequencies of sounds produced, it was possible to study and demonstrate the laws of vibrating strings.

The mixture of the musical and scientific context that Marloye bridged in nineteenth-century Paris is apparent by the three divided scales on the top. One is the "chromatic tempered scale," the second is the "chromatic physicist's scale" with the harmonic divisions of the scale, and the third is a meter stick divided into millimeters.

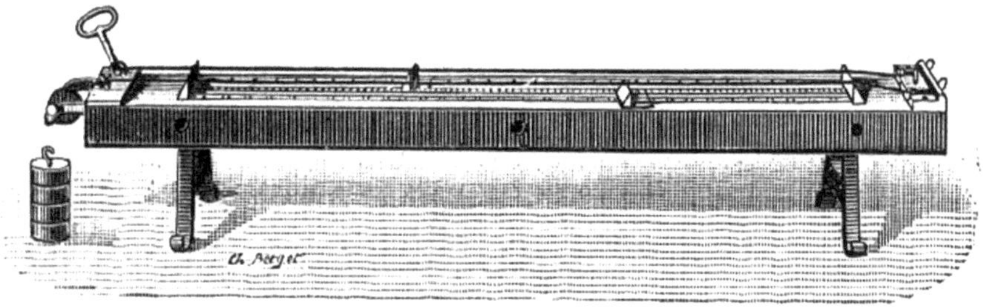

Fig. CR no. 127 Source: Koenig (1889, p. 49)

Location: Coimbra (FIS.0391; unsigned). NMAH (cat. no. 314588; unsigned). Rome (c. 1873). Teylers (1865).

Description: The unsigned instrument at the NMAH is finely varnished with a cedar resonance box, maple bridges, mahogany sides, and three sounding holes on each side. The resonance box closely resembles the wood and finish used by Koenig in his sonometers.

Measurements: (NMAH) 26.4 × 20.6 × 140 cm.

References: Barker (1892, pp. 232–233), Barnes (1898, pp. 18–32), Daguin (1867, pp. 505–506), Desains (1857a, pp. 105–106), Fau (1853, p. 369), Ganot (1893, pp. 247–248), Jamin (1868, p. 550), Loudon and McLennan (1895, p. 100), Marloye (1851, p. 50), Turner, G.L'E. (1996, p. 119), and Violle (1883, pp. 18–19).

128. Two clamp-bridges to limit the lengths of the strings. 20 fr

129. Packet of steel wires. 2 fr

130. Two brass wires, diameters 1:2. 1 fr

Reference: Marloye (1851, p. 50).

131. One iron and one platinum wire of the same diameter. 10 fr

Reference: Marloye (1851, p. 50).

132. Sonometer for the longitudinal vibrations of wires. 175 fr

Longer sonometers could be used for calculating the velocity of sound with great precision. The stretched wires are activated by rubbing them with India rubber or a resin bag. The resultant longitudinal vibrations move back and forth along the wire. From the pitch sounded, and the position of the adjustable bridge, one calculates the wavelength of the longitudinal wave, and then the velocity.

Location: NMAH (cat. no. 314601).
Description: (NMAH) There are massive cast iron clamps on the ends for holding the two wires. Like the differential sonometer, there are three divided scales – the tempered, physicist and the meter stick in millimeters. The main body is a solid oak beam.
Measurements: (NMAH) Stamped "RUDOLPH KOENIG À PARIS." 180 cm long, 21.8 cm in height and 28.6 cm wide and almost 22 kilos in weight.
Reference: Barnes (1898, p. 136).

133. Plassiart's Phonoscope for testing violin strings. 35 fr

The musical instrument market was competitive in nineteenth-century Paris. Many of the makers turned to science for an edge. Early in his career, Koenig maintained contacts with this market. This instrument tested violin strings for purity and homogeneity. He marketed it to violin players as a convenient and portable way to ensure good strings for their concerts. Plassiart, a chief engineer at Lorient, invented it and Koenig made and sold his own version. He showed it for the first time at the 1862 exhibition in London.

Variation of density and thickness of strings was a common problem. Violin makers sold strings in long segments with varying quality. Musicians, therefore, rejected dozens of string sections. According to an early review of the invention, the phonoscope allowed one to find good segments of string with a simple comparison by ear. A long segment is stretched over a wooden base. There is a sliding wooden frame on top of the base with ebony clamps for securing the string; the clamps are the same distance apart as the bridge (chevalet) and the nut (sillet) of a violin. A small hammer rests exactly in the middle of the string. One moves the frame along the string and sets it at a certain part to be tested. The string is then plucked simultaneously on both sides to compare the notes. If the notes are dissonant, the string is not of equal density and therefore not homogeneous. The frame is moved along the length of the string until a pure segment is located.

Fig. CR no.133 Source: Koenig (1889, p. 50)

References: Radau (1862a, pp. 700–701) and Pisko (1865, p. 129).

134. Barbereau's large eight-stringed sonometer for the study of scales, etc. 350 fr

The Barbereau sonometer was the most elaborate of Koenig's stringed instruments. It had eight strings, a tempered scale, physicist's scale and meter rules on each side divided into millimeters. The instrument derives from studies on the origin of scales by the French musical theorist and teacher, Auguste Barbareau (1799–1879), who taught at the Paris Conservatoire.

Location: NMAH (cat. no. 314589).
Description: The sonometer at the NMAH is one of the finest surviving examples of Koenig's wood working. It has a thinly finished spruce top, mahogany sides, walnut ends, oak bridge and steel strings. The sides have stylized lyre sound holes. The top has inlaid boxwood meter scales. The notes of the two scales (tempered and physicist's scales) are marked along with millimeters.
Markings and measurements: "RUDOLPH KOENIG À PARIS." 23 × 57 × 133 cm.
References: Barbereau (1848). Idem., 1852.

Vibrations of Rods and Bars

135. Four steel bars to illustrate the laws of transversal vibration. 30 fr

Two of the bars are the same length and thickness, but different width. The third is a different length and double thickness. The fourth is the same thickness as the first two, but its length is 1: [square root of 2].

Location: FST.
References: Giatti (2001, p. 85) and Marloye (1851, p. 47).

135a. Four brass bars. 25 fr

Reference: Marloye (1851, p. 47).

135b. Four wooden bars. 7 fr

Reference: Marloye (1851, p. 47).

136. Six bars of the same size, but different description. 14 fr

The 1889 catalogue states that "five of the bars are of different wood, one is brass, for showing the influence of material on pitch and sonority of sound."

Reference: Marloye (1851, p. 47).

137. Small support for transversally vibrating bars. 10 fr

This is a support to hold a bar so that there is no lateral movement. One arm of the support can be adjusted for different length bars.

138. Two brass rods to illustrate the law of harmonics in transversal vibrations. 12 fr

One of the rods is longer than a meter, the other is half a meter.

Reference: Marloye (1851, p. 47).

139. Two corkbridges on iron plates. 8 fr

140. Four brass bars of the same length, one straight and the others more and more bent. 18 fr

141. Four steel rods to illustrate the law of longitudinal vibrations. 45 fr

Two of these rods are cylindrical, one meter in length, with different diameters. There is also a cylindrical one which is half a meter in length. The fourth rod is prismatic and is one meter in length. These rods demonstrated that the diameter and form of the rods had no effect on the frequency of longitudinal vibrations. The rods were placed in a firm support (CR no. 142) and rubbed with resined leather. The two one meter cylindrical rods, and the rectangular rod yielded the same pitch. However, the rod that was half a meter in length produced a pitch elevated by one octave. August Zahm described these experiments as illustrations of the following law: "The number of longitudinal vibrations is inversely proportional to the lengths of the vibrating segments, or, when rods of the same material but of different lengths are employed, the number of vibrations executed per second is inversely as the lengths of the rods."

Reference: Zahm (1900, pp. 185–186).

141a. Four pine rods. 8 fr

142. Support for longitudinally vibrating rods. 40 fr

This consists of a vice that can be secured to a table. It can hold both circular and rectangular bars.

143. Four steel rods of same diameter and different length, giving the perfect chord. 60 fr

144. Apparatus to show the lengthening and shortening of a rod whilst vibrating longitudinally. 45 fr

Longitudinal vibrations are almost impossible to see in an activated rod. This apparatus, which according to Tyndall was invented by Koenig, beautifully revealed these vibrations by means of a bouncing ivory ball. The apparatus consists of a brass rod set in a wooden frame. An ivory ball hangs from the support resting just in front of the rod. When the rod vibrates the ball pushes away and continues to bounce in conjunction with the longitudinal vibrations.

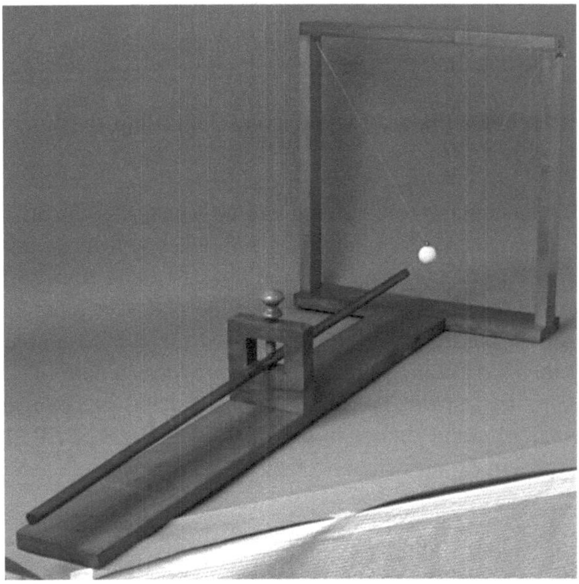

Fig. CR no. 144 Photo by author 2005. Museu de Física, University of Coimbra, Portugal. FIS.0393

Location: Coimbra (FIS.0393).
Description: Wood, ivory, brass.
Markings and measurements: Stamped "RUDOLPH KOENIG À PARIS" on top of wood frame. 47.1 × 103.6 × 44.9 cm.
References: Miller (1916, p. 4), Tyndall (1896, pp. 193–94), and Zahm (1900, pp. 179–180).

145. Apparatus to show the position of nodes on opposite sides of horse-hair vibrating longitudinally. 12 fr

Small rings move toward the nodes of the vibrating hair when set in vibration.

Location: Teylers (c. 1865).
Description: Oak and pine rectangular frame.
Reference: Turner, G.L'E. (1996, p. 113).

146. Claque-bois. [Wooden sounding bars]. 20 fr

In the English world this instrument is known as a xylophone. It consists of twelve pine bars which form one and a half octaves. The bars are supported by a straw rope (hence the alternative name of "straw fiddle") at the two nodes of the bar. When they are struck with a leather-covered wooden mallet they emit a soft bell-like tone. At the turn of the century, according to August Zahm, these instruments were "becoming more popular daily."

Reference: Zahm (1900, pp. 178–179).

Vibration of Plates

147. Stand with 6 brass plates, 3 square and 3 circular, to illustrate the law of thickness and areas. 80 fr

This is a set of Chladni plates for studying vibration patterns with different areas and thicknesses. Sand is sprinkled on the plates and, when activated with a violin bow, collects at the places of no vibration, or nodal lines. Musical notes are also produced which correspond to the size and thickness of the plates. A plate that is the same size as its neighbour but double in thickness produces a note double in frequency. A plate of half the area but the same thickness produces a note four times as high. These experiments derive from the work of Ernst Chladni, the German scientist who published original vibration studies in his 1802 book *Die Akustik* which became a foundation for modern experimental acoustics.

Fig. CR no. 147 Photo by author, 2005. Museum of Science, University of Lisbon, Portugal

Location: Vanderbilt (1875). Lisbon. Rome.
Description: The examples at the University of Lisbon and Vanderbilt University have stands with turned supports made of wood. The brass plates are painted black.
References: Auerbach in Winklemann (1909, pp. 383–401), Blaserna (1876, pp. 14–15), Chladni 1802 and 1809. Daguin (1867, p. 585), Fau (1853, p. 396), Jackson (2006, pp. 13–44), Jamin (1868, p. 593), Marloye (1851, p. 45), Tyndall (1896, pp. 168–184), Violle (1883, pp. 231–233), and Zahm (1892, p. 202).

147a. Stand with three square plates. 50 fr

Reference: Marloye (1851, p. 45).

148. Circular brass plate, diameter 30 cent. 18 fr

Reference: Marloye (1851, p. 46).

149. Square brass plate, side 30 cent. 18 fr

References: Daguin (1867, p. 583) and Marloye (1851, p. 46).

150. Triangular brass plate. 18 fr

Location: Union (c. 1875).
Reference: Marloye (1851, p. 46).

151. Pentagonal brass plate. 18 fr

152. Hexagonal brass plate. 18 fr

153. Large universal support composed of four clamps for plates. 60 fr

Location: Union (c. 1875).

153a. Support with one clamp. 15 fr

Location: Toronto. Union (c. 1875). Yale (acc. no. YPM 51298).
Markings and measurements: (Toronto). "RK" on brass collar. 21 × 11.5 cm.

154. Iron support for plates pierced at the centre. 15 fr

155. Steel rod for exciting vibrations in plates pierced at the centre. 15 fr

156. Circular wooden plate with handle. 4 fr

Fig. CR no. 153a Photo by author, 2005. Physics Department, University of Toronto, Canada

The 1889 catalogue states that the fundamental sound and the accompanying figures with sand are different depending on whether one strikes the bow on the edge of the "axes of elasticity" or between the axes.

Location: Lisbon (unsigned).

157. Apparatus to show the rotation of lycopodium in circular plates. 120 fr

Following the work of Ernst Chladni, vibrating patterns became a popular research topic for a variety of scientists. Michael Faraday, Hans Christian Ørsted and Félix Savart were fascinated by the action of lycopodium, an extremely light powder, on vibrating plates. Whereas sand gathers at the nodal lines (places of no movement) of a vibrating plate, lycopodium (spores from moss) gathers and swirls at the ventral segments (vibrating segments). The nodal lines move if one shifts the bow from right to left. The shifting can also be detected by the strengthening and weakening of sounds that are amplified by the resonator suspended over the plate. This apparatus, first sold by Albert Marloye, was developed by Savart who discovered that even without the shifting bow strokes, the vibrating plate, left to itself, demonstrated shifting nodal patterns.

Location: Dartmouth College has one made by Albert Marloye (acc. no. 2002.1.34026).

Reference: Faraday (1831, p. 314–335), Fau (1853, pp. 397–400), Jones (1937, pp. 177–180), Marloye (1851, p. 46), Ørsted (1998, p. 261), Pantalony (2005a, pp. 143–144), Savart (1827, pp. 187–208), and Zahm (1892, pp. 198–200).

158. Glass bell-shaped jar on stand with four suspended balls. 28 fr

The behaviour and sound of bells is complex and, like violins, took centuries to understand and perfect. They were a prominent part of everyday life (e.g. church bells) and bell founding was a prized art. In the same way that vibration patterns of plates came to be revealed and mapped by Chladni's methods, nodes and ventral segments of vibrating bells were also the subject of study and visual demonstrations.

In this instrument, four ivory balls are suspended near the lip of a glass jar. After stimulating the bell with a violin bow, the balls, just touching the glass, are set into vibration revealing the location of the vibrating segments. If they are near nodes, they will not move. If they are located at the vibrating segment, they will bounce.

Location: Teylers Museum (unsigned).
References: Turner, G.L'E. (1996, p. 111) and Zahm (1900, p. 204).

VII. Communications of Vibrations – Vibrations of Compound Bodies: Compound Vibrations of Simple Bodies

159. Apparatus to prove that pendular movement can excite pendular harmonic movements. 100 fr

In the nineteenth century, a clever mechanical demonstration was sometimes the most persuasive form of argument in the lecture theatre. During the dispute over combination tones, some critics suggested that Koenig's forks were not pure and emitted unwanted harmonics. He countered that his forks were pure and that the confusion was due to the fact that they stimulated harmonics in other sources. He used twelve forks (CR no. 38) based on the ut2 harmonic series to demonstrate this principle using sympathetic vibration. He also devised this graphical, pendulum apparatus to demonstrate the harmonic relations of two oscillating bodies. It was his way of showing, in mechanical terms, how one motion can excite another motion in a second body.

The main part of the apparatus is a pendulum that oscillates in time with a mercury interrupter (similar to CR no. 214). A shorter rod with a graphical stylus is attached to the axis of the pendulum and moves with its own oscillatory motion. If its oscillations are harmonically related to the main pendulum, it will exhibit movements of the combined oscillations. If it is not related, its natural vibrations are not excited and it only registers the oscillations of the main pendulum. The apparatus comes with six rods of different lengths.

Reference: Koenig (1882c, pp. 201–205).

160. Two forks ut4 on resonance boxes. 70 fr

The notion of sympathy, where two objects influence each other from a distance, had a special place in scientific and medical thinking for centuries. Tuning forks provided a particularly striking example of this phenomenon in action. In 1866 Koenig witnessed it in dramatic form while working with Victor Regnault in the sewers of Paris on speed-of-sound experiments. He was aware that a tuning fork could stimulate another (of the same natural frequency) into vibration, but was amazed when

Fig. CR no. 159 Source: Koenig (1889, p. 57)

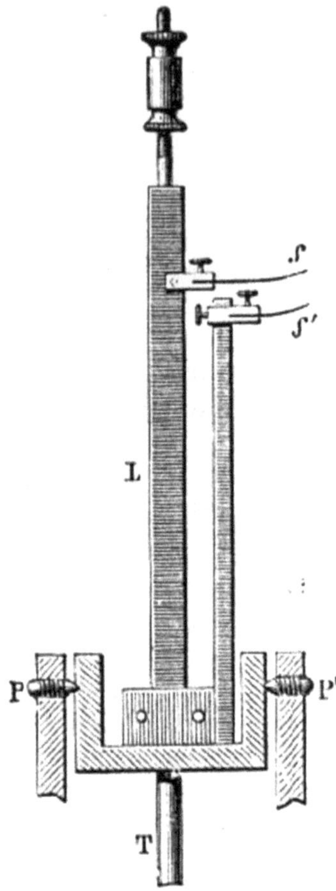

he observed this action at a distance of 1,590 m through the sewer conduit of St. Michel. For the young instrument maker, it was confirmation of just how sensitive tuning forks could be to outside influences. He used this observation to support his claims that unwanted harmonics could sometimes be activated in even the purest forks during an experiment, thus throwing off the results.

Reference: Koenig (1882c, p. 194).

161. Two similar brass plates, one with handle the other on support. 28 fr

A vibrating plate can stimulate another similarly shaped plate into sympathetic vibration. Distinctive Chladni patterns form (with sand sprinkled on the surface) when the plate with a handle is made to vibrate. If this plate is held over the second plate, the same patterns form, showing that it vibrated in sympathy with the first plate.

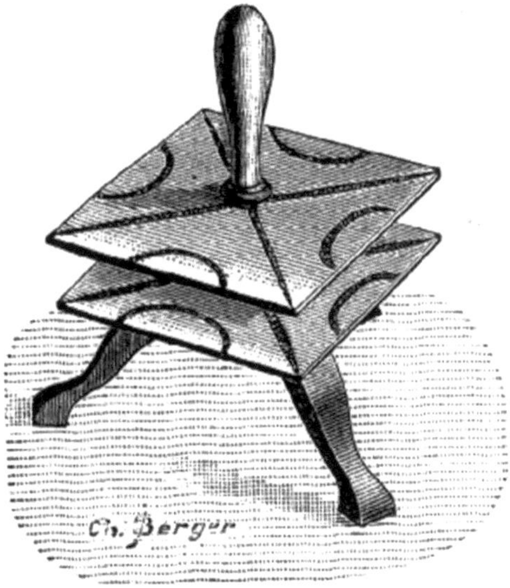

Fig. CR no. 161 Source: Koenig (1889, p. 57)

Location: CSTM (acc. no. 1998.0244). Harvard (acc. no. 1997-1-1060a).
Measurements: Both plates at the CSTM measure 15 × 15 cm.
Reference: Zahm (1900, p. 270).

162. Schaffgotsch's singing-flames apparatus. 175 fr

Glass tubes with a small gas-jet flame placed at one end produce strong, pure notes. The "singing flame" was a marvel for nineteenth century audiences and a source of fascination for scientists. It was first noticed by the Irish physician Dr. Byran Higgins in 1802. Faraday and Wheatstone studied this phenomenon and concluded that it was due to small explosions of flame that were amplified into a sound within the tube. In 1857 Franz G. Schaffgotsch developed an apparatus to test these ideas. Shortly thereafter, Koenig sold a similar apparatus with six tubes and two organ pipes.

Location: CNAM (inv. 08027; c. 1868). Coimbra (FIS.0751; date, 1881). Harvard (acc. no. 1997-1-0916). NMAH (cat. no. 315171). Teylers.
Description: Wood, brass, glass.
Markings and measurements: (NMAH) Stamped "RUDOLPH KOENIG À PARIS." 46.4 × 34.6 × 46.0 cm.
References: Ganot (1893, pp. 257–258), Guillemin (1881, pp. 666–668), Koenig (1865, pp. 27–28), Jones (1937, pp. 223–228), Pisko (1865, pp. 183–184),

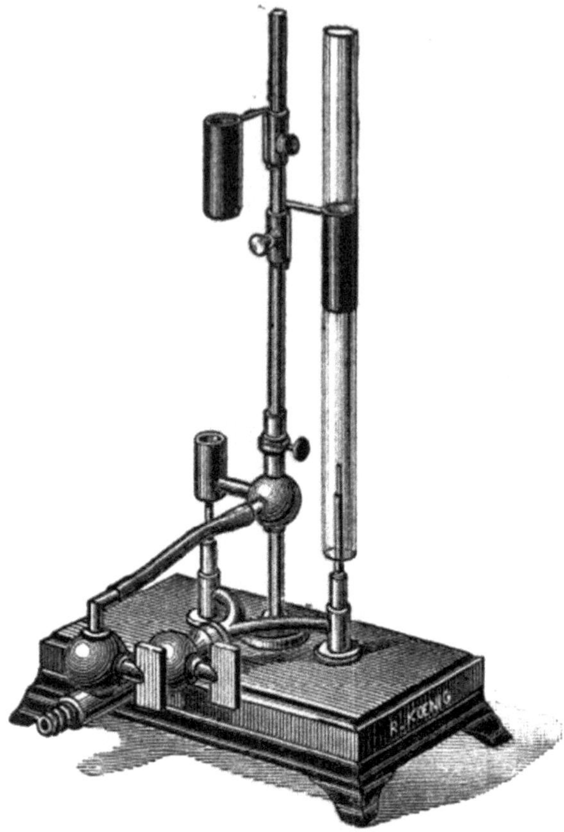

Fig. CR no. 162-1 Source: Koenig (1889, p. 58)

Schaffgotsch (1858, pp. 627–629), Turner, G.L'E. (1996, p. 128), and Tyndall (1896, pp. 244–257).

162a. The same apparatus of simpler form. 60 fr

In a simpler form of the singing flame, two similar notes could be produced to compare their pitch. Small adjustments could be made to change the frequency. The popular physics lecturer, August Zahm, used this version to compare two tones and demonstrate beat phenomena.

Reference: Zahm (1900, pp. 305–306).

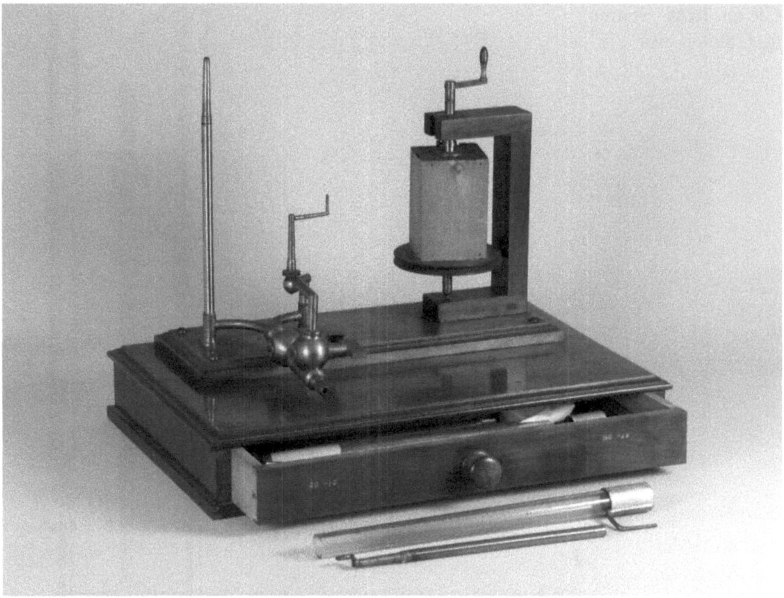

Fig. CR no. 162-2 Courtesy of the Department of the History of Science, Collection of Historical Scientific Instruments, Harvard University, USA. acc. no. 1997-1-0916

163. Sensitive flame apparatus. 25 fr

At an evening music recital in 1857 John Le Conte, a professor at South Carolina College, noticed that the flame of a gas lamp flickered rhythmically to the sounds of a violoncello. Le Conte concluded that the small vibrations in the outgoing stream of gas (at the edge of the orifice) were sympathetically amplified by the external vibrations. As the gas pressure increases, the flame becomes less stable and more sensitive to sound. This was a popular demonstration and scientists devised several means for experimenting with the effect. Koenig's version consisted of a stand and gas burners with wire gauze and a sound funnel. Gauze placed between the burner and flame greatly increased the sensitivity of the flame.

References: Auerbach in Winkelmann (1909, pp. 164, 479–484), Guillemin (1881, pp. 671–674), Jones (1937, pp. 236–238), Le Conte (1857, p. 473), Tyndall (1896, pp. 257–271), and Zahm (1900, pp. 251–254).

164. Apparatus to show the transmission of sound through solids. 60 fr

This apparatus, a music box sealed by another box to test the transmission of sound through solids, is a miniature version of a famous demonstration of the scientist and prolific inventor, Sir. Charles Wheatstone, who first played an instrument he called the "enchanted lyre" at his family music shop at Pall Mall in September 1821. In

Fig. CR no. 162a Source: Koenig (1889, p. 58)

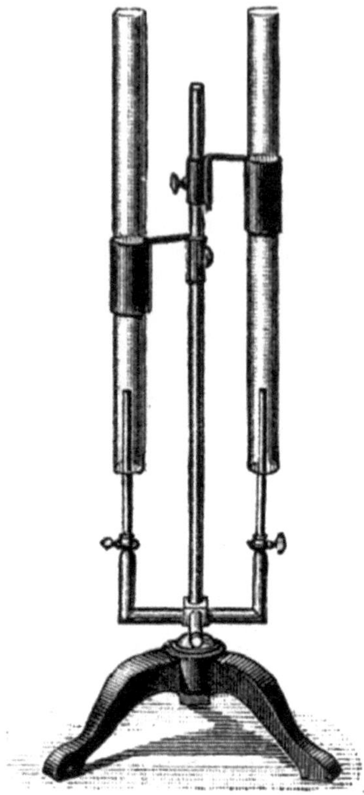

this experiment, Wheatstone suspended his lyre from the ceiling by a wire that was connected to instruments he was playing in a room above. The sound was transmitted through the wire to the lyre which had long horns pointed down toward the floor. The lyre miraculously seemed to play by itself. Even before the thought of using electricity, and amidst his studies of the vibrating properties of rods and various solids, Wheatstone promoted his findings as a possible way to transmit concerts through underground cables.

A music box is the main element of Koenig's instrument. Such musical toys were enormously popular in the nineteenth century. An entire industry developed around this clever bit of technology based on delicate clockwork. In this apparatus, a little music box (*une petite musique de Genève*) is placed in a hermitically sealed container. A long key for winding the box runs from the music box to the outside through a long tube. This sealed box is further encased in sand which smothers the sound even more. A pine rod connects the sealed box with an exterior resonating platform. The music box is activated, the sound travels along the rod, and the resonator vibrates in sympathy with the music, thus demonstrating the transmission of sound through a solid.

References: Bowers (2001). *Grove Dictionary of Music*, "Music Box." Zahm (1900, p. 172).

165. String telephone. 12 fr

The string telephone was a simple application of Wheatstone's discoveries of sound travelling through solids. Koenig's apparatus consisted of two speaking tubes with membranes attached by a fine string. The vibrations were transmitted along the string when it was held taut.

Reference: Guillemin (1881, pp. 764–765).

166. Reis telephone. 65 fr

Philipp Reis (1834–1874), a science teacher in Friedrichsdorf, invented this telephone in 1863. It was quite limited in its capability to receive and transmit speech, but was still significant enough to be considered by many contemporaries the first telephone. Reis died before the eruption of patent disputes over the telephone that followed Bell's invention, so his priority was never properly resolved.

The transmitter consisted of a drum and membrane attached to a small strip of platinum which was very lightly attached to an electrical contact finger. The receiver consisted of a bar of soft iron connected to an electrical coil, something like Whertheim's device (no. 23). Electrical constriction of the bar, therefore, produced longitudinal vibrations and then sound. Originally, Reis had used a needle hook wrapped in a coil placing it inside a violin to amplify the sound. He later built a simple resonating box to support the coil or helix, as they were called then. When one spoke into the vocal drum, vibrations were transformed into interruptions in current and transmitted to the receiver which produced faint sounds. It was these electrical interruptions which distinguished this model from later models which could transmit continuous electromagnetic variations. It could reproduce various frequencies, but not necessarily the distinct modulations and timbre of speech. Reis himself characterized his system as "make or break," the way electrical systems were understood at that time, but it is still debatable if in a limited sense it could be viewed as a device that operated in the continuous variable pressure mode which later became common in telephones and microphones. In this sense, telephone inventions were not fully understood by contemporary theory. Some of these issues, for example, spilled over into the timbre dispute between Helmholtz and Koenig.

Koenig sold this instrument in his 1865 catalogue. He stated that, "it is true that it is not of good quality, and that it stops from time to time." The membrane, he wrote, did not respond equally to all vibrations and the action of the platinum strip was "far from perfect."[31]

Location: Harvard (acc. no. 8000a-b).
Description: The transmitter consists of a mahogany box, with mica membrane, brass horn, key and coils on the side. There is a delicate strip of coiled flat metal

Fig. CR no. 166-1 Transmitter: Photo by author, 2005. Courtesy of the Department of the History of Science, Collection of Historical Scientific Instruments, Harvard University, USA. acc. no. 8000a-b

(probably platinum) under the mica diaphragm. (Receiver) Constriction coil covered by hinged pine resonator. This lies on a pine support box, similar in quality to the resonating boxes made for tuning forks. Both the cover and the box have small resonating holes. Sides are mahogany. Magnet spool is boxwood. Bridges are maple.

Markings and measurements: (Transmitter) Stamped "RUDOLPH KOENIG À PARIS" and with Harvard no. "7-49." 9.9 × 9.4 × 9.5 cm. (Receiver) stamped with Harvard no. "7-48." 6.1 × 9.1 × 24.1 cm.

References: Evenson (2000), Koenig (1865, p. 5), Pisko (1865, pp. 94–103), Shulman (2008), and Thompson (1883).

167. Apparatus to show the difference of phase between the transmitted and received sound in telephone transmission. 150 fr

Alexander Graham Bell's invention of the telephone triggered a debate about the role of phase in the quality of tone (timbre). Particular elements of the compound sound waves could be at different stages in their periodic cycle. Were there phase changes among different sound waves during electrical transmission? Did these changes affect the timbre? Emil DuBois-Reymond, Helmholtz, and Ludimar Hermann all investigated this question. Koenig, who believed timbre did change due to phase differences, invented this apparatus to show that phase shifted when transmitted through telephone transmitters and receivers. Two telephones were set up with two tuning forks, sol1 and sol1. Following the activation of one fork and

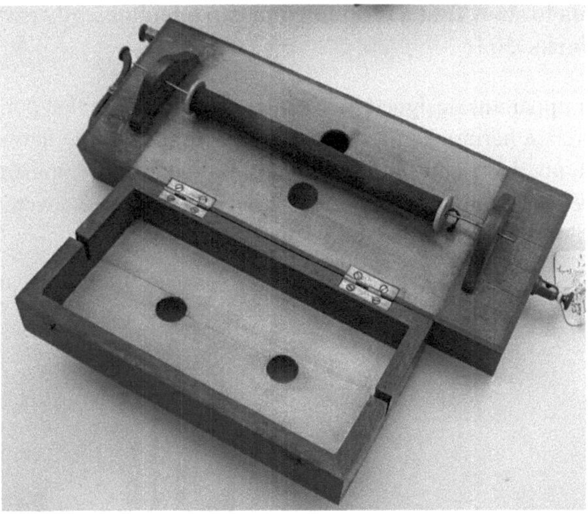

Fig. CR no. 166-2 Receiver: Photo by author, 2005. Courtesy of the Department of the History of Science, Collection of Historical Scientific Instruments, Harvard University, USA. acc. no. 8000a-b

transmission to the other, Lissajous mirrors attached to the forks enabled the experimenter to judge the phase relations. Koenig claimed that phase was off by a quarter of a vibration, in agreement with a theory of DuBois–Reymond.

Fig. CR no. 167 Courtesy of the Department of the History of Science, Collection of Historical Scientific Instruments, Harvard University, USA. acc. no. 1997-1-0999

Location: FST. Harvard (acc. nos. 1997-1-0999 and 1997-1-1001).
Descrition: The apparatus at FST has a "SOL1" tuning fork (192 vs).
Reference: Koenig (1882c, pp. 167–171).

167a. The same apparatus without the forks. 30 fr

168. Apparatus to show that a fundamental can telephonically excite vibrations in harmonic forks. 50 fr

This is another apparatus designed by Koenig to demonstrate that pure fundamental tones can stimulate harmonics in other sources. Similar to the harmonic pendular apparatus (CR no. 159) and the sympathetic tuning fork demonstration (CR no. 160) it was meant to convince critics that Koenig tuning forks were pure and that unwanted harmonics actually derived from external sources.

Fig. CR no. 168 Source: Koenig (1889, p. 60)

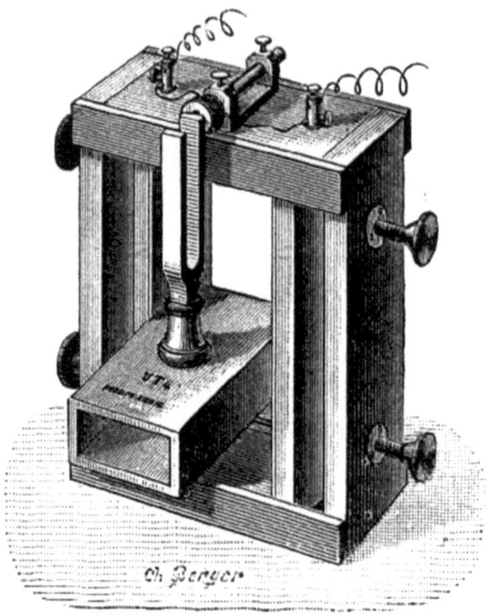

Reference: Koenig (1882c, p. 201).

169. String stretched before the slit of a resonance box of variable volume. 26 fr

The volume of this wooden resonance box can be adjusted with a piston/handle from the back. The experimenter stimulates the string and adjusts the volume until maximum resonance occurs. Likewise, the volume can be kept constant while the string tension is adjusted to achieve maximum reinforcement. Koenig introduced this simple demonstration in his 1859 catalogue.

Location: Amherst College. Teylers.
Description: Oak base.
Reference: Koenig (1859, p. 26) and Turner, G.L'E. (1996, p. 113).

Fig. CR no. 169 Source: Koenig (1889, p. 60)

170. Capsule, rod and membrane to show the transmission of sound. 16 fr

Longitudinal vibrations travel in a rod through a capsule with water and cause sand to form patterns on the surface of a membrane. Koenig introduced this instrument in 1859, saying that it was from the work of Félix Savart.

Reference: Koenig (1859, p. 25).

171. Resonance box ut3, with capsule to show the transmission of sound through liquids. 12 fr

With this apparatus one can demonstrate sound travelling through water, mercury and wood. A capsule filled with mercury is mounted on a resonant box. A glass of water is placed on top of the mercury. The experimenter then places a vibrating fork on top of the water and the resonant box begins to vibrate. This instrument originally derived from Marloye's workshop, and most probably, as with CR no. 170, the work of Félix Savart.

Location: NMAH (cat. no. 315723). Colby College.
Description: The cup appears to be mahogany with a standard pine box with mahogany veneers.
Markings and measurements: (NMAH) "UT3/ RUDOLPH KOENIG À PARIS." Cup, 7 cm diameter, 2.5 cm deep. Box has overall dimensions, 9.8 × 11.5 × 31 cm.
References: Daguin (1867, p. 623), Desains (1857a, p. 117), Fau (1853, pp. 354–355), Guillemin (1881, pp. 556–557), Jamin (1868, p. 623), Marloye (1851, p. 53), Tyndall (1896, pp. 106–108), and Violle (1883, pp. 280–281).

171a. Capsule to be placed on any resonant box. 5 fr

172. Two compound systems of vibrating wooden bars. 18 fr

This is a visual demonstration that seems to go against common sense. There are two sets of wooden bars, one pair is equal and one is off by an interval of a half-tone to a tone. Both pairs produce the same tone when vibrating, but the unequal pair displays a slight variation in nodal patterns.

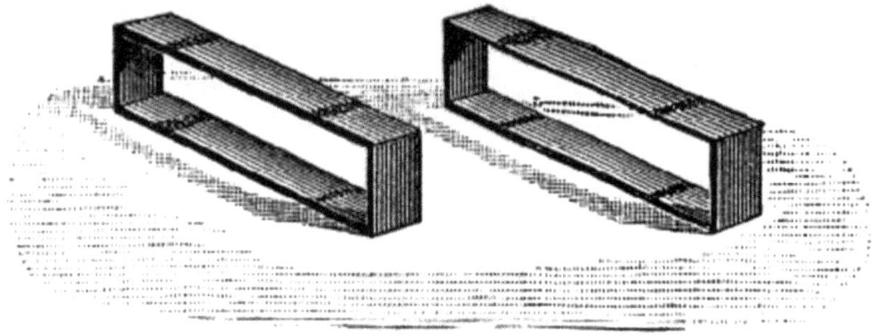

Fig. CR no. 172 Source: Koenig (1889, p. 61)

173. Three vibrating boxes. 45 fr

There are two vibrating actions in this demonstration – the air in the chamber and the vibrating boards. Even if the proportions of the boards are different, one finds that the vibrating air and boards tend to join in unison.

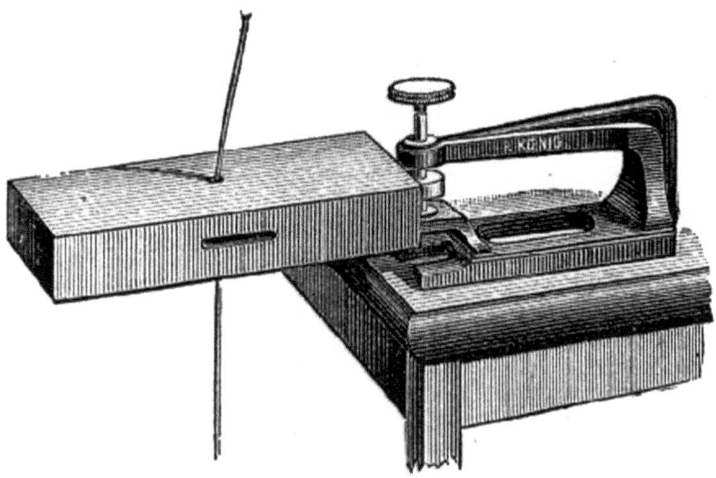

Fig. CR no. 173 Source: Koenig (1889, p. 61)

Location: Harvard (acc. no. 1997-1-0924c).

Markings: (1) "SOL 3" on the top and bottom, "SOL#2" on the side. (2) "LA#3" on top and bottom, "LA#2" on the side; (3) "SOL 3" on top, "FA3" on the bottom, "SOL2#" on the side.

174. Weber's free reed

This instrument demonstrates that pitch does not change in proportion to the changing length of a long output tube. The tongue of the free reed can be replaced and the output tube can be augmented.

Fig. CR no. 174 Photo by author, 2005. Department of Physics, University of Toronto, Canada

Location: Toronto (1878).
Description: Tube missing. Reed missing. Oak pipe.
Markings and measurements: Stamped "RUDOLPH KOENIG À PARIS." (9.7 × 9.7 × 27.4 cm) and the reed is placed on a brass spout. The wood is marked "172" in ink referring to the 1873 catalogue.

174a. The same apparatus simpler. 35 fr

175. Five parallel brass rods joined together. 50 fr

Based on the work of Félix Savart, this is a demonstration of the "law" that vibrations move in the same direction as the original oscillation. If the top branch is bowed so as to produce transversal vibrations, one finds that the other branches vibrate in the same manner. The stem (which is sometimes reinforced with putty) transmits the vibrations longitudinally (parallel to the original oscillation), while the branches vibrate transversally, which is also parallel to the original oscillation. If on the other hand, the stem is activated transversally, one finds that the branches vibrate longitudinally. The vibrations on the branches are made visible with sprinkled sand and their resultant nodal lines.

Location: University of Mississippi at Oxford.
References: Daguin (1867, p. 624), Desains (1857a, pp. 118–119), Fau (1853, p. 403), Marloye (1851, p. 52), and Savart (1820). Idem., 1824. Violle (1883, p. 282).

176. Wooden bar fixed at one end to a support and at the other to a violin string. 16 fr

Reference: Marloye (1851, p. 52).

177. Three parallel wooden bars joined together and mounted as preceding. 20 fr

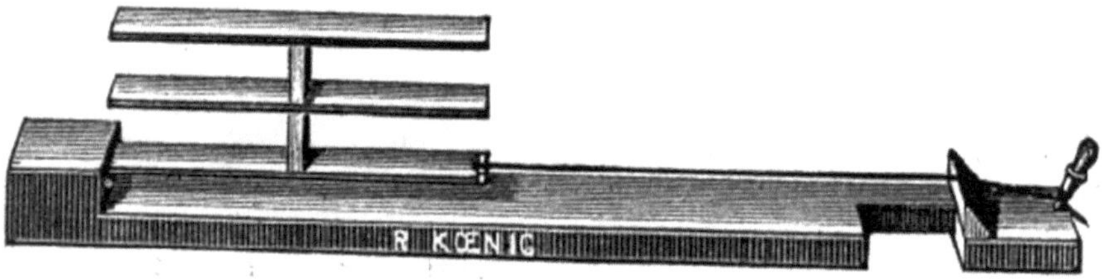

Fig. CR no. 177 Source: Koenig (1889, p. 62)

Location: Teylers (c. 1865).
Description: Oak block and three pine strips and a tension key.
Reference: Turner, G.L'E. (1996, p. 113).

178. Round wooden plate with string passing through its centre, on support. 16 fr

References: Daguin (1867, p. 624), Fau (1853, p. 402), and Marloye (1851, p. 52).

179. Round wooden plate with support, bridge and string. 20 fr

All five instruments above derive from the work of Félix Savart, who established what Albert Marloye termed "a law" that the direction of vibrations of several parts of a system is always parallel to the axis of vibration. The instruments 175 and 176–178 date back to Marloye's business, the others were added by Koenig. In the 1889 catalogue he reminded readers that Savart's law was subject to many exceptions.

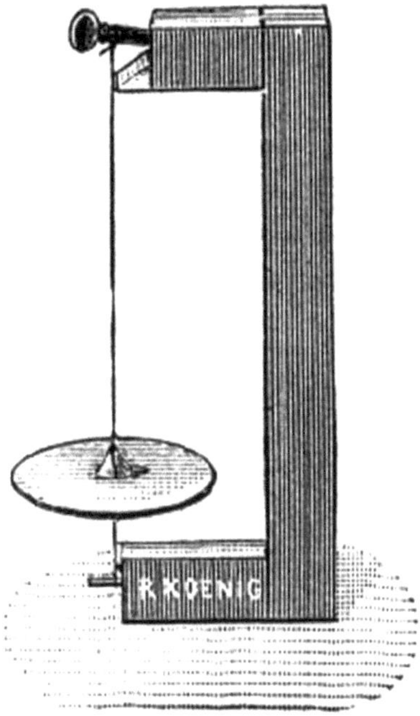

Fig. CR no. 178 Source: Koenig (1889, p. 63)

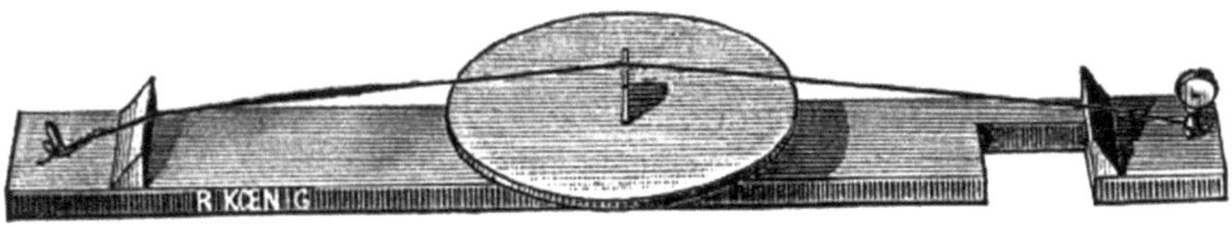

Fig. CR no. 179 Source: Koenig (1889, p. 63)

References: Marloye (1851, pp. 52–53), Koenig (1859, p. 25), and Koenig (1865, p. 29).

180. Experimental violin, trapezium shape. 200 fr

Félix Savart famously designed an experimental, trapezoidal violin in 1819. He wanted to build an improved violin based on current acoustical research, especially using the experimental techniques of Chladni for studying vibration patterns. In his announcement of the instrument, he wrote that "the efforts of scientists and those of artists are going to unite to bring to perfection an art which for so long has been limited to blind routine."[32]

Location: Harvard (acc. no. 1997-1-0949)(signed, parts missing).
References: Guillemin (1881, pp. 804–805). Savart in Hutchins (1997b, p. 18).

181. Two square brass plates, of different sizes, joined together at two angles. 15 fr

182. Two round brass plates of different sizes joined at two points of their circumference. 15 fr

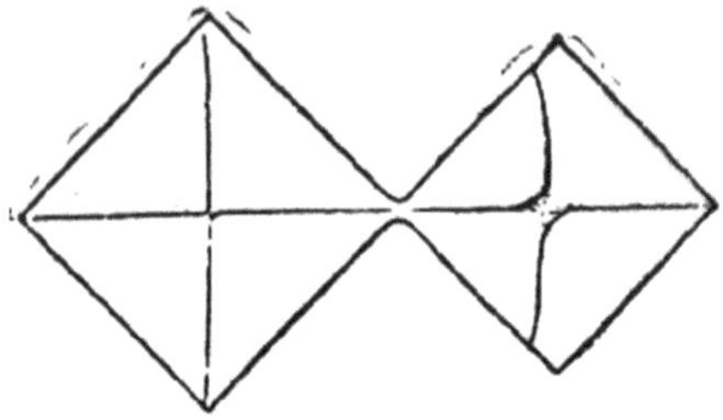

Fig. 89 (N° 181).

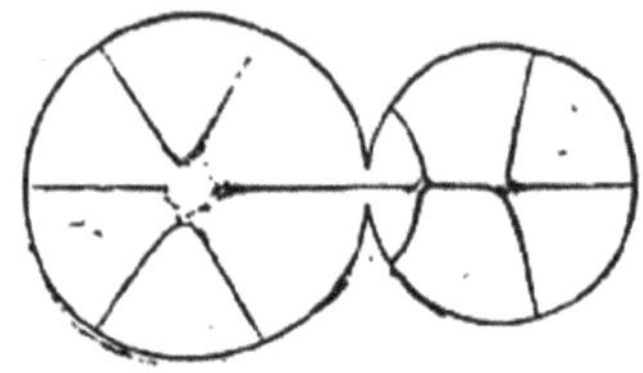

Fig. 90 (N° 182).

Fig. CR nos. 181 and 182 Source: Koenig (1889, p. 64)

183. Two square brass plates of same size, joined together. 15 fr

Location: Union (c. 1875).

184. Two round brass plates of same size, joined together. 15 fr

The above plates (181–184) simply demonstrate the communication of vibrations between plates creating identical vibration patterns.

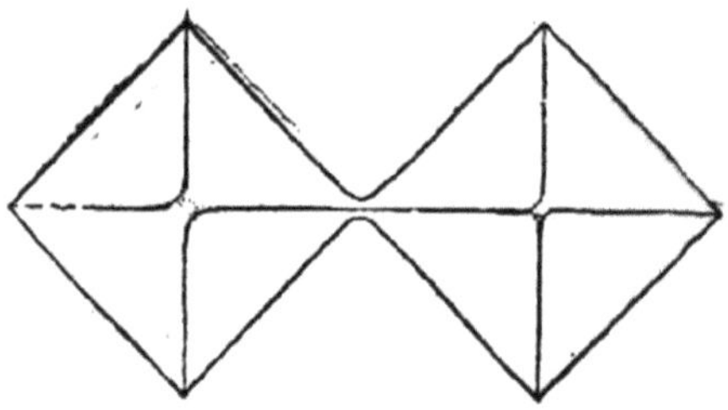

Fig. 91 (N° 183).

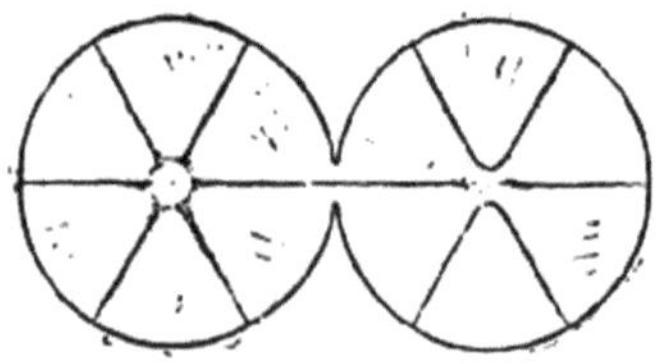

Fig. 92 (N° 184).

Fig. CR nos. 183 and 184 Source: Koenig (1889, p. 64)

185. Four brass rods for Terquem's experiments, with supports. 120 fr

Vibrating rods sometimes emit two sounds, one due to longitudinal vibrations, the other transverse vibrations. In the late 1850s, the French physicist, Alfred Terquem, performed a series of experiments on these effects. In the earliest years of his business, Koenig also studied these effects and produced a small series of Terquem's instruments for sale.

References: Terquem (1859), Koenig (1882c, pp. 32–38), and Zahm (1900, p. 190).

186. Three brass rods for Terquem's experiments on the hoarse sounds. 80 fr

These three rods demonstrate Terquem's simple rule that the "sons rauque" (hoarse sounds) derive from transverse vibrations and are one octave lower than the prime tone generated by the longitudinal vibrations.

References: Terquem (1859) and Koenig (1882c, pp. 138–139).

187. Two brass rods tuned for exciting the hoarse sound by means of the first longitudinal harmonic. 80 fr

Koenig extended Terquem's experiments by demonstrating that the first harmonic from longitudinal vibrations could produce what was called the "son rauque" or a note that is one octave lower.

Reference: Koenig (1882c, pp. 138–139).

188. Nine plates and six bars for Wheatstone's and Koenig's experiments on the formation of nodal lines. 100 fr

Charles Wheastone was one of the first scientists to seriously study Chadni's vibration plates. Based on extensive experiments, he developed a theory to explain some of the complex vibration patterns. In the early 1860s, Koenig extended Wheatstone's experiments on square Chaldni plates by doing a series of experiments on rectangular plates. Just as Wheatstone had used knowledge of the nodal positions to predict vibration patterns, Koenig did the same for rectangular plates, even those with two vibratory movements at once. It confirmed to the "highest degree," he stated, "the truth of Wheatstone's theory."[33] This set consisted of five rectangular brass plates and four wooden plates, three squares and one rectangular.

Fig. CR no. 188 Photo by author, 2005. Department of Physics, University of Toronto, Canada

Location: Toronto.

Description: There are several positions on the plate where the clamp has been applied.

Markings and measurements: One surviving brass plate marked "4:5" "RK". It measures 15.5 × 20.0 × 0.15 cm.

References: Barnes (1898, pp. 58–63), Koenig (1882c, pp. 32–38), Wheatstone (1833), and Zahm (1900, pp. 193–195).

188a. Six plates for the same experiments. 60 fr

Three plates are rectangular and brass, the other three are wood, two of which are rectangular and one square.

VIII. Phenomena Due to the Coexistence of Two or More Sounds in Air

189. Two large electrical forks ut2, one of variable pitch, mounted before resonators. 800 fr

These two forks came out of Koenig's work in the first half of the 1870s on combination tones or, as he called them, beat tones. They were made for experimenting and demonstrating beat phenomena from two powerful sound sources. Both forks

Fig. CR no. 189 Courtesy of the McPherson Collection, McGill University, Canada

are ut2 (128 Hz; C3) mounted on cast iron stands with large brass resonators. The openings and the back of the resonators are adjustable. Between the forks, there is an electromagnetic coil. The current from a battery enters the terminal at the base through copper wire, runs into a terminal at the coil, and then runs through the frame holding the coil. A fine wire brush connected to one of the prongs barely touches the coil. While in operation, vibrations cause the brush to "make or break" the circuit by continually touching the live coil. Small bluish sparks and a small amount of smoke can be seen when it is operating.[34] The current then runs down through the actual fork to the terminal at the base. One of the forks has mercury in it so that the frequency can be changed at will. Sliding brass weights had been the accepted standard for doing this, but Koenig wanted a method by which he could change the mass of the prongs (and thus the pitch) with more ease and precision. The Lissajous mirrors allow one to tune the forks to an exact frequency.

Location: McGill.
References: Koenig (1882c, pp. 84–86) and Zahm (1900, pp. 315–317).

189a. The fork ut2, of variable pitch. 480 fr

Fig. CR no. 189a Photo by author, 2005. Courtesy of the Department of the History of Science, Collection of Historical Scientific Instruments, Harvard University, USA. acc. no. 1998-1-0274

Location: Harvard (acc. no. 1998-1-0274).

189b. The fork ut2, of constant pitch. 320 fr

The large electromagnetic ut2 fork with resonator was sold on its own as a powerful, electrically driven sound source. The one at MIT probably came from the laboratory of Charles Cross, the chair of physics from 1977 to 1917.

Fig. CR no. 189b Physics Department, MIT, USA

Location: MIT.
Description: The electromagnetic coil has many windings with a small iron core.
Markings and measurements: Stamped on black resonator opening "RUDOLPH KOENIG À PARIS"; fork marked, "UT2 256 vs RK". 57 × 39 × 50 cm (depth of resonator).

190. The same apparatus as 189 with forks sol2. 700 fr
190a. The fork sol2, of variable pitch. 420 fr
190b. The fork sol2, of constant pitch. 280 fr
191. The same apparatus as 189, with forks ut3. 640 fr
191a. The fork ut3, of variable pitch. 380 fr
191b. The fork ut3, of constant pitch. 260 fr
192. Forked tube with membrane. 18 fr

This is a simple demonstration of interference. It consists of a tube that splits into two parts that can be suspended over a vibrating plate. There is a membrane at the top of the single tube. If the two branches of the forked pipe are held over two opposite, vibrating segments, sand on the membrane vibrates and forms patterns. Both segments are vibrating in the same mode therefore causing an augmentation of the resulting aerial vibrations. If on the other hand the pipes are placed over two adjacent segments that are vibrating in a contrary fashion (180 degrees out of phase), the membrane will be still. The vibrations cancel each other. Koenig stated that it worked best with high frequencies. August Zahm used this demonstration to explain the interference patterns and effects of a vibrating tuning fork, where there will be areas of quiescence and augmentation surrounding the prongs. He claimed that this experiment came from the Cambridge scientist, William Hopkins, who tutored Clerk

Maxwell and William Thomson. It also resembles aspects of Charles Wheatstone's terpsiphone, an instrument that reinforced columns of air in toroid shaped pipe.

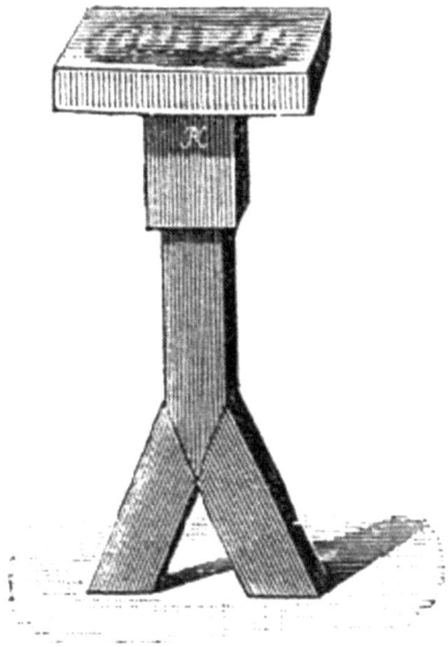

Fig. CR no. 192 Source: Koenig (1889, p. 68)

References: Blaserna (1876, pp. 77–78), Daguin (1867, p. 482), Fau (1853, p. 405), Jamin (1868, p. 588), Reid (1987), and Zahm (1900, pp. 289–290).

193. Three zinc disks, with sectors cut out, for Lissajous interference experiments. These disks are to be mounted on apparatus no. 157. 60 fr

This is an interference demonstration designed by Jules Lissajous. It is combined with the circular, vibrating plate from CR no. 157 and a suspended resonating tube. One of the plates divides into six segments (three spaces and three zinc segments), the other two disks have eight segments (four spaces and four zinc segments). In the classic experiment, the six-sector disk is held over the vibrating plate, which divides into six sectors when activated. Each alternating sector vibrates in the opposite phase – three in phase with each other, and three in another phase. The sound is not strong because the opposing sectors cancel each other's vibrations. The three zinc sectors are held over the vibrating plate and suppress pulses from three of these sectors. In this way, the sound becomes stronger because the zinc sectors remove the interference of opposing vibrations. If the zinc sectors are rotated rapidly over the vibrating plates, one hears rapid risings and fallings of volume.

Location: NMAH (cat. no. 314595).

References: Auerbach in Winkelmann (1909, pp. 600–601), Daguin (1867, p. 482), Desains (1857a, p. 47), Jamin (1868, pp. 588–589), Tyndall (1896, pp. 370–371), Violle (1883, p. 97), and Zahm (1900, p. 291).

194. Three large forks, with sliders, going from sol-1 to ut2, two large metal resonators with pistons moved by screws, and four stands. 5,000 fr

Koenig made these forks for demonstrating "beat tones" that one heard when two prime tones were played simultaneously. Combination tones, as they were more commonly known, had long been known by musicians and scientists, but Koenig resurrected the old beat-tone theory to explain the effects. These large forks were made to convince audiences of his position, and, in fact, the only known surviving set (from the South Kensington Museum, in the Science Museum) were used by Sylvanus P. Thompson in 1890 at his series of lectures for the Physical Society of London. Koenig acted as the demonstrator at these events and excited the forks with a cello bow.

In order to demonstrate his findings in a wide range of notes, these forks cover the low notes sol-1 (48 Hz) to ut2 (128 Hz). They were thick to prevent unwanted harmonics, divided precisely with sliding brass weights and reinforced with massive adjustable resonators. He developed simple mathematical rules for predicting the appearance of what he called "inferior" and "superior" beat tones. If for example one played the notes 40 and 74 vibrations (Hz) one would hear two "beat tones" – an inferior beat tone of 34 vibrations (which resulted from subtracting the lesser note 40 from the higher note 72); one would also hear a rough series of beats at 6 vibrations a second, which resulted from subtracting the higher note 74 from 80, or 2 times the lower note of 40.

Fig. CR no. 194 Photo by author, 2003. Science Museum, UK. acc. no. 1890-53

Location: Science Museum (acc. no. 1890-53).

Description: Each steel fork has a threaded stem for screwing into the cast iron stands. The resonators are a painted black metal with brass trim and an adjustable piston with a handle for changing the volume.

Markings and measurements: Fork 1: "SOL-1 – UT1 RK" Full height to end of stem = 93 cm; 11.3 cm width; 4.0 cm depth. Fork 2: "UT1 – SOL1 RK" 79 × 10.6 × 4.0. Fork 3: "SOL1 – UT2 RK" 67.3 × 10.0 × 4.0. The brass sliding weights have two screws for clamping to the prong at the gradation line marked on the steel prong. They are each marked with one of three note ranges from above, and also marked "DS & S.K. MUS." The resonators are both 100 cm long, 37 cm diameter. Two cast iron stands hold two forks, while two support the resonators. There are also three forks without the weights are marked "UT1," "SOL1," and "UT2."

References: Thompson (1891, pp. 201–202) and Zahm (1900, frontispiece and pp. 301–339).

194a. The two largest stands of preceding. 270 fr

194b. The two smallest stands of preceding. 230 fr

195. Large iron pin with female screw and handle. 40 fr

196. Same apparatus as no. 194 in simpler form. 2,500 fr

This apparatus was electrically driven allowing for prolonged demonstration of the beat tones without diminution of intensity.

197. Five large forks with sliders from ut2 to ut3, and four brass resonators with pistons moved with screws. 4,000 fr

Without the sliding weights these forks give the notes ut2, mi2, sol2, 7th harmonic of ut-1, and ut3. They are marked in double vibrations, "VD" (Hz).

Location: CSTM (acc. no. 1998.0246).

Description: The CSTM has four large resonators for low forks. Both have a rectangular metal door (black) for adjusting the size of the opening. The length and volume can be adjusted by pulling on the cylinder at the back.

Markings and measurements: (CSTM). Stamped "RUDOLPH KOENIG À PARIS." Cast iron stand is 30.5 cm h. Brass drum is 50 cm l, 17.5 cm h.

198. Nine large forks with sliders from ut3 to ut4 and six brass resonators with pistons moved by screws. 3,000 fr

These forks demonstrated Koenig's beat theory comparing the intervals from ut3 to ut4. Without the sliding weights these forks produce the notes ut3, re3, mi3, 11th harmonic of ut-1, sol3, 13th harmonic of ut-1, 14th harmonic of ut-1, si3, ut4.

Reference: Koenig (1882c, pp. 87–148).

199. Collection of 32 tracings of primary and secondary beats on glass, for projection. 100 fr

199a. Collection of 16 tracings of primary beats. 50 fr

199b. Collection of 16 tracings of secondary beats. 50 fr

200. Stopped pipe giving ut1 of feeble intensity. 30 fr

This pipe combined with the forks from no. 38 for demonstrating that the fundamental tone does not have to be strong when played with upper harmonics to produce sensible beats.

201. Twelve strong forks, ut5, ut6, ré6, mi6, fa6, 11th harmonic of ut3, la6, 14th harmonic of ut3, si6, ut7, with support. 625 fr

These are special forks with fat prongs designed for purity of sound and reducing unwanted harmonics. Two are placed on the cast iron base to produce a series of beats and beat tones in the upper octaves. Because of the high notes, they also produced what Koenig called secondary beats and beat tones, or beats that derived from combinations of the primary beats.

Fig. CR nos. 201 and 206 Photo by author, 2005. Physics Department, University of Toronto, Canada

Location: Amherst. CSTM (acc. no. 1998.0248; sol6 and ut6). Science Museum (acc. no. 1890-20). Sydney. Toronto. Yale (acc. no. YPM 50280).
Markings and measurements: (Toronto) "14/ 7168 vs RK" 7.0 cm; "LA6 6826,6 vs RK" 7.8; "13 6656 vs RK" 8.0; "SOL6 RK" 8.3; "11 5632 vs RK" 9.2; "FA6 5461,5 vs RK" 9.2; "MI6 RK" 10.0; "RÉ6" 10.6; "UT6" 11.5; "UT5" 17.0.
Reference: Zahm (1900, pp. 325–327).

201a. Eight strong forks for ut5, ut6, re6, 11th harmonic of ut3, sol6 13th harmonic of ut3, si6, ut7. 365 fr

Fig. CR no. 201 Photo by author, 2005. Physics Department, University of Toronto, Canada

Location: Harvard (acc. no. 1998-1-0141).

201b. The support of no. 201 only. 50 fr

Location: Harvard (acc. no. 1997-1-1074).

202. Apparatus for the continuous sound of beats, with 12 tuned glass tubes. 400 fr

Koenig's beat tones were difficult to hear mainly because tuning forks had a short duration. This made it a challenge to demonstrate beat tones before large audiences. In 1881 he addressed this concern with an invention for demonstrating "strong and persistent" combination tones and interference phenomena. This instrument consisted of two tuned glass tubes, a tall iron frame and a wheel covered with felt that made contact with the glass tubes. The friction of a clothed wheel rubbed against the tubes producing pure simple tones through longitudinal vibrations. Two powerful tones played simultaneously giving strong beat tones. The apparatus came with twelve glass tubes that gave different notes. As with other teaching instruments of Koenig, this instrument served as a source of information on the mechanics underlying the combination tones. Pictures of this instrument were found in several textbooks of the time, attesting to the clear way it illustrated combination phenomena.

Location: Coimbra (FIS.0969).
Description: Twelve glass tubes of differing lengths are connected to two wooden side arms that swing out and can be fastened by leather straps to the rotating wheel. The wheel has a felt-like material around the circumference which is continually dampened in a trough of water. The apparatus rests on a heavy cast iron

Fig. CR no. 202-1 Source: Koenig (1889, p. 71)

Fig. 97 (h. = 1m,05) (N° 202).

base. A smaller version of this instrument made by Lancelot can be found at the Conservatoire national des arts et métiers in Paris.

Markings and measurements: (Coimbra) Overall height, 102 cm. The 12 glass tubes vary from 45 to 104 cm. The thickness of the tubes vary from 1.7 to 2.5 cm. Each tube has a paper label with handwritten designations, e.g. "10 MI6 RKg." The notes range from ut6 to above mi7. The cast iron base has a large white plaque that reads, "RUDOLPH KOENIG À PARIS" (presumably made for exhibition).

References: Auerbach in Winklemann (1909, pp. 624–628), Koenig (1882c, pp. 163–166), and Zahm (1900, pp. 328–330).

203. Glass tubes tuned for notes between ut6 and sol7. 8 fr

204. Two locomotive whistles, one of variable pitch

The locomotive whistle produced intense sounds of high pitch. In 1881, Koenig took this simple whistle which he had offered since the 1860s, and transformed it into a research instrument to investigate the beats of higher pitches. After becoming frustrated that tuning forks did not produce a strong, continuous sounds for both research and audiences he wanted to demonstrate his beat theory in higher pitches with intense, pure and continuous sounds. He designed a model with adjustable

Fig. CR no. 202-2 Photo by author 2005, Museu de Física, University of Coimbra, Portugal. FIS.0969

mechanisms (a sliding piston and a covering tube near the wind slit) for varying the pitch and ensuring the purity of tone. But the variability of pitch could not be controlled as desired for quantitative purposes (even small changes made a big difference in beat experiments), so he developed the large apparatus with glass tubes for producing more stable longitudinal vibrations (CR no. 202).

Location: Nebraska. QUP.
Description: The University of Nebraska has two brass whistles by Koenig, one of fixed pitch (see no. 10), the other a Galton whistle.
References: Koenig (1882c, pp. 163–166) and Mollan (1990, p. 203).

205. Large wooden wheel of 128 teeth, mounted. 100 fr

In 1875 Koenig proposed that beats could be blended into a tone. Others such as Helmholtz argued that beats by nature could not be made into a tone. They were simply the by-product of the overlapping of two waves. Koenig's argument against Helmholtz's combination tones depended "beat tones" and he created a series of

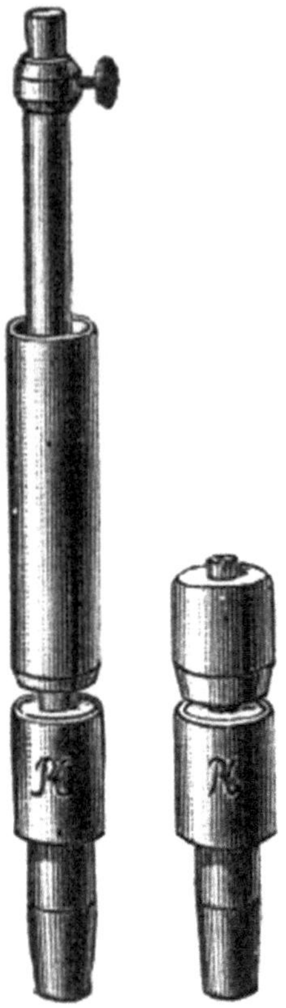

Fig. CR no. 204 Source: Koenig (1889, p. 72)

experiments for demonstrating the nature of beats. He argued, for example, that for sounds between 32 and 128 Hz one could hear beats and primary tones simultaneously. The ear, he said, could blend them into a tone and, at the same time, distinguish them as discreet pulses. Depending on the source of production, one effect could be stronger than the other; at one time only the tone would be heard, while at other times only the rattle of beats could be perceived.

He designed a wooden Savart-type wheel – 35 cm diameter, 35 mm thick, with 128 teeth – for testing this idea. If he pressed a piece of wood against it and rotated once per second he heard both a quickening succession of taps which he deemed to be 128 per second, and also a note ut2 or 128 Hz (C3). When he used a soft piece of cardboard instead, the rattle disappeared.

References: Helmholtz (1863, pp. 235–262), Helmholtz (1954, pp. 158–173, 533), and Koenig (1882c, pp. 135–136). Rayleigh in Bosanquet (1881–1882, p. 28) and Zahm (1900, pp. 330–331).

206. Eight large forks for the notes between si6 and ut7. 340 fr

Related to the debate on the nature of beats and tones (CR no. 205), Koenig investigated the lowest sound that could be produced from beats. He did this in 1875 by combining forks between si6 and ut7 with ut7 which gave the beats 256, 128, 64, 48, 40, 32 and 26. At 32 beats one could still hear a continuous tone and, by pulling away, the rattle of 32 beats. As one approached 26 beats the rolling pulses were only heard as beats. This suggested to Koenig that he had passed the lowest threshold at which beats become beat tones. In all cases one could hear the simultaneous appearance of both beats and beat tones, showing like CR no. 205, that beats and beat-tones were related to each other.

Fig. CR no. 206 Photo by author, 2005. Physics Department, MIT, USA

Location: MIT. Toronto.

Markings and measurements: MIT has three forks. (A) "8064 vs RK" "63:64/8064/8192/64 B/Ut1" (62 × 34 × 17 mm). (B) "SI6 7680 vs RK" "15:16/8192/7680/256 vs" (65 × 33 × 15 mm). (C) "UT7-8192 vs" (62 × 31 × 15 mm). (Toronto) Marked "247" on the oak box referring to an unknown catalogue. "UT7 8192 vs RK" 6.7 cm long; "8140 vs/RK/8192/8140/26" 6.3; "8128 vs/RK/127:128/8192/8128/32/UT-1" 6.3; "8112 vs/RK/507:512/8192/8112/32 UT-1" 6.3; "8096 vs/RK/253:256/8192/8096/48 SOL-1" 6.3; "8064 vs/RK/65:64/8192/8064/64 UT1" 6.4; "7936 vs/RK/31:32/8192/7936/128 UT2" 6.5; "SI6 7680 vs RK" 6.8.

Description and function: "B" and "C" differ by 512 v.s. or 256 Hz, which produces a beat tone of 256 Hz. When both forks are hit very hard with the wooden mallet the resultant beat-tone is almost as strong as that from a tuning fork. It lasts a few seconds. It is slightly lower in tone than a standard 256 Hz fork in the collection. The 64 Hz tone produced by combining "A" and "C" is strong, but very short-lived. There is a slight ruffle or flutter to both notes, which could lead one to question their nature – beat tone or true note?[35]

The 18 forks at the University of Toronto contain this set of eight forks plus the first ten forks from no. 201.

References: Koenig (1882c, p. 134), Pantalony (2005a), and Zahm (1900, pp. 331–332).

206a. The same apparatus without the forks si6 and ut7. 255 fr

206b. Five large forks for the same experiments as no. 206. 215 fr

206c. The same series without ut7. 170 fr

207. Large disk for producing a sound by the interruptions of another sound. 40 fr

Koenig studied periodic bursts of sound that themselves could blend into sound. In 1875 he took a rotating wheel with holes (siren device) and put a high pitched tuning fork beside the rotating holes. The sound travelled through the holes and he heard both the pitch of the fork and a lower note associated with the frequency of periodic bursts of the pierced disk. If the tuning fork was ut7 and the disk had 16 apertures, moving at 8 revolutions per second, one would hear ut2 (128 Hz; C3) and ut7. He tried this with other forks and got the same result. These experiments were part of Koenig's demonstrations in favour of his beat-tone theory. Periodic bursts of sound, like beats, he argued, could form their own tone. By widening the explanation of how sound was produced, Koenig hoped to persuade others that his beat tones were legitimate sound phenomena in their own right.

References: Koenig (1882c, pp. 138–140) and Zahm (1900, pp. 332–334).

208. Accessories for observing sounds of variation. 50 fr

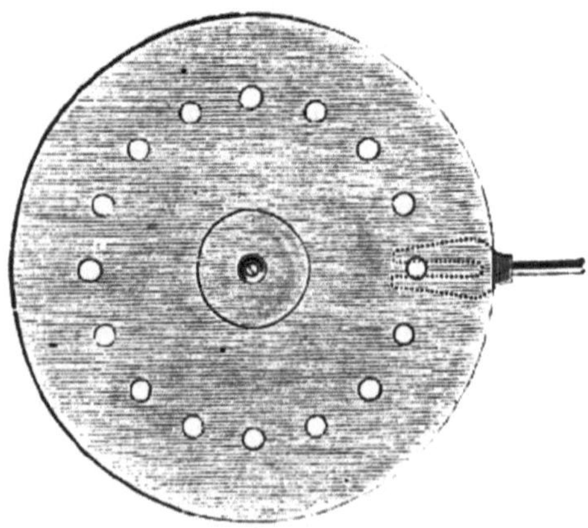

Fig. CR no. 207 Source: Koenig (1889, p. 72)

Reference: Koenig (1882c, p. 140).

209. Large siren disk for producing a sound by periodical variations of intensity of another sound. 250 fr

In his 1875 study of combination tones, Koenig started to adopt a more visual interpretation of beats and beat tones. This siren disk consisted of seven rings of 192 pierced holes. Within each ring there were periodic increases in the size of the holes. In one there were 12 maximums, in another 16, followed by 24, 32, 48, 64, and 96. When the siren rotated with a jet of air blowing at the holes, one could hear the note corresponding to the 192 holes and also the periodic maxima. Thus the beat phenomena could be seen and heard.

References: Koenig (1882c, p. 141) and Zahm (1900, pp. 334–335).

209a. The same apparatus of smaller size. 50 fr

210. Large wave siren for the sounds of beats. 1,000 fr

In 1881 Koenig developed a wave siren for demonstrating his controversial beat tones, or the third tones that were heard when two primary tones sounded together. It was an attempt to produce sounds directly from pictorial wave forms in brass. Instead of using tuning forks or traditional sirens with holes (Helmholtz double siren no. 27) he believed that the metal representations of waves would produce a purer sound. Each wave was a combination of two primary sinusoidal waveforms, (which were made from graphical inscriptions and photographs), and then cut from a brass sheet. There were eight waves rotating on an axle with a wind-slit forcing air against the curves under study. The beat tones derived from the intervals 8:9, 8:10, 8:11, 8:12, 8:13, 8:14, 8:15, and 8:16. For example, the interval 8:9 (major second) produced two primary sounds and a beat tone of 1, corresponding to what Koenig called the inferior beats. (The other set of beats, the superior beats, were too faint to hear). To make frequencies easier to hear, one could rotate the siren at such a speed to create 512 and 576 Hz (major second) and thus producing a beat tone of 64 Hz. Another traditional siren disk (with pierced holes) rotated on the top of the apparatus producing simple sounds to verify the notes heard with the wave siren. The pressure of air against the curves was supposed to be "at least 10 or 12 cent. of water."

This was a rather large instrument in a cast iron stand, with rotating axle, standing at 75 cm. It was much taller than the wave siren for timbre (no. 60) which stood at 40 cm in height.

References: Auerbach in Winkelmann (1909, pp. 266–268), Koenig (1882c, pp. 149–162), and Zahm (1900, pp. 337–338).

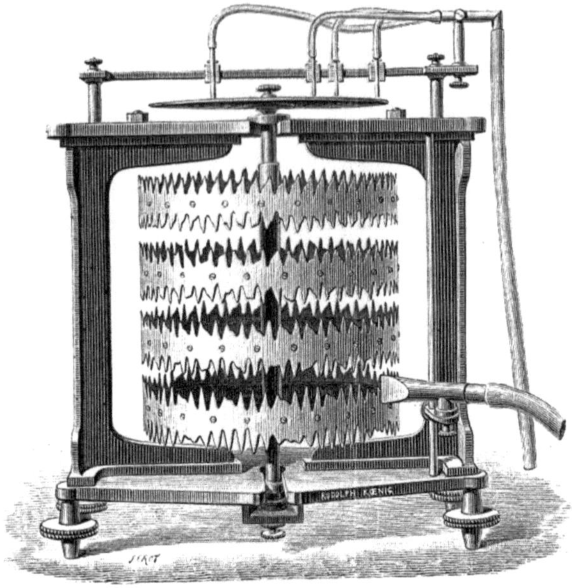

Fig. CR no. 210 Source: Koenig (1889, p. 75)

210a. Supplementary axis for preceeding, with four wheels and eight curves for the intervals of the second period from 8:17 to 8:24. 640 fr

211. The same apparatus as 210 with curves for the intervals 8:9, 8:11, 8:12, 8:13, 8:15, 8, 8:18, 8:23, 8:24. 1,000 fr

212. Collection of 16 wave-siren disks with air tube for the sounds of beats. 1,280 fr

With this instrument, Koenig attempted to produce complex sounds from brass wave patterns. The disks are the individual representations of two combined sounds that, similar to the above apparatus, produced beats and beat-tones. They rotated on a Savart wheel combined with an air jet (like the basic wave siren disk, CR no. 62). Koenig cut the edge of the disk in the exact shape of a waveform that had been produced by two combined, pure tones. In one of the first examples, he used a waveform that combined 120 simple sinusoidal waves with 64, which together formed a slightly mistuned major seventh (ratio 8:15). The combined waveform ran the circumference of the disk. When it was sounded, one heard the two prime tones and a resultant beat-tone. For comparisons of these components, he created two concentric rings of 120 and 64 holes. If these holes sounded at one revolution a second, 8 beats would result (the superior beat frequency being 128 minus 120). He added a series of eight holes in the interior for comparison with the beats and beat-tones produced by the wave component. When the wave disk increased in speed, the beats from the wave component blended into the beat tone. Then the series of eight holes were played for comparison, and it resulted in the same tone. He therefore claimed to recreate a beat tone from an artificial metal waveform.

This set covers a range of intervals from 8:9 to 8:21. The pressure of air "must not be less than 10 or 12 cent. of water."

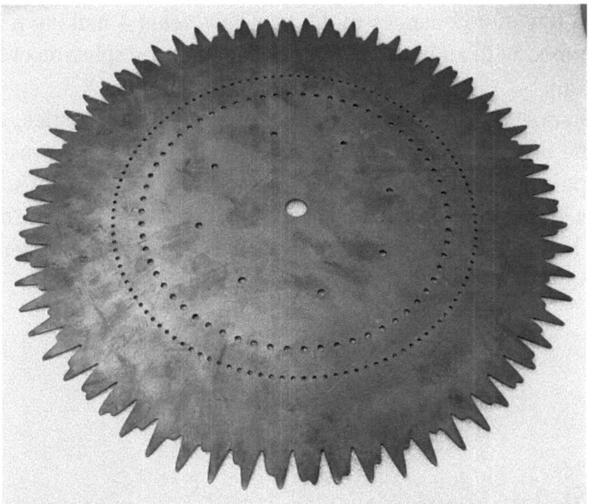

Fig. CR no. 212 Photo by author, 2005. Courtesy of the Department of the History of Science, Collection of Historical Scientific Instruments, Harvard University, USA. acc. no. 1997-1-1010

Location: Harvard (acc. no. 1997-1-1010; two disks).
References: Koenig (1882c, pp. 157–162) and Zahm (1900, pp. 335–336).

212a. Collection of 10 disks for the intervals 8:9, 8:11, 8:12, 8:13, 8:15, 8:16, 8:18, 8:20, 8:23, 8:24, with air tube. 800 fr

212b. Collection of 5 disks for the intervals 8:9, 8:12, 8:13, 8:15, 8:23, with air tube. 400 fr

212c. One disk for any interval from 8:9 to 8:24. 80 fr

IX. Methods of Studying Sonorous Vibrations Without the Assistance of the Ear

213. The Phonautograph

The phonautograph was the first mechanical instrument to record sounds from the air. It consisted of a collecting chamber to receive the sound, a writing stylus that was connected to a sensitive membrane, and a rotating drum with paper that recorded the movements of the vibrating stylus. Édouard-Léon Scott patented the first version of this instrument in 1857. It had a simple collecting chamber and a stylus that rested on a moving piece of paper that was connected to a steadily falling weight. In 1859 he approached Koenig for help with the design. They signed a contract and the young instrument maker changed the shape of the collecting chamber, added a rotating writing drum, and improved the efficiency of the membrane. Koenig made a few more changes in the next few years – making a zinc parabolic collecting chamber, with an improved membrane and a graphic recording device on the rotating drum.

Koenig, who quickly cornered the market on graphical acoustics, went on to sell and promote the instrument as a centerpiece of his business. This instrument, along with other graphical instruments, radically transformed the study of sound making it more reliant on vision. In an essay attached to his 1859 catalogue, the author claimed that acoustics before the phonautograph was like "astronomy before the invention of the telescope."[36]

Fig. CR no. 213-1 Photo by Gilberto Pereira. Museu de Física, University of Coimbra, Portugal

Locations: Coimbra (FIS.0403 and FIS.0909; date, 1867). NMAH (acc. no. 215,518). MCQ (acc. no. 1993.13267). Teylers (1865).

Description: (Coimbra) Cast-iron base with a series decorative curves. Wooden bar under front of drum rests in leather padding for adjusting height. Sound collector appears to be a tin alloy, painted brass or copper colour. Membrane is a thin sheet of parchment attached tightly to frame with thin string.

Markings and measurements: (Coimbra). Collecting chamber and stand, 53 × 52 × 54 cm. Rotating cylinder and stand, 36 × 96 × 22 cm.

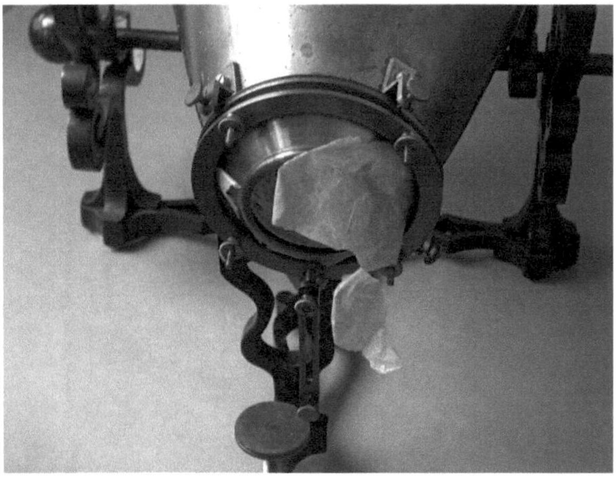

Fig. CR no. 213-2 Photo by Gilberto Pereira. Museu de Física, University of Coimbra, Portugal

References: Auerbach in Winkelmann (1909, pp. 155–156), Blaserna (1876, pp. 156–158), Daguin (1867, pp. 495–496), Donders (1864), Ganot (1893, pp. 269–270), Guillemin (1881, pp. 655–656), Helmholtz (1863, pp. 34, 248), Jamin (1868, pp. 508–509), Koenig (1859, appendix), Miller (1916, pp. 71–73), Pisko (1865, pp. 71–82), Scott de Martinville (1878), Turner, G.L'E. (1996, pp. 135–136), Violle (1883, pp. 22–23), and Zahm (1900, pp. 70–73).

213a. The cylinder of the preceding instrument on a support. 200 fr

Location: Harvard (acc. no. 1996-1-0351).

214. Clock with interrupting pendulum and electric signal. 300 fr

Reference: Auerbach in Winkelmann (1909, pp. 154–155).

214a. Electric signal. 80 fr

Location: Harvard. Old version at MCQ.
Reference: Auerbach in Winkelmann (1909, pp. 154–155).

215. Iron support for fixing vibrating bodies before the cylinder. 50 fr

The iron support would be placed in front of a rotating cylinder and used as a timing device. A vibrating tuning fork of known frequency acted as a "chronograph," while the electric signal marked the start and finish of a measurable event. Earlier in his career, Koenig had employed a small escapement chronometer that marked the roller every six seconds, but he found that the act of marking retarded the movement of the roller thus throwing off the measurements.

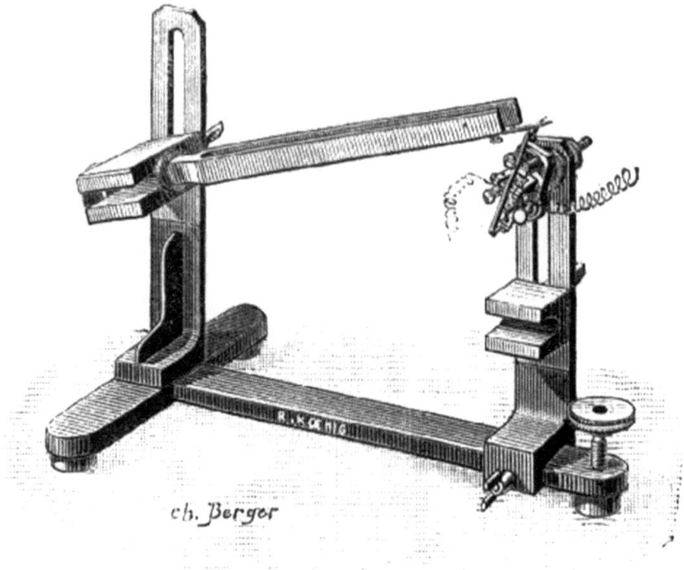

Fig. CR no. 214a Source: Koenig (1889, p. 78)

Reference: Koenig (1882c, pp. 2–6).

216. Regnault's chronograph with tracing forks of 100, 200, and 120 s.v. 1,000 fr

In 1866 Koenig collaborated with the celebrated experimentalist Victor Reganult to measure the speed of sound. They did their experiments underground in the sewers of Paris during the Haussman renovations when long pipes were available as sound carriers. Regnault invented the electrical chronograph in order to measure very small intervals of time, such as the short period of time that sound travelled in the pipes. The "Regnault chronograph" (as it came to be known after Koenig began making and selling it) consisted of an electromagnetic tuning fork held upright in a heavy rigid frame. A small brass stylus attached to the end of one prong made contact with a roll of smoked paper drawn continuously by a handle at the rear. With a tuning fork of known vibration one could easily calculate a time interval by making electrical marks on the paper and counting the vibrations between marks. In order to do this, two styluses rested on either side of the tuning fork writer; both styluses were hooked up to an electric circuit and ran continuously unless their circuit was broken, at which instant a small mark was recorded on the paper. In Regnault's experiments, he attached one of these styluses to a seconds-pendulum in order to calibrate the potential errors of the tuning fork. The other stylus recorded the events under study. A break in the circuit registered the original report (a trumpet blast) and after travelling through a series of reflections in the pipes (to make distances of up to 20 km), the sound wave activated a membrane that broke the circuit. The chronograph, being connected to the circuit, recorded all of these events on the blackened roller paper.

The reels and bobbin rollers behind the frame are designed for the smooth functioning of the inscription process. There is a rotating handle, with three different-sized grooves (for different speeds) that attach through a leather strap to an electrical rotation machine. On the reel and bobbins, there are fine adjustments and pressure screws to regulate speed. The central bobbin (where the inscriptions take place) is wood. The whole frame pivots also, allowing changes in speed due to pressure put on a weighted brass pivot (with two toothed wheels that grip the paper) that rest on the last rubber bobbin.

Making the black recording paper was an art in itself. Koenig wrote detailed instructions to James Loudon on how to operate the apparatus. A smoking coal oil lamp covers the white paper as it winds itself into the reel to be used for the chronograph. The feeding reel and the receiving reel have to be set in the right position so as to ensure that the paper rolls in a smooth and regular fashion. The paper slides under a brass cylinder which is just above the smoking lamp. To prevent the paper from burning, the cylinder is filled with three quarts of water for cooling. The water, however, should be kept at 40 to 50°C, so that water droplets do not form on the cylinder and mark the paper. The operation should take place in a room with no air currents. "If all the arrangements are properly taken," Koenig wrote to Loudon, "one can easily reel in 50 or 60 m of paper, very uniformly blackened, in less than a half hour." After the paper is marked, it can be fixed with a mixture of 1 g of "gommelaque" (shellac) in a litre of alcohol.[37]

Fig. CR no. 216 Courtesy of the McPherson Collection, Physics Department, McGill University, Canada

Locations: Case. CNAM (inv. 12593). (both Case and CNAM have the blackening apparatus as well). McGill. NMAH (cat. no. 314597; date, 1877).

References: Loudon and McLennan (1895, pp. 117–118). Koenig, "Chronographe d'áprés Regnault catal. No. 205a," in letter of Dec. 1878, UTA-JLP. Koenig (1882c, pp. 11–12), Regnault (1868), and Violle (1883, p. 64).

216a. The same apparatus with fork of 200 s.v. 900 fr

217. Chronographic electric fork of 100 s.v. 110 fr

218. Similar fork of 200 s.v. 110 fr

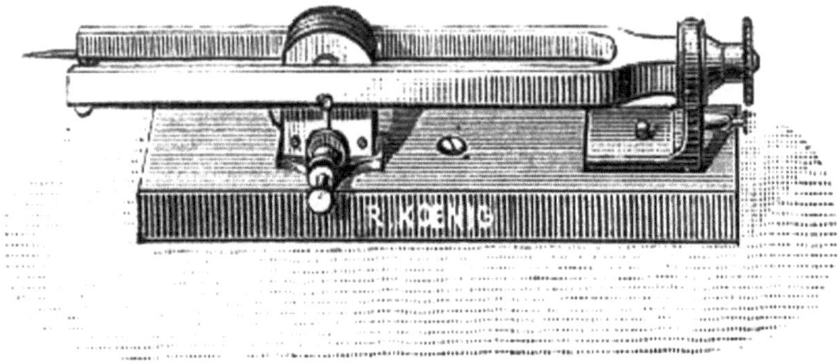

Fig. CR no. 218 Source: Koenig (1889, p. 79)

219. Similar fork of 500 s.v. 100 fr

220. Similar fork of 1000 s.v. 100 fr

Location: CNAM (inv. 12592).

221. Similar fork of 2,000 s.v. 110 fr

222. Similar fork of 128 s.v. 110 fr

223. Similar fork of 256 s.v. 100 fr

224. Similar fork of 512 s.v. 100 fr

The above forks (CR nos. 217-224) can work through auto-interruption (the vibrations make and break the circuit thus continuously pulling and releasing the fork). They can also operate via an interrupting current from another interrupter fork in unison (similar to the one found in CR no. 56). The latter method could be used to avoid the loud noise of auto-interruption.

225. Chronographic fork without electrical mounting of 100 s.v. 50 fr

226. Similar fork of 200 s.v. 40 fr

227. Similar fork of 500 s.v. 35 fr

228. Similar fork of 128 s.v. 50 fr

229. Similar fork of 256 s.v. 40 fr

230. Similar fork of 512 s.v. 35 fr

Koenig noted that these forks (CR nos. 225-230) could be activated with the stroke of a violin bow. He could also make forks of other frequencies.

231. Large fork of 20 s.v. with transmitting capsule. 180 fr

This fork was 1.25 m in length.

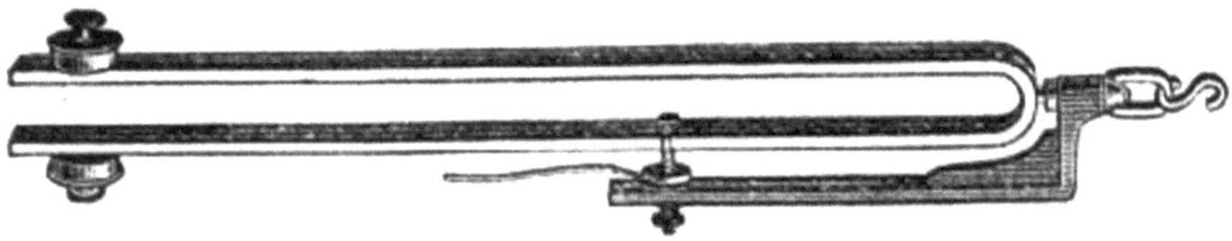

Fig. CR no. 231 Source: Koenig (1889, p. 80)

232. Marey's membrane capsule [Tambour] with tracer. 45 fr

This was an inscription device that connects to CR no. 231. It recorded subtle physiological movements with a membrane and writing stylus. The inventor, Etienne Jules Marey, was a pioneer of graphical recording in mid nineteenth century Paris. He based his work on the pneumatic devices developed by Charles Buisson.

Reference: Marey 1878.

232a. Attachment for fixing Marey's capsule upon the support no. 215. 10 fr

233. Apparatus for graphically compounding two vibratory movements at any inclination. 1,100 fr

Shortly after the introduction of the phonautograph, Koenig applied graphical techniques to numerous acoustical phenomena thus producing beautiful visuals of harmony on paper. He developed a set of instruments to display graphically the Lissajous patterns produced when two vibrating bodies were combined through one apparatus. In his book, he described Lissajous and Desains' first attempts to do this in 1860, and his subsequent improvement of this method 2 years later. Aside from the beautiful tracings, these instruments were a marvel themselves. They consisted of a large and very heavy cast iron frame (1 m in length) with two adjustable steel mounts for the tuning forks. One fork held a blackened glass plate on its prong with a counter balance on the other prong; the other fork had a small writer on the end of the prong that moved slowly and smoothly backwards as it rested on the vibrating glass plate of the adjacent fork. The combined movements created distinctive graphical curves on the glass plate. For more elaborate geometric patterns, the writing fork was placed at different angles to the fork in relation to the glass plate. The apparatus came with two electrical mountings for maintaining the vibrations of the forks.

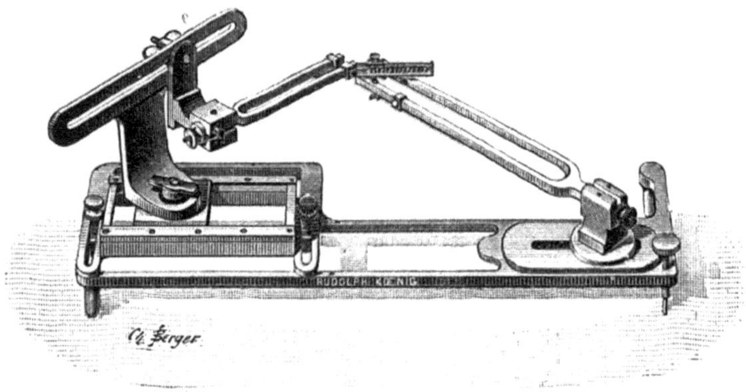

Fig. CR no. 233-1 Source: Koenig (1889, p. 80)

Location: CSTM (acc. no. 1998.0263; ut1 fork). Nebraska. NMAH (cat. no. 314592; date, 1877). Sydney (forks only). Teylers (1875). Toronto (1878) (forks only). Vanderbilt (1875) (forks only).

Description: Two large tuning forks, ut-1 and ut1, hold the glass plate. Another eight forks, made of highly quality, highly polished steel, ut1 to ut3, carry the writing stylus. The Toronto forks have black needles (strips of soft lead) on the end of the prongs. The largest fork has ivory washers at the stem. There are also brass sliding weights for adjusting the frequency.

The apparatus at the Teylers museum comes with iron frame with wooden base plates. A mahogany box contains 12 smoked glass plates with traces, signed "RK." It also comes with a photograph of the apparatus with an instruction manual. Seven tuning forks survive

.

Markings and measurements: (Toronto) Marked "208a" on the oak box referring to the 1873 catalogue. "1:4 UT3 512 vs RK" 16.1 cm long; "2:7 448 vs RK" 17.2; "1:3 SOL2 384 vs RK" 18.4; "2:5 MI2 320 vs RK" 19.9; "LA1 – UT2 RK" 21.9;

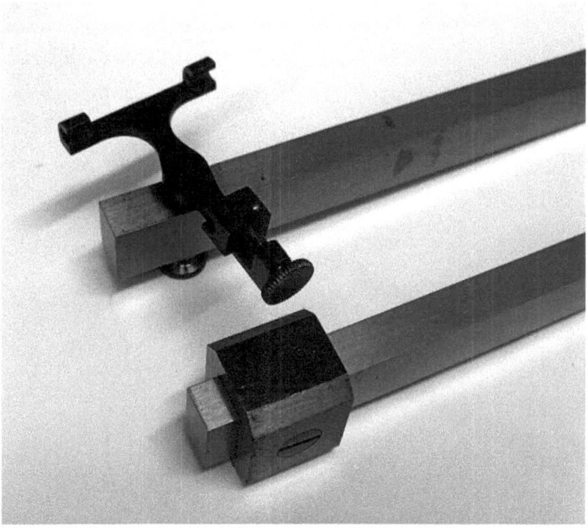

Fig. CR no. 233-2 Photo by author, 2005. Physics Department, University of Toronto, Canada

"– SOL1 – LA1 RK" 23.4; " – M11 – FA1 – RK" 25.5; "UT1 – RÉ1 – RK" 27.0; "UT1 – MI1 – RK" 32.0; "1:2 UT-1 64 vs RK" 49.0.

References: Desains (1857b), Guillemin (1881, pp. 656–657), Koenig (1882c, pp. 12–18), Koenig (1865, pp. 40–41), Loudon and McLennan (1895, pp. 108–109), and Pisko (1865, pp. 64–65). Rudolph Koenig to Joseph Henry in SIA, *Record Unit 26*, vol. 166, 269–275. Turner (1977), Turner, G.L'E. (1996, p. 130), and Zahm (1900, pp. 420–422).

233a. The same apparatus less complete. 750 fr

This graphical device has no electrical mounting and the base for the inscription forks is made of wood.

233b. The same apparatus very simple. 250 fr

Similar to CR no. 233b, this instrument has no electrical mounting and the base for the inscription forks is made of wood.

Location: Coimbra (FIS.0405; date, 1867). Teylers (1875).

Description: The Teylers instrument comes with signed examples of smoked plates by Koenig. There are two forks, ut1 holds the glass plate, and the inscription fork covers the range from "UT1" to "MI1" (128–160 v.s.) giving the intervals 1:1–4:5.

Reference: Turner, G.L'E. (1996, p. 130).

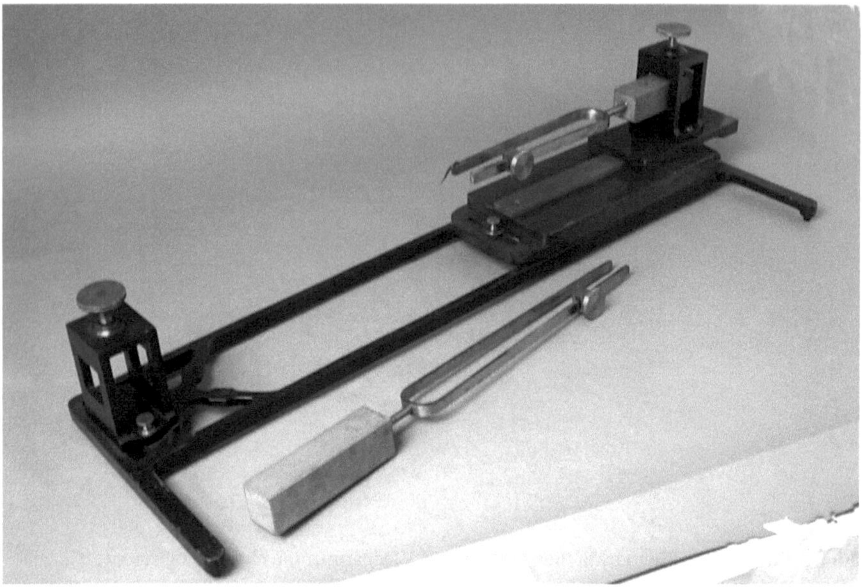

Fig. CR. no. 233b Photo by author 2005. Museu de Física, University of Coimbra, Portugal. FIS.0405

Optical Method

234. Large apparatus for compounding two vibratory movements by Lissajous' optical method. 1,800 fr

In the 1850s a professor of physics at the Lycée Saint-Louis in Paris, Etienne Jules Lissajous, developed an optical method for comparing the frequency of two tuning forks. It was based on the relations of standard musical intervals – octave, third, fifth etc. The forks had mirrors attached to the end of the prongs which were used for projecting a beam of light on a screen. The vibrations of one fork were so rapid that they appeared as a still line of light on the screen (opposed to a dot of light if the fork were not vibrating). Lissajous combined two such motions at perpendicular angles to each other. He bounced a light beam off the prong of one fork (vibrating up and down) and then directed it at the other vibrating mirror (vibrating sideways). The combined motions of the light beam fell on a screen creating a pattern representing the two motions. For example, if the forks had the same frequency, the pattern would be a circle, or a combination of vertical and horizontal movements. A pair of forks that were an octave apart would create a figure-eight pattern. This technique offered a dramatic improvement for precision tuning because the forks had to be exactly tuned to create the characteristic figures.

The most precise application of Lissajous's method came in the form of what was called the vibration microscope, or "comparateur," that allowed one to study the vibrations of strings, tuning forks or any vibrating bodies using a microscope assembly attached to a known fork. The objective lens rested on the end of a tuning fork and vibrated in one direction, and the object under study would be illuminated and vibrate in the opposite direction. Thus the vibrations combined to form characteristic Lissajous patterns as seen through the microscope lens. This was Lissajous's mosts significant precision instrument for tuning and observing vibrating bodies. Helmholtz used it in his studies of violin strings.

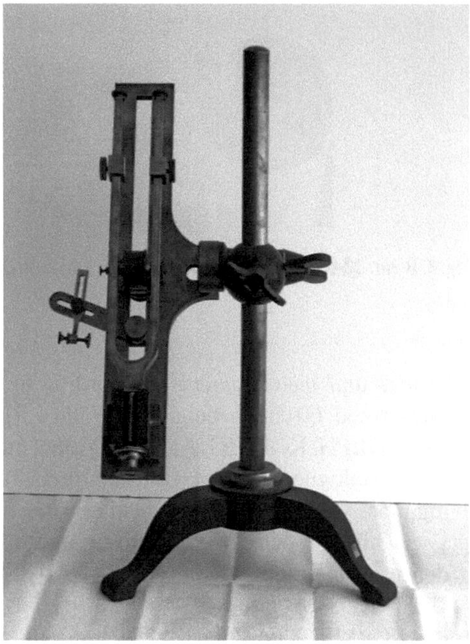

Fig. CR no. 234-1 Courtesy of the McPherson Collection, McGill University, Canada

Location: Cornell. FST. McGill. NMAH (cat. no. 314591). Oxford (Clarendon Laboratory; 8 forks, 1865 catalogue no. 206). Rome. Science Museum (acc. no. 1968-634; eight forks, 1865 catalogue no. 206). Teylers (1875). Toronto (1878, only forks). Union (8 forks, 1865 catalogue no. 206). Vanderbilt (c. 1900).

Description: (Toronto) Ten large tuning forks made of highly quality, highly polished steel with adjustable brass handles and sliding brass weights on each prong; one prong has a polished steel mirror on the end; the other has a connection for a microscope objective. Some forks have mirrors on both prongs. They range from ut1 (64 Hz) to ut3 (256 Hz), graduated with a specific frequency range. They are stored in an oak box. There are two supports of steel and cast iron that are extremely robust so as to prevent unwanted vibrations.

The set at the fondazione scienza e tecnica (FST) is from the earlier workshop, c. 1865. Eight forks are mounted in square wooden bases, which are secured in wooden supports and vices. Koenig later used cast iron supports. The apparatus at McGill and Cornell have electromagnetic coils between the prongs which was an adaptation made by Lord Rayleigh.

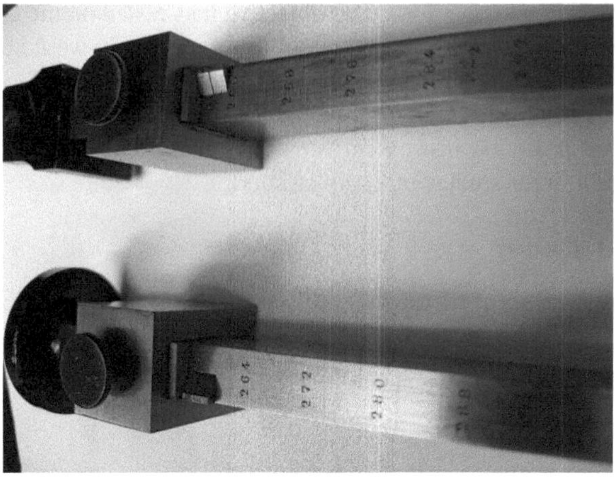

Fig. CR no. 234-2 Photo by author, 2005. Physics Department, University of Toronto, Canada

Markings and measurements: According to the 1889 catalogue the whole apparatus stood 60 cm in height. (Toronto) The forks rest in an oak box stamped "RUDOLPH KOENIG À PARIS" and marked in ink "209a," referring to the 1873 catalogue. The forks are stamped with the "RK" monogram and have sliding brass weights with graduated divisions from ut1 to ut2 and divided every two v.s., while the ones from ut2 to ut3 are divided every four v.s. There is no evidence of fine tuning (filing) at the base of the yoke. The forks include: "– SI2 – UT3 RK" 17.6 cm long; "– LA2 – RK" 18.2 cm long; "– SOL2 – RK" 19.5 cm long; "– MI2 – FA2 – RK" 21.2 cm long; "UT1 – RÉ1 – RK" 23.2 cm long; "– SOL1 – RK" 27.4 cm long; "– MI1 – FA1 – RK" 30.0 cm; "– UT1 – RÉ1 – RK" 32.0 cm long; "1 UT1 128 vs RK" 35 cm long; "1 UT1 128 vs RK" 35.7 cm long, with mirror on end. Stand on cast iron tripod – 60 cm high.

References: Auberbach in Winkelmann (1909, p. 162), Daguin (1867, p. 520–521), Deschanel (1877, pp. 852–854), Giatti (2001, pp. 101–103), Gregory (1889), Guillemin (1881, pp. 720–721), Helmholtz (1863, p. 138), Helmholtz (1954, p. 81), Koenig (1865, pp. 40–41), Ku (2006), Lissajous (1857, p. 10), Loudon and McLennan (1895, pp. 107–108), Thompson (1886), Turner, G.L'E. (1996, p. 132), and Turner, S. (1996).

234a. Apparatus for compounding two vibratory movements by Lissajous' optical method consisting of six forks with steel mirrors attached, and two iron stands. 540 fr

Location: Amherst. Sydney.

References: Auerbach in Winklemann (1909, p. 162) and Deschanel (1877, pp. 850–852).

234b. Same apparatus with smaller forks and wooden stands. 360 fr

234c. The same apparatus with four forks. 250 fr

234d. The two stands of no. 234. 150 fr

234e. The two stands of 234a. 40 fr

234 f. The two stands of 234b. 20 fr

234 g. Optical comparator consisting of five forks with sliders from ut2 to ut3. 700 fr

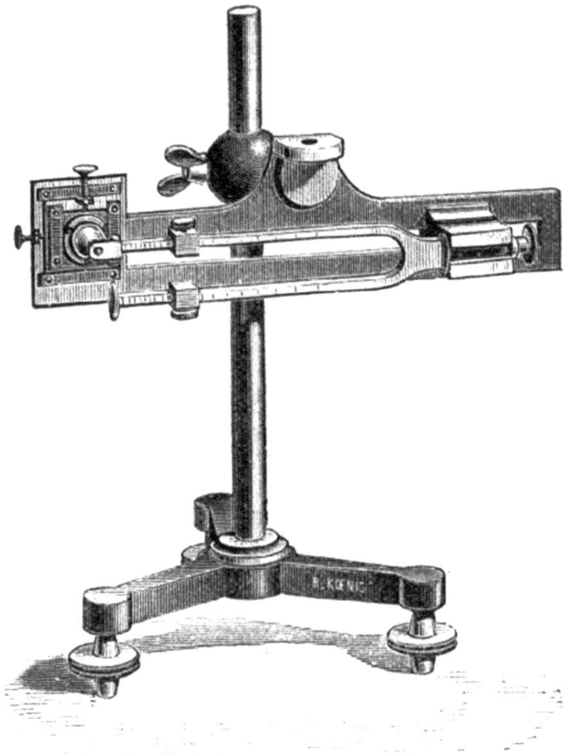

Fig. CR no. 234 g Source: Koenig (1889, p. 82)

234 h. Optical comparator ut2. 90 fr
234i. The same apparatus mounted electrically. 140 fr

Location: Amherst. Case. Coimbra (FIS.1040). Harvard (acc. no. 1997-1-0883). FST. Lisbon (both instruments). Nebraska. NMAH (cat. no. 315724). Teylers. Vermont (with metal frame). Wesleyan. Yale (acc. no. YPM 50532).

Fig. CR no. 234 h Photo by author, 2005. Physics Department, University of Toronto, Canada

Fig. CR no. 234i Photo by author, 2005. Museu de Física, University of Coimbra, Portugal. FIS.1040

Description: There were two kinds of these instruments. One had two coils that straddled the fork and microscope. The other had a coil in the middle of the prongs (developed by Lord Rayleigh). A mahogany support holds the steel tuning fork, microscope and electromagnetic coils. The whole microscope can be moved up and down the steel stand, mounted on a sturdy cast iron base. There is a little eyepiece ("Huygenian" [Turner]) at the end of a short body-tube.

Markings and measurements: The apparatus at the NMAH (two coils) measures 43.5 × 24 × 22 cm.

References: Auerbach in Winkelmann (1909, pp. 152–153), Giatti (2001, p. 106), Ku (2006), Thompson (1886), Turner, G.L'E. (1996, pp. 127–128), Violle (1883, pp. 273–274), and Zahm (1900, p. 418).

235. Apparatus for the same experiments as no. 234a, b, c, consisting of two large electrical forks with steel mirrors attached and sliders, mounted on iron stands. 300 fr

236. Kundt's polarization vibroscope. 200 fr

Manometric Flame Method

237. Organ pipe with three manometric flames. 45 fr

The manometic capsule made sound visible through a flickering flame. The cinematic, silent dance of flame viewed in a rotating mirror became an icon of nineteenth-century acoustics. The manometric capsule and a whole family of related optical instruments were developed between 1862 and 1866.

A thin membrane divided the capsule into two parts: one part was open to the sound vibrations under study; the other was closed to a flow of gas that came in through an input and exited through a gas jet, which was lit creating a tiny candle-sized flame. The membrane picked up vibrations in the air and transferred these vibrations to the gas, which caused the flame to flicker. A rotating mirror spreads this flickering flame across the surface of the mirror through persistence of vision. The pattern of flame flickerings resemble a saw-tooth pattern of ups and down that would otherwise be imperceptible to the viewer.

Koenig first applied this technique in observing the fluctuations of air in an organ pipe. At the 1862 London exhibition he displayed his manometric pipe with three capsules at three nodal positions along the length of the pipe. (A node of vibration corresponds to a place where there is changing density or pressure, yet no longitudinal vibration. For example, at the centre of the pipe two longitudinal segments compress into each other creating a dead zone in the middle. The continuous squeezing and pulling create pressure changes, and cause the flame to vibrate). The middle capsule corresponded to the node of the fundamental and the outer two capsules corresponded to the nodes of the octave. When the pipe sounds with the fundamental note, the middle capsule vibrates strongly, since it is located at the node of vibration, while the other two vibrate less strongly, being halfway between the node and the ventral sections. When the higher octave sounds there is a strong response at the two outer capsules, as they are at the nodes of vibration, while the middle capsule does not vibrate, being at a ventral segment.

The membranes varied in material from "a very thin membrane of india rubber," gold-beater's skin or a thin sheet of caoutchouc," "a flexible membrane of oiled silk," "India rubber from a toy balloon," and a "membrane of parchment or thin rubber."

Location: Coimbra (FIS.0375). CSTM (acc. no. 1998.0250.2). Dartmouth (acc. no. 2002.1.34055). Harvard (acc. no. 1997-1-9076). NMAH (acc. no. 315727). Teylers. Toronto.

Markings and measurements: (Toronto) Stamped "RUDOLPH KOENIG À PARIS" and measures 7.8 × 7.8 × 80 cm. There is a glass window on one of the sides. The nodal and ventral segments are marked top to bottom, "V, N2, V2, N1, N2, V." (NMAH) Three original membranes were examined under the microscope with ultraviolet light to reveal a thin layer of gelatin or rabbit's glue applied to a thin piece of paper.

References: Daguin (1867, p. 533), Barnes (1898, p. 186), Deschanel (1877, pp. 847–848), Dolbear (1877, p. 64), Ganot (1893, p. 253), Guillemin (1881, p. 722), Helmholtz (1954, p. 374), Jamin (1868, p. 539), Koenig (1865). Idem., 1864b. Pisko (1865, p. 197), Richardson (1947, p. 185), Turner, G.L'E. (1996, p. 121), Tyndall (1896, p. 215), Violle (1883, p. 128), and Zahm (1900, pp. 230–231).

238. Stopped organ pipe with three manometric flames. 45 fr

In a closed organ pipe there is always a node of vibration (place of no longitudinal movement, but large changes in density) next to the closed end. If the fundamental is sounded, the manometric capsule at this node will be greatly agitated. The position in the middle and the one closest to the opening (where there is a ventral segment) will be less agitated respectively. If the next octave is sounded, the two outside capsules, located at nodes, will be agitated, while the middle capsule, being at a ventral segment (no changes in density), will remain motionless.

Fig. CR no. 238 Photo by author, 2005. Physics Department, University of Toronto, Canada

Location: CSTM (acc. no. 1998.0250.1). Toronto.

Markings and measurements: (Toronto) Stamped "RUDOLPH KOENIG À PARIS" and measures 7.8 × 7.8 × 81 cm. It is marked in ink "214" referring to the 1873 catalogue. From mouthpiece to the closed end the nodal and ventral markings read "V, N3, V3, [illegible]"

References: Daguin (1867, p. 533), Deschanel (1877, pp. 847–848), Ganot (1893, p. 253), Guillemin (1881, p. 722), Jamin (1868, p. 539), Koenig (1864b), Pisko (1865, p. 197), Tyndall (1896, p. 215), Violle (1883, p. 128), and Zahm (1900, pp. 230–231).

239. Apparatus for compounding and comparing the vibrations of two air columns by the method of manometric flames, with 9 pipes. 300 fr

Vibrating flames were convenient for demonstrating relations of musical intervals based on the optical-tuning methods of Lissajous. Instead of using the vibrations of two tuning forks for making comparisons, Koenig used two adjacent manometric pipes. Both pipes rested vertically in a wind-chest and each had a capsule attached to the middle of the pipe. Each capsule had a rubber gas input tube and an output tube that connected to a stand for the burners, which were placed one on top of the other. A rotating mirror sat adjacent to the stand in order to pick up the signal from the burners. Two ut3 pipes, for example, displayed identical flame signals. Other combinations demonstrated the differences between octaves, thirds, fifths, etc.

Fig. CR no. 239 Photo by author 2005. Museu de Física, University of Coimbra, Portugal. FIS.0406

Koenig also built a burner that combined the output lines from both pipes. Such a set-up created specific harmonic patterns, for example, a unique pattern of an octave, or third, etc. He saw this not only as a demonstration of basic musical intervals, but as very good at tuning.[38]

This apparatus comes with nine manometric pipes, each with a capsule in the middle. Two of the pipes are ut3 with small mahogany sliding doors for adjusting the note by a half tone. The other seven pipes form the scale, re3, mi3, fa3, sol3, la3, si3, and ut4.

Location: CNAM (inv. 12609). CSTM (acc. no. 1998.0251 and 0252). Coimbra (FIS.0406; FIS0756; FIS.1349). Dartmouth (acc. no. 2002.1.34051; 54; 55; 57; 71; 77). Harvard (acc. no. 1997-1-0933). MCQ (old version on wood base similar to one pictured in 1865 catalogue no. 215). NMAH (acc. no. 315170 and 315727). Rome. Teylers. Toronto.

Description: Pine pipes with a mahogany lip. The two ut3 pipes have a sliding door at open end.

Markings and measurements: (Toronto) Marked "215" in ink referring to the 1873 catalogue. Each pipe is stamped "RUDOLPH KOENIG À PARIS." "UT3" pipes = 8.3 × 7.8 × 59.7 cm; "MI3" has a lead flap at the open end, pipe = 6.7 × 6.0 × 45.5 cm; "SOL3 "has lead flap at the open end, pipe = 6.0 × 5.5 × 38.4 cm. The windchest measures, 13.3 × 28.0 × 17.5 cm.

References: Auerbach in Winkelmann (1909, pp. 158–161, 169), Blaserna (1876, pp. 23–25), Guillemin (1881, p. 725), Jamin (1868, pp. 540–541, 590–591), Koenig (1882c, pp. 50–52). Idem., 1873, pp. 4–7. Loudon and McLennan (1895, pp. 126–127), Pisko (1865, p. 199), Turner, G.L'E. (1996, p. 119), Violle (1883, pp. 99–101), and Zahm (1900, pp. 292–293).

239a. The same apparatus with five pipes. 240 fr

239b. The revolving mirror of no. 239. 150 fr

Location: Amherst. Coimbra. Harvard (acc. no. 2000-1-0014). Lisbon. Teylers.
Reference: Turner, G.L'E. (1996, p. 133).

239c. The stand of the gas burners of no. 239. 6 fr

∗(Note: The preceding organ pipes were made of unvarnished pine. Koenig varnished CR nos. 237 and 238 for an extra 5 fr, and the ones from 239 for an extra 4 fr.)

240. Manometric capsule with tube and mouthpiece. 20 fr

Vibrating flames produced beautiful figures when applied to violin and vocal sounds. Koenig's apparatus consisted of a small rotating mirror resting on a cast iron frame, a capsule attached to a stand and gas input, and a rubber tube connected to either a stethoscope (for pressing against the sounding-post of a violin) or a hand-held funnel speaker tube into which one sings vowels or musical notes. He used it for a series of experiments to determine the characteristic pitch of the five vowels sounds, OU, O, A, E, and I. Each vowel was sung at fifteen different pitches from ut1 to ut3, producing distinctive flame pattern.

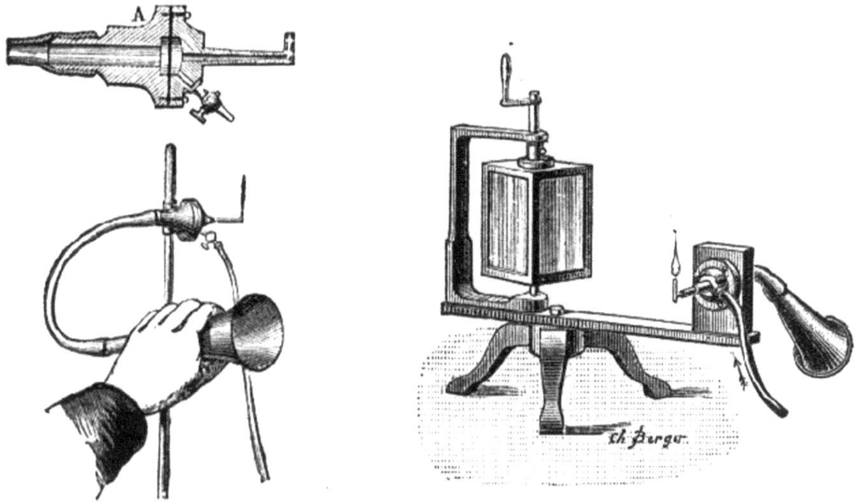

Fig. CR nos. 240 and 241 Source: Koenig (1889, p. 85)

Location: Case (c. 1894). NMAH (cat. no. 325951; only capsule, c. 1865). Teylers (1889).
Description: The capsule at Case University is metal, has an "s" shaped gas inlet tube, and is mounted on a cast iron stand. The capsule at the NMAH is mahogany (c. 1865) with a turned inlet for the vibration tube. The membrane is rubber.
References: Auerbach in Winkelmann (1909, p. 167), Ganot (1893, pp. 271–274), Koenig (1882c, pp. 56–67), Loudon and McLennan (1895, p. 125), Miller (1916, pp. 73–74), Turner, G.L'E. (1996, p. 133), Violle (1883, pp. 298–301), and Zahm (1900, pp. 358–360).

241. Small revolving mirror, manometric capsule, tube and mouthpiece. 60 fr

The mouthpiece can be replaced by a resonator to display the pattern of a specific frequency.

241a. Small revolving mirror. 50 fr

Location: Case (c. 1894).
Reference: Miller (1916, pp. 73–74).

242. Manometric flame analyser for the timbre of sounds, with 14 universal resonators. 650 fr

Koenig's flame analyser was, next to the sound synthesiser, one of the clearest expressions of Hermann von Helmholtz's theory that complex sounds were made up of a spectrum of elemental or pure tones. The adjustable resonators covering a range of 65 notes from sol1 to mi5 (96–1,280 Hz), could each be rendered visible with a connection to a manometric flame capsule. The resonators connected to a gas-filled capsule with a rubber tube. If activated, the distinctive pattern would appear in the rotating mirror. A human voice, for example, would activate a series of capsules revealing its rich harmonic structure. A tuning fork, representing a pure, elemental tone, would only activate one resonator and capsule. Koenig invented this analyser for his vowel studies between 1865 and 1872. It was flexible with a wide range compared to the earlier model with eight fixed resonators (CR no. 242a).

The analyser at the University of Toronto is still in operation although the membranes (not all original) in the capsules require a considerable amount of adjustment. They must be a certain material and tightness in order to respond adequately. The kind of gas and its pressure must also be taken into consideration. In the nineteenth century the membranes could be rubber or thin paper painted with animal glue. Coal gas was commonly used in laboratories.

Location: Amherst. Case (c. 1896). FST (c. 1890). Liceo Visconti. Rennes. Toronto (1878). Vanderbilt (1875).

Description: The capsules are wooden. Metal gas tubes with small holes protrude from each capsule. The bottom tubes are longer, tapering to a smaller size toward the top (smaller resonators). The black screens are tin. Rubber tubes connect resonators to the back of the capsules. The front of the capsules connect via rubber tubes to eight stop cocks, which in turn connect to common wooden gas reservoir. The mirror consists of four glass mirrors (painted silver on glass) on wooden rectangular prism. The glass panes are held in place with black tape. The analyser at Case University has metal capsules. It was most likely bought in 1893, 1894 or 1896 by D.C. Miller.

Markings and measurements: (Toronto) Overall dimensions (91 × 86 × 33 cm). The resonators, each stamped "RK," are as follows (same as no. 55): (1) SOL1 – SI1, (2) SI1 – RE2, (3) RE#2 – F#2, (4) FA#2 – LA2, (5) LA2 – UT3, (6) UT3 – MI3, (7) MI3 – LA3, (8) LA#3 – RE4, (9) UT4 – MI4, (10) RE4 – FA4, (11) MI4 – SOL#4, (12) FA4 – LA4, (13) SOL#4 – UT5, (14) UT5 – MI5. The mirror measures 39.5 × 11.0 × 11.0 cm.

References: Ganot (1893, pp. 238–239), Giatti (2001, pp. 96–98), Koenig (1882, pp. 70–74), Loudon and McLennan (1895, pp. 123–124), Pantalony (2001), and Zahm (1900, pp. 352–356).

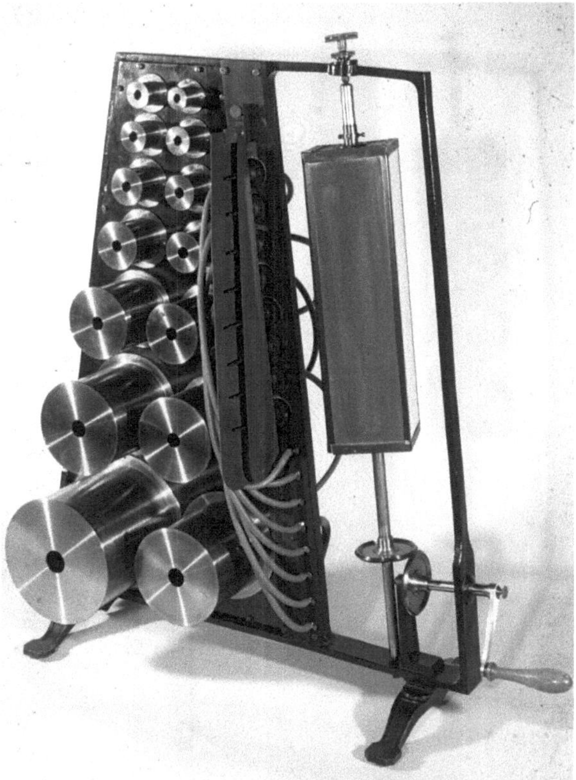

Fig. CR no. 242 Photo by Louisa Yick. Courtesy of the Physics Department, University of Toronto, Canada

242a. Analyser for one sound (ut2) with eight resonators. 325 fr

This was the first form of analyser developed c. 1865. It was based on the fundamental ut2, with seven other harmonics for demonstrating timbre in this limited range. The example at Dartmouth has a number of original membranes made of thin paper covered with as thin coating of rabbit glue. The cast-iron frame and the cast-iron parts that hold the resonators into the frame all have matching manufacturing marks in the form of dots. These markers are evidence of the production techniques used in Koenig's workshop.

Locations: Barcelona. CNAM (inv. 12605). Dartmouth (acc. no. 2002.1.34112). Dublin. Geneve. Duke. Harvard (acc. no. 1998-1-1606). Lisbon. NMAH (cat. no. 314583). QUP. Rome. Porto. Science Museum (acc. no. 1947–126). Sydney. Western.

Markings and measurements: (NMAH) The resonators from large to small, bottom to top are each stamped "RK" "1" to "8" and, UT3, SOL3, UT4, MI4, SOL4, 7,

Fig. CR no. 242a Photo courtesy of the National Museum of American History, Smithsonian Institution, Washington, DC, cat. no. 314583, neg. 76-1827

UT5. The arrangements of the gas chamber, capsules and rotating mirror are the same as no. 242. Overall dimensions are 66 cm wide × 91.5 cm high. (Dartmouth) Manufacturing marks on the bottom of the cast iron frame and stand.

References: Blaserna (1876, pp. 171–173), Deschanel (1877, p. 856), Guillemin (1881, pp. 735–736), Jamin (1868, pp. 632–633), Koenig (1882c, pp. 70–74), Mollan (1990, pp. 194, 322), Pantalony (2001), Pantalony et al. (2005, pp. 137–138), Violle (1883, pp. 292–295), and Zahm (1900, pp. 352–356).

243. Manometric flame interference apparatus. 250 fr

Koenig invented the manometric interference apparatus to provide an optical method for showing and studying beats and interference phenomena. A spherical resonator and tuning fork produced a known frequency that was sent along two parallel sets of brass tubing. One of the tubes could be adjusted like a trombone to extend or shorten its length by a measurable amount. Two sound vibrations met in a joined capsule to produce a combined flame signal. If, for example, the waves met while in perfectly opposite phase, they would cancel each other and produce an unmoving band of flame. This apparatus was also used as a precision instrument for measuring the wavelength of certain notes in different gases, and for calculating the velocity of sound. The wave length could be measured accurately by slowly adjusting the tubing until it was "visibly" out of phase; at such a point the tube had been moved half a wave length, which in turn could be used to calculate the speed of sound. The one at the NMAH could also be sounded using a Koenig monochord (NMAH cat. no. 314587).

Fig. CR no. 243 Photo by Louisa Yick. Courtesy of the Physics Department, University of Toronto, Canada

Locations: CNAM (inv. 12606; c. 1894). Coimbra (FIS.0701). Dublin. Harvard (acc. no. 1997-1-0902). FST (c. 1865). Lisbon. NMAH (cat. no. 314594; c. 1865). Naples. Rome. Teylers (1889). Toronto (1878).Vanderbilt (1875).

Description: The graduated scale on the instrument at Toronto reads from 0 to 35 cm, with every mm marked. The one at the University of Lisbon (formerly the Polytechnical school) has four capsules, perhaps for further comparison. The metal scale in the middle reads 0 to 50, with divisions of 10 between each number. The tubes on the instruments at Florence and at the NMAH lie horizontal on a wooden table. They date from around 1865.

Markings and measurements: (Toronto) Stamped "RUDOLPH KOENIG À PARIS." 85 × 45 × 31 cm.

References: Koenig (1882c, pp. 76–83), Giatti (2001, pp. 94–96), Loudon and McLennan (1895, p. 128), Mollan (1990, p. 322), Turner, G.L'E. (1996, p. 134), Violle (1883, p. 104), Zahm (1900, pp. 295–299), and Zoch 1866.

244. Wheatstone's kaleidophone with twelve rods. 100 fr

In 1825 Charles Wheatstone invented this simple demonstration related to musical intervals. It consisted of rectangular rods set vertically in a cast-iron frame, each capped with shiny brass beads. Vibrating figures were determined by the proportions of their sides. For example, a rod with two sides of 1:1 executed a circular figure. A rod with sides 1:2 produced a figure eight, as one of the sides vibrated twice as easily as the other.

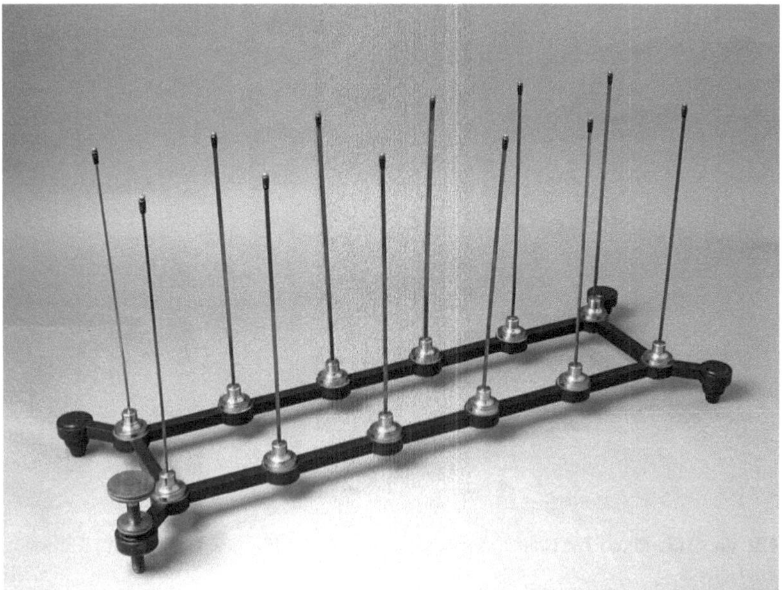

Fig. CR no. 244 Photo by author 2005, Museu de Física, University of Coimbra, Portugal. FIS.0755

Locations: Coimbra (FIS.0755; date, 1881). Lisbon. Vanderbilt (1875).

Description: (Coimbra) This instrument has twelve rods with the following musical intervals marked at each brass base: row 1: 1:1 (square rod), 3:4, 3:5, 4:5, 5:6,

6:7. Row 2: 1:1 (round rod), 1:2, 1:3, 1:4, 2:3, 2:5. The numbers are separated by the "RK" monogram.

Measurements: (Coimbra) 72.4 × 34.7 × 26.5 cm.

References: Auerbach in Winkelmann (1909, p. 163), Holland (2000), Jamin (1868, pp. 614–615), Marloye (1851, p. 55), Pisko (1865, pp. 117–119), Wheatstone (1827), and Zahm (1900, pp. 411–412).

244a. The same apparatus with six rods. 60 fr

Location: CNAM (inv. 12262). Teylers (1863). Sydney. Dublin. Wesleyan University in Middletown. Connecticut.

Description: (Sydney) The bars read 1:3, 2:3, 1:2, 3:4, 4:5, and 1:1. Mirrors are attached to the end of the rods, instead of shiny metal balls. There is no frame/bass like the standard instrument. The set at Teylers Museum has two 1:1 rods, one circular, the other square. Each rod has polished steel balls at the end.

References: Turner, G.L'E. (1996, p. 129) and Mollan (1990, p. 321).

245. The same apparatus with twelve rods for projection, and stand. 220 fr

The rods carry small mirrors instead of brass balls. Each rod is secured into a cast iron support stand which swivels to any desired angle.

References: Zahm (1900, pp. 411–412).

245a. The same apparatus with six rods, without a stand. 85 fr

Location: Sydney.

245b. The stand for no. 245. 50 fr

246. Four long kaleidophone rods to produce figures with incandescent charcoal points. 50 fr

These long rods with heated, glowing tips are used with the stand in no. 245b. In a darkened room they dramatically illustrate acoustic figures.

Location: Harvard.

247. Compound rod to show the composition of parallel vibrations. 20 fr

Mounted on no. 245b.

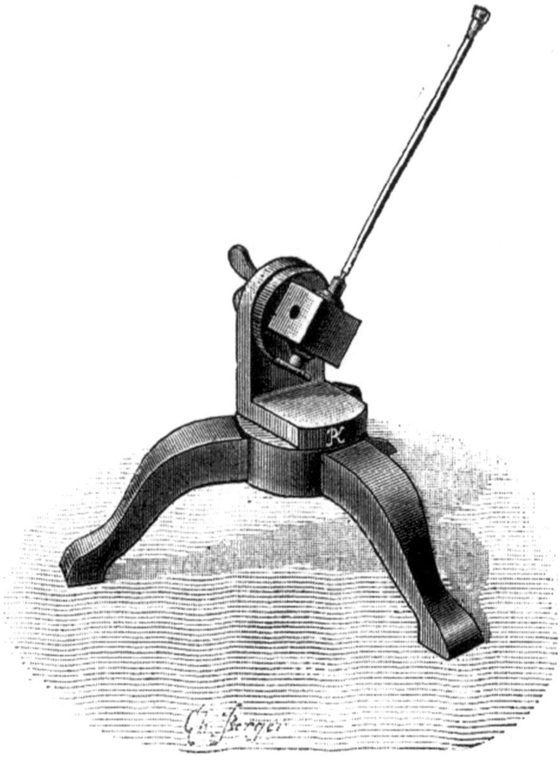

Fig. CR no. 245 Source: Koenig (1889, p. 89)

248. Melde's apparatus for studying simple and composed vibrations of strings, with five forks. 350 fr

Franz Emil Melde, a professor of physics at the University of Marburg, designed a way to test the behaviour of vibrating strings using mounted tuning forks. With this version including two forks, one can study combinations of vibrating strings.

Locations: Coimbra (FIS.0408; date, 1867). ISEP. Teylers (1876). Toronto (1878). Vanderbilt (1875).

Description: (Coimbra) Fine silk strings are attached to the forks through brass hooks on the end of the prongs. The forks are secured to mahogany plates on a sturdy, cast-iron frame. They are driven by electrical coils. The forks at Teylers Museum stand vertical.

Markings and measurements: (Coimbra) Overall dimensions 11.2 × 90.0 × 15.5 cm; (Toronto) Base missing. Includes five forks: "UT1/128 vs/RK" 29.8 cm long; "SOL1/192 vs/ RK" 24.8 cm long; "SOL1/192 vs/RK" 23.1 cm long; "UT2/256 vs/RK" 21.5 cm long; "SOL2/384 vs/RK" 17.6 cm long. The oak box

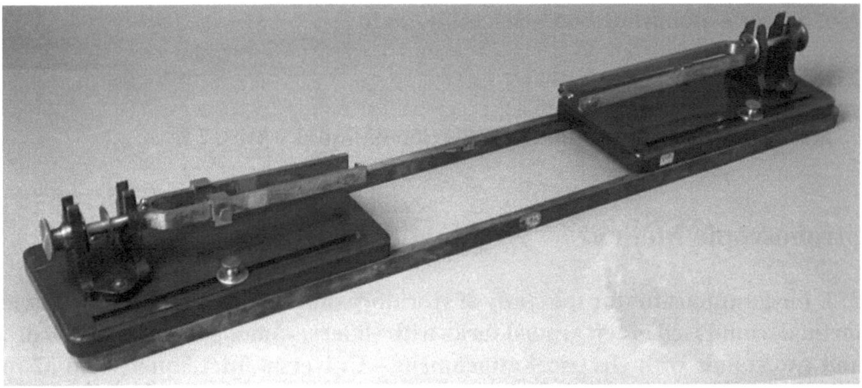

Fig. CR no. 248 Photo by author 2005, Museu de Física, University of Coimbra, Portugal. FIS.0408

is stamped "RUDOLPH KOENIG À PARIS" and marked "223" in ink, referring to the 1873 catalogue.

References: Auerbach in Winkelmann (1909, pp. 150–152), Loudon and McLennan (1895, pp. 120–121), Melde (1860a,b), Pisko (1865, pp. 129–132), Turner, G.L'E. (1996, p. 123), Violle (1883, pp. 170–74), and Zahm (1900, pp. 157–159).

249. Melde's electrical monochord. 200 fr

This apparatus consists of an electrically driven tuning fork connected to a fine, silk thread held taught by a suspended weight. The wooden frame is approximately 1.5 m in height. There is a scale on the frame for marking nodal points and measuring wave-length. By varying the frequency of the fork, tension and length of the string, one could test Mersenne's laws of vibrating strings – the number of vibrations of a string is inversely proportional to the length of the string, and, proportional to the square root of its tension. This apparatus was also a striking visual demonstration of the nodes and ventral segments of a vibrating string.

Location: McGill.

References: Barnes (1898, pp. 74–75), Jones (1937, pp. 204–208), Melde (1860a,b), Miller (1916, pp. 64–66), and Zahm (1900, pp. 157–160).

249a. Small apparatus of Melde, with one electrical fork. 120 fr

249b. The same apparatus with fork without electrical attachments. 60 fr

References: Auerbach in Winkelmann (1909, p. 315) and Tyndall (1896, pp. 133–139).

250. Savart's monochord on black table. 20 fr

251. Weber's wave-canal. 100 fr

252. Elliptical vase to exhibit the reflection of liquid waves. 7 fr

Stroboscopic Method

253. Large apparatus for the study of vibratory movements by the stroboscopic method, composed of ten graded forks with sliders, ranging from 32 to 256 d.v. and two stands with electrical attachment – Universal interrupter from 32 to 256 interruptions. 1,400 fr

This apparatus was used to produce stroboscopic images of vibrating tuning forks. It did this by means of aluminum screens with tiny slits or windows, attached to the end of the tuning fork prongs. The subsequent stroboscopic effect was used to study other vibrating bodies. The apparatus consisted of two stands, approximately 60 cm in height, with ten graduated forks and sliding weights. The forks ranged from ut-1

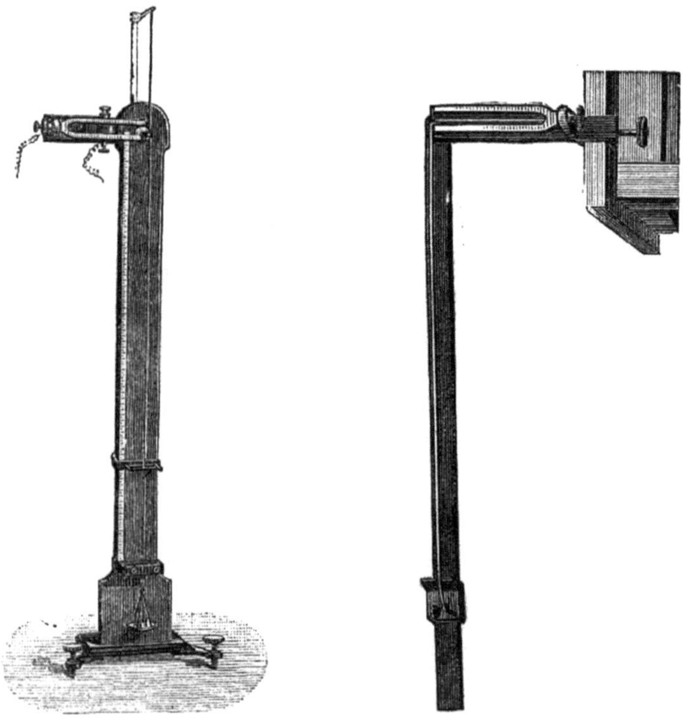

Fig. 122 (h. — 1m,50) (No 249). Fig. 123 (No 249 b).

Fig. CR no. 249 Source: Koenig (1889, p. 91)

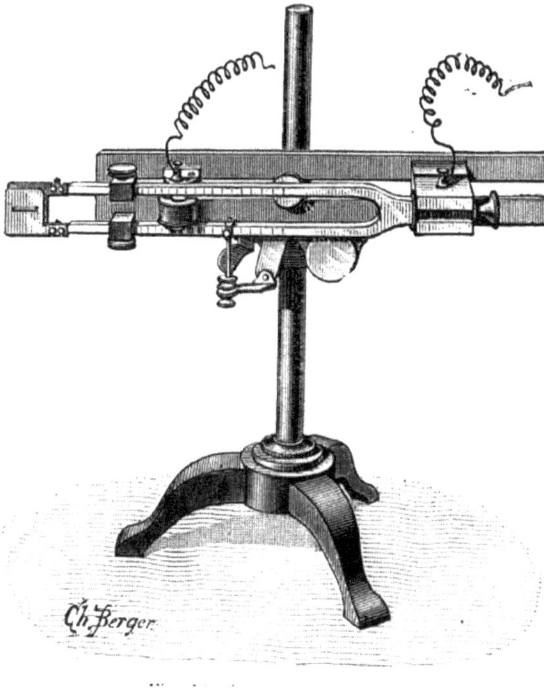

Fig. CR no. 253 Source: Koenig (1889, p. 92)

to ut3 (32–256 Hz). From ut-1 to ut1 the forks are divided by one v.s., from ut1 to ut2 by two v.s., from ut2 to ut3 by four v.s. The screens are larger for the lower notes and smaller for the higher notes. Jospeh Henry reputedly used the instrument preserved at the NMAH.

Location: NMAH (cat. no. 314733; c.1875). Toronto (1878).

Markings and measurements: The University of Toronto has a set of ten graduated forks. The oak box is stamped "RUDOLPH KOENIG À PARIS," and marked in ink, "218a" referring to the 1873 catalogue. The forks include: "SI2 – UT2 RK" 18.8 cm long; "– SI2 RK" 19.7 cm long; "– LA2 – RK " 21.0 cm long; "MI2 – SOL RK," 21.6 cm long; "UT2 – RÉ2 – RK" 21.5 cm long; "LA1 – UT2 RK" 25.0 cm long; "– FA1 – SOL1 – RK" 28 cm long; "– SOL-1 – UT1 RK" 36 cm long; "50 VD RK" 38.7 cm long; "UT-1 – FA-1 – RK" 41.7 cm long. The two cast iron supports at the NMAH are 58.5 cm high.

253a. The same apparatus with five forks giving from 32 to 128 interruptions. 950 fr

253b. The same apparatus with two forks giving from 32 to 64 d.v. 600 fr

253c. Support of preceding with electrical attachment. 150 fr

254. Accessory pieces to adapt 253, 253a and b, for graphical and optical composition and comparison of two vibratory movements and for Melde's experiments. 200 fr

255. Electrical interrupter with three forks ut-1, ut1, and ut2. 225 fr

256. Toepler and Boltzmann's pipe for studying the vibrations of an air column by the stroboscopic method. 250 fr

Fig. CR no. 256 Source: Koenig (1889, p. 93)

Fig. 125 (N° 256).

References: Toepler (1866) and Toepler and Boltzmann (1870).

257. Mach's organ pipe for representing stroboscopically the vibrations of an air column. 60 fr

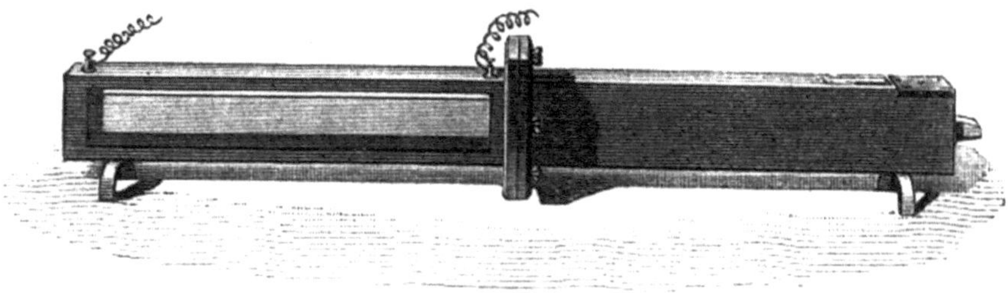

Fig. 126 (N° 257).

Fig. CR no. 257 Source: Koenig (1889, p. 94)

258. Kundt's apparatus for producing dust figures in an air column. 100 fr

In 1866 August Kundt, while assistant to Professor Heinrich Gustav Magnus in the University of Berlin, invented a method for measuring the speed of sound in different gases. This is done by making nodes of vibrations visible with very light powder (cork dust or lycopodium) in a closed glass tube. In this apparatus, a glass piston is rubbed and longitudinal vibrations are communicated into the glass chamber. The length between the nodes is then measured and used to calculate the speed of sound in that medium. Koenig's apparatus is a glass tube with brass fittings, a glass piston for exciting longitudinal vibrations, and two stop-cocks for filling the main tube with various gasses. It rests on a wooden, horizontal support.

In the late 1890s, Koenig used Kundt's technique to determine and visibly prove the frequency of his highest forks at 45,000 Hz, well above the threshold of hearing. He photographed the wave patterns next to a millimeter bar.

Fig. CR no. 258 Source: Koenig (1889, p. 95)

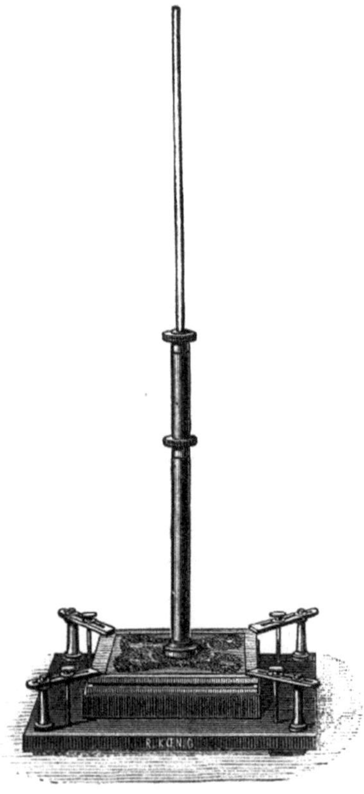

References: Blaserna (1876, pp. 22–23), Ganot (1893, pp. 256–257), Jones (1937, pp. 208–210), Kundt (1866a,b), Koenig (1899), Tyndall (1896, pp. 229–238), and Zahm (1900, pp. 256–259).

258a. The same apparatus without the cocks. 80 fr

259. Kundt's apparatus for producing dust figures in plates of air. 100 fr

The vertical rod transmits longitudinal vibrations to an enclosed plate of air producing figures.

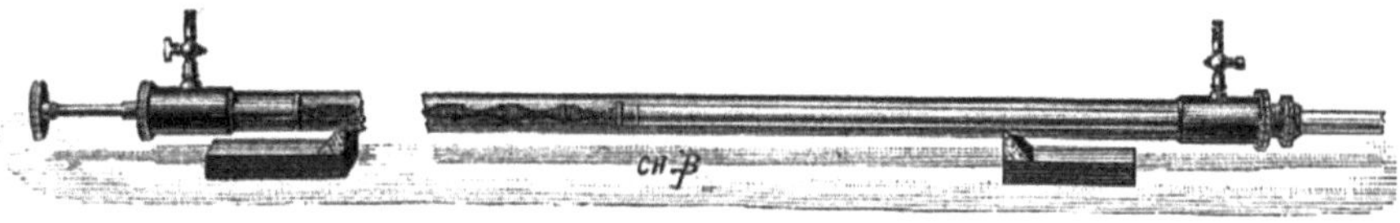

Fig. 127 (N° 258).

Fig. CR no. 259 Source: Koenig (1889, p. 95)

X. Apparatus for the Mechanical Representation of Vibrations and Wave Movements

260. Mach's apparatus, large model. 300 fr

260a. The same apparatus, smaller model. 150 fr

261. Eisenlohr's apparatus to show the molecular movement of liquid waves. 100 fr

262. Crova's apparatus for representing vibratory movements on the screen, with eight disks. 400 fr

Black glass disks, with circular wave patterns carefully inscribed on the surface, project various wave formations on a screen – propagation of a wave pulse, reflection of a wave pulse, propagation of a sound wave, reflection of continuous vibratory movement, fundamental tone of sound pipes, first harmonic of sound pipes, vibrations of ether, and interference of two vibratory movements. The French scientist Andre Prosper Crova commissioned Koenig to make this instrument and it was first shown at the 1867 World Fair in Paris.

Locations: Union. Wesleyan.
References: Auerbach in Winkelmann (1909, pp. 106–107), Crova (1867), and Lissajous (1868, pp. 480–484).

262a. The same apparatus with 7 disks without the lenses for showing interference. 250 fr

Fig. CR no. 262a Photo by author 2005. Museu de Física, University of Coimbra, Portugal. FIS.1282

Location: Coimbra (FIS.1282).

Markings and measurements: (Coimbra) Each disk (36 cm diam) has a distinctive pattern of waves. A white label identifying the wave is pasted to the black glass disk and handsigned "R.Kg." The labels read "Vibration d'ether," "Son fundamental d'un tuyau ouvert," "Deuxiéme son d'un tuyau ouvert," "Reflexion d'un movement vibratoire continue," "Reflexion d'une onde isolée," "Propogation d'une onde isolée," "Propogation des ondes sonores."

263. Wheatstone's wave apparatus. 1,000 fr

In the late 1840s Charles Wheatstone developed dynamic mechanical models for demonstrating wave properties of light and sound. It had both vertical and horizontal waves with the ability to show varying differences of phase. "Sliders" in the shape of a sinusoidal wave move along the axis causing bead-wire units (with both vertical and horizontal components) to move as a wave.

References: Holland (2000), Loudon and McLennan (1895, pp. 112–114), and Secchi 1850.

263a. Iron stand for preceding. 150 fr

Approximately 1.5 m in height, this stand could be adjusted to display the wave machine at an angle for classroom demonstrations.

263b. Wheatstone's wave apparatus, small model (original model). 600 fr

Locations: Union (1867–1874). Minnesota. Vanderbilt (1875).

Fig. CR no. 263b-1 Photo by author, 2005. Physics Department, Union College, USA

Fig. CR no. 263b-2 Photo by author, 2005. Physics Department, Union College, USA

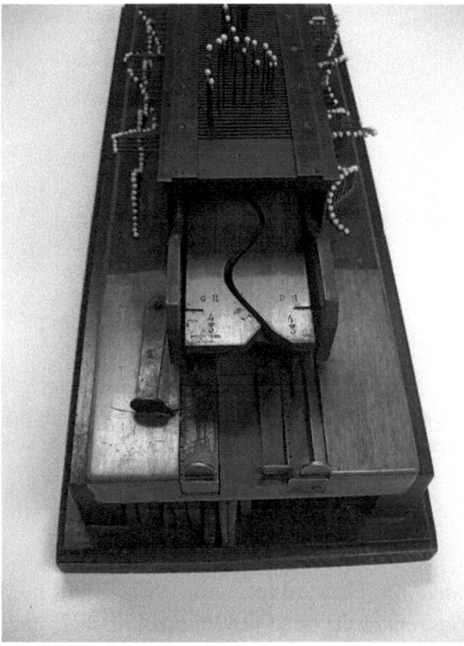

Description: Although most of the English models have wooden sliders, Koenig's version at Vanderbilt University comes with 20 metal sliders. It also comes with a table for reproducing figures.

Measurements: The one at Union College, which has 16 wooden sliders, measures, 25 × 27 × 69 cm.

264. Apparatus, which shows only the theoretic curves resulting from two systems of waves in the same plane. 100 fr

265. Apparatus, which shows the theoretical curves resulting from two systems of waves, equal and perpendicular to each other. (Circular and elliptical polarization). 50 fr

266. Wheatstone's apparatus for mechanically compounding two rectangular vibratory movements. 200 fr

This is a clever mechanical demonstration of harmonic relations. A rod with a point of light is moved by two motions, creating a combined harmonic movement. The rod moves in a ball socket in any direction. The lower end of the rod is connected to two arms that move back and forth and are set in motion through the central wheel. Owing to the gearing arrangement, they move in different relative motions.

Fig. CR no. 266 Courtesy of the Department of the History of Science, Collection of Historical Scientific Instruments, Harvard University, USA. acc. no. 1997-1-0901

Location: Harvard (acc. no. 1997-1-0901).
References: Tyndall (1896, pp. 420–422) and Pisko (1865, pp. 123–126).

267. Apparatus for mechanically and optically compounding two rectilineal movements, a rectilineal with a circular movement, and two circular movements. 500 fr

268. Lymann's apparatus for graphically compounding two pendular movements. 250 fr

XI. Acoustic Apparatus for Practical Use

269. Stethoscope with one tube. 10 fr

Koenig developed this stethoscope for listening to sounding bodies, suchs as violins or a piano. In 1864, Koenig claimed that it had the potential to transmit sounds better than a resonator, "because all the sounds produced before the membrane appear with astonishing force to the ear."[39] He used this instrument to study the sounds of a violin in his famous set of manometric experiments. It has two caoutchouc sides which were inflated to form a double convex lens shape. These sheets rest in a metal frame with an input stop-cock. A rubber hose connected the inflated lens to the ear.

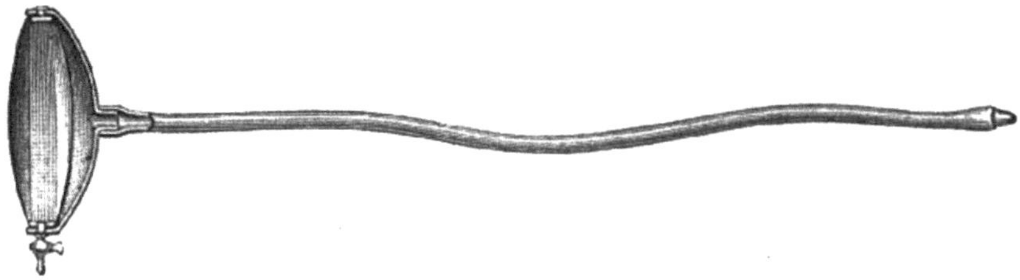

Fig. 132 (N° 269).

Fig. CR no. 269 Source: Koenig (1889, p. 100)

Reference: Koenig (1882c, pp. 39–40, 58) and Ganot (1893, pp. 222–223).

270. Stethoscope with 5 tubes. 20 fr

With this instrument, five people could simultaneously study the sounds of a body.

271. Ear trumpet. 10 fr

272. Speaking trumpet. 15 fr

Miscellaneous Instruments Not Found in Koenig Catalogues

Visible sound flame apparatus for showing reflection of gases, vapours, and heated air.
(Harvard acc. no. 1998-1-0919) (engraving in Tyndall 1896, p. 317)

Simple wave machine, (Yale, acc. no. YPM 50322)
Wooden triangle and square, (Yale, acc. no. YPM 40206a and b)
Pipe marked in ink "253" 13.8 × 13 × 56.5 cm. Perhaps no. 253 from 1882 catalogue, (Toronto)
Thirteen brass organ pipes, Rome
Pine monochord, (NMAH cat. no. 314, 587)
Unusual Mercury Interrupters (Harvard)

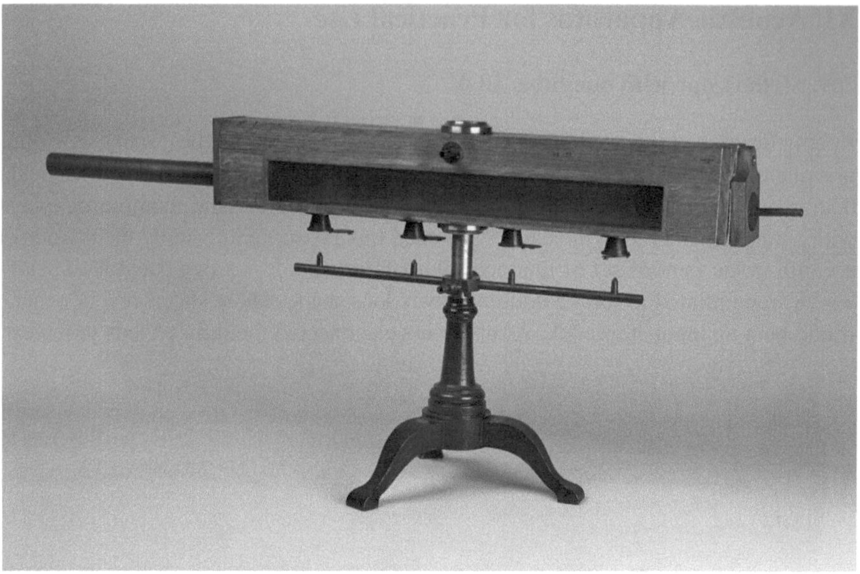

CR Fig. HU1997-1-0919 Courtesy of the Department of the History of Science, Collection of Historical Scientific Instruments, Harvard University, USA. HU1997-1-0919

"Boxwood Flageolet," (Harvard, acc. no. 1997-1-2054)
Set of 17 glass tubes, probably CR no. 202 (Harvard, acc. no. 2001-1-0049)
Boxwood pipe (Harvard, acc. no. 1998-1-009)
Early wave machines (MCQ; acc. nos. 1993.13264; 1993.13266; 1993.13265; 1993.12456)
Three unusual resonance boxes (Harvard acc. no. 1997-1-0924)
Resonance box and electromagnetic telegraph device (Harvard)
Bar with various notes marked "RK" used to demonstrate combination tone effects, see Fig. 7.11 (CSTM; acc. no. 1998.0273.12)
8 large forks from Sol-1 to Ut4, see Fig. 6.8 (Toronto).

Notes

1 The 1889 catalogue has been published on-line by the National Museum of American History, Smithsonian Institution, *Instruments for Science, 1800–1914.*
2 Turner, G.L'E. 1996.
3 Gogh 1991, April 21, 1889 and Feb. 20, 1890.
4 From the official web site of the Eiffel tower.
5 Miller (1916, pp. 22–24).
6 Ibid., p. 23.
7 Helmholtz (1863, pp. 241–242).
8 Ibid., p. 243. "...der Klang voll, stark und weich wie ein schooner Hornton." English translation from Helmholtz (1954, p. 163).
9 Koenig (1882c, p. 173).

10 Helene Neumann to Ernst Christian Neumann, Oct. 22, 1901. NFA.
11 George Barker "Memorandum" on Koenig collection, c.1882, UARCUP.
12 Over thirty forks were measured from each group, entailing six measurements of different dimensions from each fork. I gratefully acknowledge the help and insights of Smithsonian research specialist, Roger Sherman, in the examinations of the tonometer on May 8 and 9, 2003 at the NMAH.
13 Scheibler (1834, p. 53).
14 Helmholtz (1863, p. 301).
15 Ellis in Helmholtz (1954, pp. 443–446). Idem., 1968, pp. 17–18. Miller (1935, pp. 55–56).
16 Brooke (1863, p. 33).
17 Ibid.
18 Radau (1862a, p. 112).
19 I would like to thank Professor Sam Allen of the Department of Material Science and Engineering at MIT for providing laboratory time and equipment for this study. Throughout the summer of 2004, I prepared the sample (U of T fork 512 vs) and Allen's laboratory technician, Yinlin Xie, took the micrographs and performed the hardness tests. Hardness HV, the ferrite area, she obtained 134, 117, 117, 122.5 and 112.4 for an average of 120.58 w/25 g. Hardness HV, the pearlite area, she obtained 146.3, 147.9, 139.9, 152.2, and 136.1 for an average of 144.48 w/25 g. The sample was micrographed in three areas – at the base of the U, on length of the prong, and near the elbow. The micrographs revealed a sample of 0.55% annealed carbon steel (hypoeutectoid).
20 Koenig (1889, p. 23).
21 Edwin G. Boring, "The Construction and Calibration of Koenig Cylinders," *The American Journal of Psychology* 38(1927): 125–127.
22 Henri Chamoux, Inventaires des instruments scientific ancient dans les établissements publics, http://www.inrp.fr/she/instruments/instruments.htm
23 In 1830 the German physicist Ernst Chladni adopted the physicist's scale for his research in acoustics. Zahm (1900, p. 79).
24 September 18, 2001. Dept. Physics, University of Toronto.
25 Loudon and McLennan (1895, pp. 122–123).
26 Bell Papers, Library of Congress. Ganot (1893, pp. 239–240).
27 Rudolph Koenig to James Loudon, Jul. 15, 1897. UTA-JLP.
28 Roger Sherman, Steve Turner and I tried the above experiment on Aug. 25, 1999 at the National Museum of American History.
29 Koenig (1882c, p. 208).
30 Barnes (1898, pp. 18–32).
31 Koenig (1865, p. 5).
32 Savart in Hutchins (1997b, p. 18).
33 Koenig (1882c, pp. 32–38).
34 Based on experiments done with the instrument at MIT (no. 189b) in 2005 with Elizabeth Cavicchi.
35 Based on experiments done with the tuning forks at MIT in 2005 with Elizabeth Cavicchi.
36 Koenig (1859, appendix).
37 Koenig, "Chronographe d'áprés Regnault catal. No. 205a," in letter of Dec. 1878, UTA-JLP.
38 Koenig (1882c, p. 53).
39 Koenig (1882c, pp. 39–40, 58).

Bibiliography

Alexander, W. J. (Ed.). (1906). *The University of Toronto and its Colleges, 1827–1906*. Toronto: University Library.

Allin, E. J. (1981). *Physics at the University of Toronto: 1843–1980*. Toronto: University of Toronto Department of Physics.

Ames-Lewis, F., Wright, J., Nottingham University Art Gallery, Arts Council of Great Britain, & Victoria and Albert Museum. (1983). *Drawing in the Italian Renaissance workshop: An exhibition of early Renaissance drawings from collections in Great Britain held at the University Art Gallery, Nottingham, 12 February to 12 March 1983 in association with the Arts Council of Great Britain... and at the Victoria and Albert Museum, London, 30 March to 15 May 1983*. London: Victoria and Albert Museum.

Ash, M. G. (1995). *Gestalt Psychology in German Culture, 1890–1967: Holism and the Quest for Objectivity*. Cambridge, New York, USA: Cambridge University Press.

Auerbach, F. (1878). Untersuchungen über die Natur des Vocalklange. *Annalen der Physik, 8*, 177–225.

Baird, D. (2004). *Thing Knowledge: A Philosophy of Scientific Instruments*. Berkeley, California: University of California Press.

Ballot, C. H. D. B. (1845). *Annalen der Physik, 66*, 321.

Barbereau, A. (1848). *Traité d'harmonie théorique et pratique*. Paris: Lemoine.

Barbereau, A. (1852). *Etudes sur l'origine du système musical*. Paris: Bachelier

Barker, G. F. (1892). *Physics: Advanced Course*. New York: H. Holt and company.

Barnard, F. A. P. (1869). *Machinery and Processes of the Industrial Arts, and Apparatus of the Exact Sciences* (Vol. 3). Washington: Government Printing Office.

Barnard, F. A. P. (1870a). Physics. In W. P. Blake (Ed.), *Reports of the United States Commissioners to the Paris Universal Exposition*, 1867 (Vol. 3, pp. 482–575). Washington: Government Printing Office.

Barnard, F. A. P. (1870b). Exact sciences. In W. P. Blake (Ed.), *Reports of the United States Commissioners to the Paris Universal Exposition*, 1867 (Vol. 3, pp. 469–481). Washington: Government Printing Office.

Barnes, C. L. (1898). *Lessons in Elementary Practical Physics*. London: MacMillan and Co., Limited.

Bedini, S. A. (1961). The optical workshop equipment of Giuseppe Campani. *Journal of the History of Medicine and Allied Sciences, 16*, 18–38.

Bedini, S. A. (1966). Lens making for scientific instrumentation in the seventeenth century. *Applied Optics, 5*, 687–694.

Beare, C. (1980). Vuillaume, Jean-Baptiste. In S. Sadie (Ed.), *The New Grove Dictionary of Music and Musicians*. London; Washington, DC: MacMillan Publishers; New Grove's Dictionaries of Music.

Bell, J. F. (1980). Koenig, (Karl) Rudolph. In S. Sadie (Ed.), *The New Grove Dictionary of Music and Musicians*. London; Washington, DC: MacMillan Publishers; New Grove's Dictionaries of Music.

Benjamin, W., & Arendt, H. (1968). *Illuminations*(1st ed.). Edited and with an introd. by Hannah Arendt. Translated by Harry Zohn. New York: Harcourt.

Benjamin, W., & Demetz, P. (1978). *Reflections: Essays, Aphorisms, Autobiographical Writings* (1st ed.). New York: Harcourt Brace Jovanovich.

Bennett, J. A. (1983). *Science at the Great Exhibition*. Cambridge: Whipple Museum of the History of Science.

Bennett, J. A. (1985). Instrument makers and the 'decline of science in England': The effect of institutional change on the Elite Makers of the early nineteenth century. In P. R. de Clercq (Ed.), *Nineteenth-Century Scientific Instruments and Their Makers* (pp. 13–27). Amsterdam: Rodopi.

Bennett, J. A. (2002). Shopping for instruments in London and Paris. In P. S. P. Findlen (Ed.), *Merchants & Marvels: Commerce, Science, and Art in Early Modern Europe* (pp. 370–395). New York, London: Routledge.

Benshop, R., & Draaisma, D. (2000). In pursuit of precision: The calibration of minds and machines in late nineteenth-century psychology. *Annals of Science, 57*(1), 1–25.

Berg, R. E., & Stork, D. G. (1995). *The Physics of Music* (2nd ed.). Englewood Cliffs, NJ: Prentice Hall.

Bergsagel, J. (1980). Ole Bull. In S. Sadie (Ed.), *The New Grove Dictionary of Music and Musicians*. London; Washington, DC: MacMillan Publishers New Grove' s Dictionaries of Music.

Beyer, R. T. (1998). *Sounds of Our Times: Two Hundred Years of Acoustics*. New York: Springer-Verlag.

Blackbourn, D. (1997). *Fontana History of Germany, 1780–1918: The Long Nineteenth Century*. London: Fontana Press.

Blaserna, P. (1876). *The Theory of Sound in its Relation to Music*. London: H.S. King & Co.

Blondel, C. (1997). Electrical instruments in 19th century France, between makers and users. *History and Technology, 13*, 157–182.

Blondel, C., & Dörries, M. (1994). *Restaging Coulomb: usages, controverses et réplications autour de la balance de torsion*. Firenze: L.S. Olschki.

Bonelli, M. L. R., & van Helden, A. (1981). *Divini and Campani – A Forgotten Chapter in the History of the Accademia del Cimento*. Florence: Istituto e museo di storia della scienza.

Boring, E. G. (1929). *A History of Experimental Psychology*. New York, London: The Century Co.

Boring, E. G. (1942). *Sensation and Perception in the History of Experimental Psychology*. New York, London: D. Appleton-Century Company.

Boring, E. G. (1957). *A History of Experimental Psychology* (2nd ed.). New York: Appleton-Century-Crofts.

Bosanquet, R. H. M. (1879). On a mode of producing continuous notes from resonators. *Proceedings of the Musical Association, 6*, 15–21.

Bosanquet, R. H. M. (1881). On the beats of consonances of the Form h: 1. *Philosophical Magazine, 2*, 420–436; 492–506; 270–282; 434–436.

Bosanquet, R. H. M. (1881–1882). On the beats of mistuned harmonic consonances. *Proceedings of the Musical Association, 8*, 13–27.

Bosscha, J. (1854). Ueber ein Mittel, die Schallgeschwindigkeit in einem eingeschlossenen Raume geradezu zu messen. *Annalen der Physik, XCII*, 485–494.

Bowers, B., & Science Museum (Great Britain). (2001). *Sir Charles Wheatstone FRS, 1802–1875* (2nd ed.). London: The Institution of Electrical Engineers in association with the Science Museum.

Boyer, J. (1901). Rudolph Koenig. Revue Universelle, 1054.

Brain, R. (1993). *Going to the Fair: Readings in the Culture of Nineteenth-Century Exhibitions*. Cambridge, UK: Whipple Museum of the History of Science.

Brain, R. (1998a). Kymograph. In R. Bud & D. J. Warner (Eds.), *Instruments of Science: An Historical Encyclopedia*. London; Washington, DC: The Science Museum, London and the National Museum of American History in association with Garland Publishing, Inc.

Brain, R. (1998b). Standards and semiotics. In T. Lenoir (Ed.), *Inscribing Science: Scientific Texts and the Materiality of Communication* (pp. 249–283). Stanford, California: Stanford University Press.
Braun, M. (1992). *Picturing Time: The Work of Etienne-Jules Marey (1830–1904).* Chicago, London: University of Chicago Press.
Bregman, A. S. (1990). *Auditory Scene Analysis: The Perceptual Organization of Sound.* Cambridge, MA: MIT Press.
Brenni, P. (1991). L'industria degli strumenti scientifici in Francia nel XVIII e XIX secolo. In G. L'E. Turner (Ed.), *Gli strumenti.* Torino: Giulia Einaudi.
Brenni, P. (1993). The Chevalier Dynasty. *Bulletin of the Scientific Instrument Society, 39,* 11–14.
Brenni, P. (1993–1996). 19th century french scientific instrument makers. *Bulletin of the Scientific Instrument Society, 38–51.*
Brenni, P. (1994a). Lerebours and Secretan. *Bulletin of the Scientific Instrument Society, 40,* 3–6.
Brenni, P. (1994b). Heinrich Daniel Ruhmkorff (1803–1877). *Bulletin of the Scientific Instrument Society, 41,* 4–8.
Brenni, P. (1994c). Jules Carpentier (1851–1921). *Bulletin of the Scientific Instrument Society, 43,* 12–15.
Brenni, P. (1995a). The triumph of experimental acoustics: Albert Marloye (1795–1874) and Rudolph Koenig (1832–1901). *Bulletin of the Scientific Instrument Society, 44, 13–17.*
Brenni, P. (1995b). Paul Gustave Froment (1815–1865). *Bulletin of the Scientific Instrument Society, 45,* 19–24.
Brenni, P. (1995c). Gli strumenti di fisica dell'Istituto Tecnico Toscano: Ottica. Fondazione scienza e tecnica, Firenze: Giunti.
Brenni, P. (1996). Soleil, Duboscq, and their successors. *Bulletin of the Scientific Instrument Society, 51,* 7–16.
Brenni, P. (2000). *Gli Strumenti di Fisica dell'Istituto Tecnico Toscano, Elettricita e Magnetismo.* Firenze: La Lettere.
Brenni, P. (2002). 19th century scientific instrument advertising. *Nuncius, 16(2),* 497–514.
Brenni, P. (2004). Mystery object answered: The Helmholtz myograph. *Bulletin of the Scientific Instrument Society, 82,* 34–36.
Brenni, P. (2006). Dumotiez and Pixii: The transformation of french philosophical instruments. *Bulletin of the Scientific Instrument Society, 89,* 10–16.
Brenni, P. (2007). Silbermann's didactic paintings in the University of Mississippi Museum. *Bulletin of the Scientific Instrument Society, 95,* 2–5.
British Association for the Advancement of Science. (1831). *Report of the annual meeting.* 1st–108th; 1831–1938 (pp. 107 v). London.
British Association for the Advancement of Science. (1860). *Report of the twenty-ninth meeting of the British Association for the Advancement of Science; held at Aberdeen in September 1859.* London: John Murray.
Broadhouse, J. (1890). *How to Make a Violin.* London: William Reeves.
Brock, W. H. (1997). *Justus von Liebig: The Chemical Gatekeeper.* Cambridge: Cambridge University Press.
Brock-Nannestad, G. (2008). *Prosody in French theatrical declamation traced backwards in time.* Conference paper at Acoustics'08, June 29–July 4, 2008, Paris, France.
Brock-Nannestad, G., & J. M. Fontaine (2008). *Early use of the Scott-Koenig phonautograph for documenting performance.* Conference paper at Acoustics'08, June 29–July 4, 2008, Paris, France.
Brock-Nannestad, G. (2007). *150 years of time-base in acoustic measurement and 100 years of audio's best publicity stunt – 2007 as a commemorative year.* Conference paper at Audio Engineering Society 2007, May 5–8, 2007, Vienna, Austria.
Brooke, C. (1863). *Class XIII, philosophical instruments and processes depending on their use. Acoustic instruments.* In London. International Exhibition, 1862. Reports by the Juries on the Subjects in the Thirty-six Classes into which the Exhibition was Divided (pp. 32–33). London: Printed for the Society of Arts by William Clowes & Sons.

Buchwald, J. Z. (1994). *The Creation of Scientific Effects: Heinrich Hertz and Electric Waves*. Chicago: University of Chicago Press.

Buchwald, J. Z., & Franklin, A. (Eds.). (2005). *Wrong for the Right Reasons*. Dordrecht: Springer.

Bud, R., & Warner, D. J. (Eds.). (1998). *Instruments of Science: An Historical Encyclopedia*. London; Washington, DC: The Science Museum, London and the National Museum of American History in association with Garland Publishing, Inc.

Burr, S. J., Burr, S. D. V., Hollingsworth, C. T., Sherman, M. L., & Gay, M. (1877). *Memorial of the International Exhibition*. Hartford: L. Stebbins.

Cadogan, J. K., & Ghirlandaio, D. (2000). *Domenico Ghirlandaio: Artist and Artisan*. New Haven, Conn.: Yale University Press.

Cahan, D. (1989). *An Institute for an Empire: The Physikalisch-Technische Reichsanstalt, 1871–1918*. Cambridge [Cambridgeshire]; New York: Cambridge University Press.

Cahan, D. (1996). The Zeiss werke and the ultramicroscope: The creation of a scientific instrument in context. In J. Z. Buchwald (Ed.), *Scientific Credibility and Technical Standards in 19th and Early 20th Century Germany and Britain*. Dordrecht, Boston, London: Kluwer.

Cahan, D. (1998). Helmholtz resonator. In R. Bud & D. J. Warner (Eds.), *Instruments of Science: An Historical Encyclopedia*(pp. 308–310). London; Washington, DC: The Science Museum, London and the National Museum of American History in association with Garland Publishing, Inc.

Cahan, D. (Ed.). (1993). *Hermann von Helmholtz and the Foundations of Nineteenth Century Science*. Berkley: University of California Press.

Cahan, D. (Ed.). (1995). *Science and Culture: Popular and Philosophical Essays*. Chicago: University of Chicago Press.

Campos, R. (Ed.). (1998). *Violins, Vuillaume (1785–1875): A Great French Violin Maker of the 19th Century*. Paris: Cité de la Musique.

Canales, J. (2001). Exit the frog, enter the human: Physiology and experimental psychology in nineteenth-century astronomy. *British Journal for the History of Science, 34*, 173–197.

Carle, P. (1986). *Le Cabinet de Physique et l'Enseignement des Sciences au Canada Francais: Le Cas du Séminaire de Québec et de l'Université Laval (1663–1920)*.Montreal: Université de Montreal.

Cattermole, M. J. G., & Wolfe, A. F. (1987). *Horace Darwin's shop: a history of the Cambridge Scientific Instrument Company, 1878 to 1968*. Bristol, Boston: A. Hilger.

Champfleury. (1872). *Souvenirs et portraits de jeunesse*. Paris: E. Dentu.

Charbon, P. (1981). *La Machine Parlante*: J.P. Gyss.

Child, T. (1889). Characteristic Parsian Cafés. *Harper's New Monthly Magazine, LXXVII*, 687–703.

Chladni, E. F. F. (1802). *Die Akustik*. Leipzig: Breitkopf & Härtel.

Chladni, E. F. F. (1809). *Traité d'Acoustique*. Paris: Courcier.

Chute, W. J. (1978). *Damn Yankee!: The First Career of Frederick A. P. Barnard, Educator, Scientist, Idealist*. Port Washington, NY: Kennikat Press.

Cohen, I. B. (1970). Alfred M. Mayer. In C. C. Gillespie (Ed.), *Dictionary of Scientific Biography*. New York: Scribner.

Coningsby, R. (1867). The condition and habits of the French working classes. Special report. In *Reports of the artisans selected by a committee appointed by the Council of the Society of Arts to Visit the Paris Universal Exhibition, 1867* (pp. 428–451). London: Bell and Daldy.

Core, T. H. (1877). Modern discoveries in sound. In *Science Lectures for the People, Science Lectures Delivered in Manchester, 1877-78-79*. London & Manchester: Manchester University Press.

Courbet, G., & Chu, P. t.-D. (1992). *Letters of Gustave Courbet*. Chicago: University of Chicago Press.

Crookes, W. (1870). *Quarterly Journal of Science*, July, p. 316.

Crova, A. P. (1867). Description d'un appareil pour la projection mécanique des movements vibratoires. *Annales de chimie et de physique, 12*, 288–308.

Daguin, P. A. (1867). *Traité élémentaire de physique théorique et expérimentale avec les applications à la météorologie et aux arts industriels à l'usage des facultés, des établissements d'enseignement secondaire et des écoles spéciales du gouvernement* (3rd éd.). Toulouse: É. Privat; [etc.].

Darrigol, O. (1998). From organ pipes to atmospheric motions: Helmholtz on fluid mechanics. *HSPS, 29*, 1–51.

David M. Stewart Museum, Pyenson, L., & Gauvin, J.-F. (2002). *The Art of Teaching Physics: The Eighteenth-Century Demonstration Apparatus of Jean Antoine Nollet*. Sillery, Québec: Septentrion.

Davis, A. B., & Merzbach, U. C. (1975). *Early Auditory Studies: Activities in the Psychology Laboratories of American Universities*. Washington, DC: Smithsonian Institution.

De Clercq, P. R. (1997). *At the Sign of the Oriental Lamp: The Musschenbroek Workshop in Leiden, 1660–1750*. Rotterdam [The Netherlands]: Erasmus Publishing.

Delvau, A., Courbet, G., Flameng, L., & Rops, F. (1862). *Histoire anecdotique des cafés & cabarets de Paris: avec dessins et eaux-fortes de Gustave Courbet, Léopold Flameng et Félicien Rops*. Paris: E. Dentu.

den Tonkelaar, I., Henkes, H. E., & van Leersum, G. K. (1996). *Eye and Instruments. Nineteenth-Century Ophthalmological Instruments in the Netherlands*. Amsterdam: Batavian Lion International.

Desains, P. Q. (1857a). *Leçons de physique*. Paris: Dezobry E. Magdeleine et cie.

Desains, P. Q., & Lissajous, J. (1857b). *Annals de Chimie et Physique, 3rd ser., LI*.

Deschanel, A. P. (1877). *Elementary Treatise on Natural Philosophy... Translated and edited, with extensive additions, by J. D. Everett* (4th ed.). London.

Deschanel, P. (1867). Class XII, Mathematical instruments and apparatus for teaching science. By Authority of the Imperial Commission (Ed.), *Paris – Universal Exhibition, 1867. Complete Official Catalogue, Including the British and all other Sections. English Version, translated from the French Catalogue, Published by the Imperial Commission* (p. 187), J.M. Johnson & Sons.

Dolbear, A. E. (1877). *The Art of Projecting, A Manual of Experimentation in Physics, Chemistry and Natural History*. Boston: Lee & Shepard.

Donders, F. C. (1864). Zur Klangfarbe der Vocale. *Annalen der Physik* (Poggendorff's), *123*, 527–528.

Doppler, C. (1843). Über das farbige licht der Doppelsterne und eininger anderer Gestirne des Himmels. *Abhandlungen der königlichen böhmischen Gesellschaft der Wissenschaften, 2*, 465–482.

Doppler, C. J. (1846). Bemerkungen zu meiner Theorie des farbigen Lichtes der Doppelsterne, mit vorzüglicher Rücksicht auf die von Herrn Dr. Ballot in Utrecht dagegen erhobenen Bedenken. *Annalen der Physik, 68*.

Dörries, M. (1998). Easy transit: Crossing boundaries between physics and chemistry in mid-nineteenth-century France. In C. Smith & J. Agar (Eds.), *Making Space for Science. Territorial Themes in the Shaping of Knowledge* (pp. 246–262). Basingstoke: MacMillan.

Dörries, M. (2001). Purity and objectivity in nineteenth-century metrology and literature. *Perspectives on Science, 9*(2), 233–250.

Dostrovsky, S. (1970). Jean-Marie-Constant Duhamel (1797–1872). In Charles C. Gillispie (Ed.) *Dictionary of Scientific Biography*. New York: Scribners.

Dostrovsky, S., Bell, J. F., & Truesdell, C. (1980). Physics of music. In S. Sadie (Ed.), *The New Groves Dictionary of Music and Musicians*. London; Washington, DC: MacMillan Publishers; New Grove's Dictionaries of Music.

Drovàk, V. (1876). Ueber die akustische Anziehung und Abstossung. *Annalen der Physik, 157*, 42–73.

Drovàk, V. (1878). Ueber die akustische Abstossung. *Annalen der Physik, 3*, 328–338.

Drummeter, L. F. J. (1989). Philosophical instruments: Free of duty! *Rittenhouse, 3*, 113–119.

Dunn, R. (2006). Made to measure: Some thoughts on the design of scientific instruments. In L. Taub & F. Willmoth (Eds.), *The Whipple Museum of the History of Science: Instruments and*

Interpretations, to Celebrate the Sixtieth Anniversary of R. S. Whipple's Gift to the University of Cambridge. Cambridge, UK: University of Cambridge.

Edmonson, J. M. (1997). *American Surgical Instruments: The History of Their Manufacture and a Directory of Instrument Makers to 1900*. San Francisco: Norman Pub.

Ellis, A. J. (1877a). Koenig's tuning-forks and the French Diapason Normal. *Nature, 16*(85), 227.

Ellis, A. J. (1877b). Appunn and Koenig. *Nature, 17*, 26.

Ellis, A. J. (1968). On the history of musical pitch: A paper read before the society of arts, 3 March 1880. In A. Mendel (Ed.), *Studies in the History of Musical Pitch: Monographs by Alexander J. Ellis and Arthur Mendel*. Amsterdam: Frits Knuf.

Ellis, R. (1851). *Official Descriptive and Illustrated Catalogue. Great Exhibition of the Works of Industry of all Nations 1851*. London: Spicer Brothers.

Evenson, A. E. (2000). *The Telephone Patent Conspiracy of 1876: The Elisha Gray-Alexander Bell Controversy and its Many Players*. Jefferson, NC: McFarland.

Fabre et Kunemann. (1855). *Fabre et Kunemann. Constructeurs d'instruments de physique de chimie et de mathématique, Maison Pixii fondée en 1780 Paris*.

Faraday, M. (1831). On a peculiar class of acoustical figures. *Philosophical Transactions, 121*, 314–335.

Fau, J., & Chevalier, C. (1853). *Nouveau manuel complet du physicien-préparateur; ou, Description d'un cabinet de physique*. Paris: Roret.

Feffer, S. M. (1996). Ernst Abbe, Carl Zeiss, and the transformation of microscopical optics. In J. Buchwald (Ed.), *Scientific Credibility and Technical Standards in the 19th and early 20th century Germany and Britain*. Dordrecht, Boston, London: Kluwer Academic Publishers.

Feldmann, H. (1995). Die Galton-Pfeife und die Entdeckung der Altersschwerhörigkeit. *Laryngo-Rhino-Otologie, 74*, 329–334.

Feldmann, H. (1997). Die Geschichte der Stimmgabel. *Laryngo-Rhino-Otologie, 76*, 116–122.

Fétis, F.-J. (1864). *Anthony Stradivari, the Celebrated Violin-Maker, Known by the Name of Stradivarius* (J. Bishop, Trans.). London: William Reeves.

Feynman, R. P. (1963). *Lectures in Physics* (Vol. 1). Reading, MA: Addison-Wesley Publishing Company.

Feynman, R. P. (1995). *Six Easy Pieces: Essentials of Physics, Explained by Its Most Brilliant Teacher*.New York: Basic Books.

Finkelstein, G. (2003). M. du Bois-Reymond goes to Paris. *BJHS, 36*(3), 261–300.

Fleming, E. M. (1982). Artifact study: A proposed model. In T. J. Schlereth (Ed.), *Material Culture Studies in America* (pp. 162–173). Nashville: The American Association for State and Local History.

Fletcher, H. (1929). *Speech and Hearing*. London: MacMillan.

Fletcher, N. H., & Rossing, T. D. (1991). *The Physics of Musical Instruments*. New York: Springer-Verlag.

Fostle, D. W. (1995). *The Steinway Saga: An American Dynasty*. New York: Scribner.

Fox, R. (1971). *The Caloric Theory of Gases from Lavoisier to Regnault*. Oxford: Oxford Press.

Fox, R. (1973). Scientific enterprise and the patronage of research in France, 1800–70. *Minerva, 11*, 442–473.

Fox, R. (1980). The savant confronts his peers: Scientific societies in France, 1815–1914. In R. Fox & G. Weisz (Eds.), *The Organization of Science and Technology in France, 1808–1914* (pp. 241–282). Cambridge: Cambridge University Press.

Fox, R. (1992). *The Culture of Science in France, 1700–1900*. Aldershot, UK: Variorum.

Fox, R., & Guagnini, A. (1998–1999). Laboratories, workshops, and sites. Concepts and practices of research in industrial Europe, 1800–1914. *Historical Studies in Physical & Biological Sciences, 29*(1–2), 55–139; 191–294.

Franklin, U. (1994). Silence and the notion of the commons. *Musicworks, 59*, 39–42.

Freeman, I. M. (1974). What is Trevelyan's rocker. *Physics Teacher, 12*, 382.

Fulton, J. (1896). *Memoirs of Frederick A. P. Barnard*. London: Macmillan and Co.

Galeries nationales du grand palais (2007: Paris), The metropolitan museum of art, & Musée Fabre. (2007). *Gustave Courbet: [Catalogue de l'exposition] galeries nationales du grand palais,*

Paris, 13 octobre 2007–28 janvier 2008, the metropolitan museum of art, New York, 27 février-18 mai 2008, musée Fabre, Montpellier, 14 juin-28 septembre 2008. Paris: Réunion des musées nationaux.
Galison, P. (1988). History, philosophy, and the central metaphor. *Science in Context, 2*(1), 197–212.
Galison, P. L. (1987). *How Experiments End*. Chicago: University of Chicago Press.
Ganot, A. (1869). *Elementary Treatise in Physics* (E. Atkinson, Trans.). New York: William Wood and Co.
Ganot, A. (1893). *Elementary Treatise in Physics* (E. Atkinson, Trans., 14th ed.). New York: William Wood and Co.
Gause, F. (1965). *Die Geschichte der Stadt Königsberg in Preussen*(Vol. 2). Köln: Böhlau.
Gee, B. (1990). Joseph Henry's trade with instrument makers in London and Paris. *Bulletin of the Scientific Instrument Society, 25*, 19–24.
Giatti, A., & Miniati, M. (Eds.). (2001). *Acoustics and its Instruments: The Collection of the Istituto Tecnico Toscano*. Firenze.
Gibbs, W., Pickering, E. C., & Trowbridge, J. (1879). *List of Apparatus Available for Scientific Researches Involving Accurate Measurements, and Contained in Different American Laboratories*. Cambridge, MA: University Press.
Giberti, B. (2002). *Designing the Centennial: A History of the 1876 International Exhibition in Philadelphia*. Lexington: University Press of Kentucky.
Gingras, Y. (1991). *Physics and the Rise of Scientific Research in Canada* (P. Keating, Trans.). Montreal & Kingston: McGill-Queen's University Press.
Gogh, V. (1991). *The Complete Letters of Vincent van Gogh: With Reproductions of all the Drawings in the Correspondence*. Boston: New York Graphic Society.
Gooday, G. (2004). *The Morals of Measurement: Accuracy, Irony, and Trust in late Victorian Electrical Practice*. Cambridge, England: Cambridge University Press.
Gooding, D., & Pinch, T. (Eds.). (1989). *The Uses of Experiment: Studies in the Natural Sciences*. Cambridge [England]: Cambridge University Press.
Goodway, M. (1987). Phosphorous in antique iron music wire. *Science, 236*(4804), 927–932.
Gordon, R. B. (1996). *American Iron, 1607–1900*. Baltimore, MD: Johns Hopkins University Press.
Gordon, R. B., & Malone, P. M. (1994). *The Texture of Industry: An Archaeological View of the Industrialization of North America*. New York: Oxford University Press.
Gourden, J.-M. (1992). *Le peuple des ateliers: les artisans du XIXe siècle*. Paris: Créaphis.
Gray, A. (1897). Famous scientific workshops. 1. Lord Kelvin's laboratory in the University of Glasgow. *Nature, 44*(1430), 486–492.
Greenslade, T. B. (1981). The rotating mirror. *The Physics Teacher, 19*, 253–255.
Greenslade, T. B. (1992). The acoustical instrumentation of Rudolph Koenig. *Physics Teacher, 30*, 518–522.
Gregory, W. G. (1889). On a method of driving tuning-forks electrically. *Philosophical Magazine, 28*, 490–491.
Guillemin, A. (1875). *Le son; notions d'acoustique physique et musicale. Ouvrage illustré de 70 figures gravées sur bois*. Paris: Hachette.
Guillemin, A. (1881). *Le monde physique: Tome premier* (Vol. 1). Paris: Hachette et cie.
Hacking, I. (1983). *Representing and Intervening*. Cambridge [England]: Cambridge University Press.
Hacking, I. (1991). The self-vindication of the laboratory sciences. In A. Pickering (Ed.), *Science as Practice and Culture* (pp. 29–64). Chicago: The University of Chicago Press.
Hackmann, W. D. (1985). The nineteenth-century trade in natural philosophy instruments in Britain. In P. R. de Clercq (Ed.), *Nineteenth-Century Scientific Instruments and their Makers* (pp. 53–91). Leiden, Amsterdam: Rodopi.
Haine, M. (1998). Jean-Baptiste Vuillaume: Innovator or conservationist? In R. Campos (Ed.), *Violins, Vuillaume (1785–1875): A Great French Violin Maker of the 19th Century*. Paris: Cité de la Musique.

Hankins, T. J., & Silverman, R. J. (1995). *Instruments and the Imagination*. Princeton, NJ: Princeton University Press.

Hatfield, G. (1993). Helmholtz and classicism: The science of aesthetics and the aesthetics of science. In D. Cahan (Ed.), *Hermann von Helmholtz and the Foundations of Nineteenth Century Science* (pp. 522–558). Berkley: University of California Press.

Helden, A. V. (1994). The telescope and authority from Galileo to Cassini. In T. L. Hankins & A. V. Helden (Eds.), *Orisis* (Vol. 9, pp. 9–29).

Helmholtz, H. (1856). Ueber Combinationstöne. *Annalen der Physik* (Poggendorff's), *49*, 497–540.

Helmholtz, H. (1857). Ueber die Vocale. *Archive für Holländischen Beiträge zue Natur- und Hiekunde, 1*, 354–355.

Helmholtz, H. (1859). Ueber die Klangfarbe der Vocale. *Annalen der Physik* (Poggendorff's), *108*, 280–290.

Helmholtz, H. (1860a). On the motion of the strings of a violin. *Proceedings of the Philosophical Society of Glasgow, 5*, 17–21.

Helmholtz, H. (1860b). On vowel sounds. *Philosophical Magazine, 19*, 81–88.

Helmholtz, H. (1860c). Theorie der Luftschwingungen in Röhren mit Offenen Enden. *Journal für die reine und angewandte Mathematik, 57*, 1–72.

Helmholtz, H. (1863). *Die Lehre von den Tonempfindungen als Physiologische Grundlage für die Theorie der Musik* (1st ed.). Braunschweig: Friedrich Vieweg und Sohn.

Helmholtz, H. (1865a). *Die Lehre von den Tonempfindungen als Physiologische Grundlage für die Theorie der Musik* (2nd ed.). Braunschweig: Friedrich Vieweg und Sohn.

Helmholtz, H. (1865b). Die Physiologishen Urachen der Musikalischen Harmonie. In *Populäre Wissenschatliche Vorträge*. Braunschweig: Friedrich Vieweg und Sohn.

Helmholtz, H. (1870). *Die Lehre von den Tonempfindungen als Physiologische Grundlage für die Theorie der Musik* (3rd ed.). Braunschweig: Friedrich Vieweg und Sohn.

Helmholtz, H. (1877). *Die Lehre von den Tonempfindungen als Physiologische Grundlage für die Theorie der Musik* (4th ed.). Braunschweig: Friedrich Vieweg und Sohn.

Helmholtz, H. (1879). Telephones and timbre. *Journal de Physique, 8*, 168–175.

Helmholtz, H. (1882). *Wissenschaftliche Abhandlungen* (Vol. 1). Liepzig: Barth.

Helmholtz, H. (1995). On the physiological causes of harmony in music. A public lecture. In D. Cahan (Ed.), *Science and Culture: Popular and Philosophical Essays* (pp. 46–75). Chicago: University of Chicago Press.

Helmholtz, H. V., & Ellis, A. J. (1954). *On the Sensations of Tone as a Physiological Basis for the Theory of Music* (2nd English ed.). New York: Dover Publications.

Helmholtz, H. V., Guéroult, G., & Wolff, A. D. B. (1868). *Théorie physiologique de la musique, fondée sur l'étude des sensations auditives*. Paris: V. Masson et fils.

Henry, J. (1847). Letter to Alexander Dallas Bache, March 31, 1847. In M. Rothenberg (Ed.), *The Papers of Joseph Henry* (Vol. 7). Washington and London: Smithsonian Institution Press.

Henry, J. (1850). *Fourth Annual Report of the Board of Regents of the Smithsonian Institution to the Senate and House of Representatives*. Washington, DC: United States Senate.

Henry, J. (1856). *Research in acoustics*. Annual Report, Smithsonian Institution (pp. 221–234), Washington, DC.

Hensen, V. (1863a). Studien über das Gehörogan der Decapoden *Zeitschrift für Wissenschaftliche Zoologie*, September, pp. 319–412.

Hensen, V. (1863b). Zue Morphologie der Schnecke des Menschen und der Säugethiere. *Zeitschrift für Wissenschaftliche Zoologie*, December, pp. 481–512.

Hentschel, K. (2007). Gaus, Meyerstein and Hanoverian metrology. *Annals of Science, 64* (1), 41–75.

Her Majesty's Commissioners. (1862). *International Exhibition. 1862. Official Catalogue of the Industrial Department* (3rd ed.). London: Printed for her Majesty's Commissioners, by Trusscott, Son, & Simmons.

Hermann, L. (1889). Phonophotographische Untersuchungen. *Pflüger's Archiv für Physiologie, 45*, 582–592.

Hermann, L. (1890). Phonophotographische Untersuchungen. *Pflüger's Archiv für Physiologie, 47*, 44–53, 347–391.

Hermann, L. (1891a). Ueber die Prüfung von Vocalcurven mittels der König'schen Wellensirene. *Pflüger's Archiv für Physiologie, 48*, 574–577.

Hermann, L. (1891b). Zur Theorie der Combinationstöne. *Pflüger's Archiv für Physiologie, 49*, 499–518.

Hermann, L. (1893). Phonophotographische Untersuchungen. *Pflüger's Archiv für Physiologie, 53*, 1–51.

Hermann, L. (1894). Phonophotographische Untersuchungen. *Pflüger's Archiv für Physiologie, 58*, 264–279.

Heron-Allen, E. (1885). *Violin-Making, as it was and is: Being a Historical, Theoretical, and Practical Treatise on the Science and Art of Violin-Making, for the Use of Violin Makers and Players, Amateur and Professional* (2nd ed.). London, New York: Ward, Lock, & co.

Hiebert, E., & Hiebert, E. (1994). Musical thought and practice: Links to Helmholtz's Tonempfindungen. In V. L. Krüger (Ed.), *Universalgenie Helmholtz: Rückblick nach 100 Jahren* (pp. 295–311). Berlin: Akademie Verlag.

Hoff, H. E., & Geddes, L. A. (1959). Graphic Registration before Ludwig; The antecedents of the Kymograph. *Isis, 50* (1), 5–21.

Holland, J. (1993). Relations between Scientific Instrument Manufacturers. *Bulletin of the Scientific Instrument Society, 38*, 15–16.

Holland, J. (2000). Charles Wheatstone and the Representation of Waves, (Part I & II). *Rittenhouse, 13 &14*, 86–106; 127–146.

Holland, J. (2003). A Testing Machine for Adelaide. *Rittenhouse, 17* (2), 99–106.

Holmes, F. L., & Levere, T. (Eds.). (2000). *Instruments and Experimentation in the History of Chemistry*. Cambridge, MA: MIT Press.

Holmes, F. L., & Olesko, K. M. (1995). The Images of Precision: Helmholtz and the Graphical Method in Physiology. In M. N. Wise (Ed.), *The Values of Precision*, (pp. 198–221). Princeton, N.J.: Princeton University Press.

Hörz, H. (Ed.). (1997). *Brückenschlag zwischen zwei Kulturen: Helmholtz in der Korespondenz mit Geisteswissenschaftlern und Künstlern*. Marburg: Basilisken-Presse.

Hosler, D. (1994). *The Sounds and Colors of Power: The Sacred Metallurgical Technology of Ancient West Mexico*. Cambridge, MA.: MIT Press.

Hounshell, D. A. (1984). *From the American system to Mass Production, 1800–1932: The Development of Manufacturing Technology in the United States*. Baltimore: Johns Hopkins University Press.

Hunt, F. V. (1992). *Origins of Acoustics: The Science of Sound from Antiquity to the Age of Newton*. Woodbury, N.Y.: Acoustical Society of America.

Hutchins, C. M. (1997b). 350 Years of Violin Research: Violin Development from the 16th through the 19th Century. In C. M. Hutchins (Ed.), *Research Papers in Violin Acoustics: 1975–1993*. (pp. 3–12). Washington D.C.: Acoustical Society of America.

Hutchins, C. M. (Ed.). (1975). *Musical Acoustics, Part 1: Violin Family Components* (vol. 5). Stroudsburg, Pennsylvania: Dowden, Hutchinson & Ross, Inc.

Hutchins, C. M. (Ed.). (1976). *Musical Acoustics, Part II: Violin Family Functions*, (vol. 6). Stroudsburg, Pennsylvania: Dowden, Hutchinson & Ross, Inc.

Hutchins, C. M. (Ed.). (1997a). *Research Papers in Violin Acoustics: 1975–1993*. Washington D.C.: Acoustical Society of America.

Ianniello, M. G. (2003). *La storia dell' istituto di fisica della sapienza attraverso le sue collezioni di strumenti*. Roma: Museo di Fisica di Roma.

Jackson, M. W. (2000). *Spectrum of Belief: Joseph von Fraunhofer and the Craft of Precision Optics*. Cambridge: MIT press.

Jackson, M. W. (2006). *Harmonious Triads: Physicists, Musicians, and Instrument Makers in Nineteenth-Century Germany*. Cambridge: MIT Press.
James, W. (1950). *The Principles of Psychology* (Authorized ed.). New York: Dover.
Jamin, J. (1868). *Cours de physique de l'école polytechnique*. Paris: Gauthier-Villars.
Jeans, J. (1937). *Science and Music*. London: Cambridge University Press.
Joerges, B., & Shinn, T. (2001). *Instrumentation: Between Science, State, and Industry*. Dordrecht; Boston: Kluwer Academic Publishers.
John T. Morse. (1896). *Life and Letters of Oliver Wendell Holmes*Boston and New York Houghton, Mifflin and Company.
Johnston, I. (1989). *Measured Tones: The Interplay Between Physics and Music*. Bristol: Hilger.
Jones, A. T. (1937). *Sound: A Textbook*. New York: D. Van Nostrand.
Jordan, D. P. (1995). *Transforming Paris: The Life of Baron Haussmann*: Free Press.
Jordan, S. (1889). Notes on the Iron and Steel Manufacture in France in 1887. *The Journal of the Iron and Steel Institute*, 10–48.
Kevles, B. (1997). *Naked to the Bone: Medical Imaging in the Twentieth Century*. New Brunswick, NJ: Rutgers University Press.
Kevles, D. J. (1970). Miller, Dayton Clarence. In *Dictionary of Scientific Biography*. New York: Scribners.
Kielhauser, E. A. (1907). Die Stimmgabel: Ihre Schwingunsgesetze und Anwendungen in der Physik. Leipzig.
Klein, U. (1996). The chemical workshop tradition and the experimental practice: Discontinuities within continuities. *Science in Context, 9*, 251–287.
Koenig, R. (1859). *Catalogue des Principaux Appareils d'Acoustique*. Paris: Bailly, Divry et Cie.
Koenig, R. (1862). Appareil pour la mesure de la vitesse du son. *Comptes Rendus de l'Académie des Sciences, 55*, 603–605.
Koenig, R. (1863). Apparat zur Messung der Geschwindigkeit des Schalls. *Annalen der Physik und Chemie,118*, 610–614.
Koenig, R. (1864a). Theorie der Klangfiguren von Wheatstone. *Annalen der Physik und Chemie, 122*, 238–242.
Koenig, R. (1864b). Ueber ein Mittel den wechselnden Dichtigkeitszustand der Luft in tönenden Orgelpfeifen sichtbar darzustellen. *Annalen der Physik und Chemie, 122*, 242–245.
Koenig, R. (1864c). Neuer Apparat um Schwingungen mit möglichst geringem Verluste ihrer Intensität vom tönenden Körper zum Ohre zu leiten. *Annalen der Physik und Chemie, 122*, 473–476.
Koenig, R. (1864d). Nouvelles recherches sur les plaques vibrantes. *Comptes Rendus de l'Académie des Sciences, 58*, 562–565.
Koenig, R. (1865). *Catalogue des appareils d'acoustique*. Paris: Simon Raçon et Comp.
Koenig, R. (1870). Sur les notes fixes caractéristiques des diverses voyelles. *Comptes Rendus Hebdomadaires des Seances de l'Académie des Sciences, 70*, 931–933.
Koenig, R. (1872). Die manometrischen Flammen. *Annalen der Physik und Chemie, 146*, 161–199.
Koenig, R. (1873a). *Catalogue des appareils d'acoustique*. Paris.
Koenig, R. (1873b). Flammes manométriques. *Journal de Physique, 2*, 182–189.
Koenig, R. (1873c). On Manometric Flames. *Philosophical Magazine, 45*(297), 1–18; 105–114.
Koenig, R. (1876a). On the Simultaneous Sounding of Two Notes. *Philosophical Magazine, 1*(6), 417–446;511–525.
Koenig, R. (1876b). Ueber den Zusammenklang zweier Töne. *Annalen der Physik und Chemie, 157*, 177–237.
Koenig, R. (1876c). Stimmgabel mit veränderlichem Tone. *Annalen der Physik, 157*, 621–623.
Koenig, R. (1877). Koenig's Tuning Forks. *Nature, 16*, 162.
Koenig, R. (1879). Recherches sur la différence de phase qui existe entre les vibrations de deux téléphones associés. *Journal de Physique, s. 1; t. 8*, 175–179.

Koenig, R. (1880a). Untersuchungen über die Schwingungen einer Normalstimmgabel. *Annalen der Physik, 9*, 394–417.

Koenig, R. (1880b). Ueber die Erregung harmonischer Töne durch Schwingungen eines Grundtones. *Annalen der Physik, 11*, 857–870.

Koenig, R. (1881a). Uber den Ursprung der Stösse und Stosstöne bei harmonischen Intervallen. *Annalen der Physik, 12*, 335–349.

Koenig, R. (1881b). Beschreibung eines Stosstöneapparates für Vorlesungsversuche. *Annalen der Physik, 12*, 350–353.

Koenig, R. (1881c). Ueber die Beobachtung der Luftschwingungen in Orgelpfeifen. *Annalen der Physik, 13*, 569–582.

Koenig, R. (1881d). Bemerkungen über die Klangfarbe. *Annalen der Physik, 14*, 369–393.

Koenig, R. (1881e). Recherches sur les vibrations d'un diapason normal. *Journal de Physique, 10*, 214–215.

Koenig, R. (1881f). Sur l'origine des battements et des sons résultants des intervalles harmoniques. *Journal de Physique, s. 1; t. 10*, 408–409.

Koenig, R. (1882a). *Catalogue des Appareils d'Acoustique*. Paris.

Koenig, R. (1882b). *Note concerning the large acoustical instruments exhibited by Dr. Rudolph Koenig, of Paris, at the Centennial Exhibition of Philadelphia, 1876*Paris.

Koenig, R. (1882c). *Quelques Expériences d'Acoustique*. Paris: A. Lahure.

Koenig, R. (1889). *Catalogue des appareils d'acoustique*. Paris.

Koenig, R. (1890a). Ueber Stösse und Stosstöne zweier in demselben Körper erregten Schwingungsbewegungen. *Annalen der Physik, 39*, 395–402.

Koenig, R. (1890b). Ueber Klänge mit ungleichförmigen Wellen. *Annalen der Physik, 39*, 402–411.

Koenig, R. (1891a). Battements et sons résultants de deux mouvements vibratoires produits simultanément dans le même corps. *Journal de Physique, s. 2, t, 10*, 527–528.

Koenig, R. (1891b). Timbres produits par des ondes irrégulière. *Journal de Physique, s. 2, t. 10*, 528–529.

Koenig, R. (1896a). Die Wellensirene. *Annalen der Physik, 57*, 339–388.

Koenig, R. (1896b). Zur Frage über den Einfluss der Phasendifferenz der harmonischen Töne auf die Klangfarbe. *Annalen der Physik, 57*, 555–566.

Koenig, R. (1899). Ueber die höchsten hörbaren und unhörbaren Töne. *Annalen der Physik, 69*, 626–660; 721–738.

Koenig, R. (1900). Sur les sons très aigus perceptibles ou non. *Journal de Physique, s. 3, t, 9*, 158–160.

Koenig, R. (1901). *Quelques notes. UTA-JLP B72-0031/017(05).*Unpublished manuscript, Toronto.

Koenigsberger, L. (1902). *Hermann von Helmholtz*. Braunschweig: F. Vieweg und Sohn.

Koenigsberger, L. (1965). *Hermann von Helmholtz* (F. A. Welby, Trans.). New York: Dover Publications, Inc.

Kohl, M. (1909). *Catalogue of Physical Apparatus*. Chemnitz.

Kremer, R. L. (1993). Innovation through Synthesis: Helmholtz and Colour Research. In D. Cahan (Ed.), *Hermann von Helmholtz and the Foundations of Nineteenth Century Science* (pp. 205–257). Berkley: University of California Press.

Kremer, R. (1992). From Psychophysics to Phenomenalism: Mach and Hering on Color Vision. In M. J. Nye (Ed.) *The Invention of Physical Science*,(pp. 147–173). Dordrecht: Kluwer.

Ku, J. H. (2005). J.W. Strutt, Third Baron Rayliegh, The Theory of Sound, First Edition (1877–1878). In I. Gratton-Guinness (Ed.), *Landmark Writings in Western Mathematics 1640–1940*, (pp. 588–599). Amsterdam: Elsevier.

Ku, J. H. (2006). British Acoustics and its Transformation from the 1860s to the 1910s. *Annals of Science, 63* (4), 395–423.

Kuhn, T. S. (1977). *The Essential Tension: Selected Studies of Scientific Tradition and Change*. Chicago: University of Chicago Press.

Kundt, A. (1866a). Ueber eine neue Art akustischer Staubfiguren und über die Anwenddung derselben zur Bestimung der Schallgeschwindikeit in Gesten Körpern und Gasen. *Annalen der Physik, 127*, 497–523.

Kundt, A. (1866b). Ueber die Erzeugung von Klangfiguren in Orgelpfeifen und über die Wirkung tönender Luftsäulen auf Flammen. *Annalen der Physik, 128*, 337–355.

Kursell, J. (2006). Physiology of one ear: Herman von Helmholtz's research on phase differences. *Unpublished conference paper. Acoustics Conference, May 4–6, 2006*. Cambridge, MA: Dibner Institute, MIT.

Ladis, A., Wood, C. H., Eiland, W. U., Museo Horne (Florence Italy), Dixon Gallery and Gardens., & Georgia Museum of Art. (1995). *The Craft of Art: Originality and Industry in the Italian Renaissance and Baroque Workshop*. Athens: University of Georgia Press.

Lagemann, R. T. (1983). *The Garland Collection of Classical Physics Apparatus at Vanderbilt University*. Nashville, Tenn.: Folio Publishers.

Latour, B., & Woolgar, S. j. a. (1979). *Laboratory Life: The Social Construction of Scientific Facts [by] Bruno Latour, Steve Woolgar; introd. by Jonas Salk*. Beverly Hills: Sage Publications.

Le Conte, J. (1857). On the Influence of Musical Sounds on the Flame of a Jet of Coal-Gas. *Philosophical Magazine, 8* (4), 473.

Le Conte Stevens, W. (1890). Sketch of Rudolph Koenig. *The Popular Scientific Monthly, 37*, 545–550.

Le Conte Stevens, W. (1901). Rudolph Koenig. *Scientific American Supplement, 1353*, 21692.

le Phonautographe, appareil pour la fixation graphique des bruits, des sons, de la voix, inventé par M. Édouard-Léon Scott et construit par M. Rudolph Koenig. (1859). (1859). In *Catalogue des principaux appareils d'acoustique*. Paris: Bailly, Divry et Ce.

Lecoeuvre. (1867). Class LIV, Machine Tools. In By Authority of the Imperial Commission (Ed.), *Paris – Universal Exhibition, 1867. Complete Official Catalogue, Including the British and all other Sections. English Version, translated from the French Catalogue, Published by the Imperial Commission* (pp. 682–683): J.M. Johnson & Sons.

Lenin, V. I. (1938) *Selected Works, Vol. XI*. New York: International Publishers.

Lenoir, T. (1994). Helmholtz and the Materialities of Communication. *ORISIS, 9*, 185–207.

Lenoir, T. (1997). The Politics of Vision: Optics, Painting, and Ideology in Germany 1845–95. In T. Lenoir (Ed.), *Instituting science: the cultural production of scientific disciplines* ,(pp. 131–178). Stanford, California: Stanford University Press.

Lerebours, & Secretan. (1853). *Catalogue et prix des instruments d'optique, de physique, de chimie, de mathématiques, d'astronomie et de marine*. Paris.

Levere, T. (1994). Balance and Gasometer in Lavoisier's Chemical Revolution. In T. Levere (Ed.), *Chemists and Chemistry in Nature and Society, 1770–1878* (pp. 313–332). Aldershot: Variorum.

Levere, T. (2000). Measuring Gases and Measuring Goodness. In F. L. olmes & T. Levere (Eds.), *Instruments and Experimentation in the History of Chemistry* (pp. 105–136). Cambridge, MA, London, England: MIT Press.

Lieberman, R. K. (1995). *Steinway & Sons*. New Haven, London: Yale University Press.

Lindsay, B. R. (Ed.). (1973). *Acoustics: Historical and Philosophical Development*. Stroudsburg, Pa.: Dowden, Hutchingson & Ross.

Lindsay, J. (1973). *Gustave Courbet: his life and art*. Bath; London (167 Hermitage Rd., N. 4): Adams and Dart; Distributed by Jupiter Books (London) Ltd.

Lissajous, J. (1857). *Instruction pratique relative à l'emploi des instruments d'acoustiques*. Paris: Secretan.

Lissajous, J. (1868). Section III. Instruments d'astronomie, de géodésie, de topographie, de marine, d'optique et d'acoustique. In *Exposition Universelle de Paris 1867, Rapports du Jury International, publiés sous la direction de M. Michel Chevalier, membre de la Commission Impériale* (vol. 2, pp. 446–484). Paris: Paul Dupont.

Loewenherz, L. (1888). Ueber die Herstellung von Stimmgabeln. *Zeitschrift für Instrumentenkunde, 8*, 261–267.

Loudon, J. (1901a). *A Century of Physics in Acoustics: Presidential Address to Section*. Toronto: Copp-Clark.
Loudon, J. (1901b). *Rudolph Koenig in UTA-JLP, B72 0031/016/05*. Unpublished manuscript, Toronto.
Loudon, J. (1902). A Century of Progress in Acoustics. *Scientific American Supplement*,1362, 21826–21827.
Loudon, J. (1916). *Memoirs*. Unpublished manuscript, UTA-JLP, B72-0031/016.
Loudon, W. J., & McLennan, J. C. (1895). *A Laboratory Course in Experimental Physics*. New York, London: MacMillam and Co.
Lourenço, M. C. (2005). *Between Two Worlds: The Distinct Nature and Contemporary Significance of University Museums and Collections in Europe*. Paris: Conservatoire National des Arts et Métiers.
Lubar, S. D., & Kingery, W. D. (1993). *History from Things: Essays on Material Culture*. Washington: Smithsonian Institution Press.
Mach. (1861). Annalen der Physik, CXII, 66–68.
Mach. (1862).*CXVI*, 335–336.
Mack, G. (1970). *Gustave Courbet*. Westport, Conn.: Greenwood Press.
Maley Jr., V. C. (1990). *The Theory of Beats and Combination Tones*. New York, London: Garland.
Maniguet, T. (1998). Savart and Vuillaume (M. de Mazières & B. Camurat, Trans.). In R. Campos (Ed.), *Violins, Vuillaume (1785–1875): A Great French Violin Maker of the 19th Century*, (pp. 60–65). Paris: Cité de la Musique.
Marey, E.-J. (1878). *La Méthode Graphique en Sciences Expérimentales*. Paris: G. Masson.
Marloye, A. (1840). *Catalogue des Principaux Appareils d'Acoustique et Autres Objets*. Paris: Ducessois.
Marloye, A. (1845). *Catalogue des Principaux Appareils d'Acoustique et Autres Objets* (2nd ed.). Paris: Ducessois.
Marloye, A. (1851). *Catalogue des Principaux Appareils d'Acoustique et Autres Objets* (3rd ed.). Paris: Bonaventure et Ducessois.
Martins, D. R. (2001). A ciência em Coimbra no *século XIX, (2001, pp. 333–345). In Actas do 1°Congresso Luso-Brasileiro de História da Ciência e da Técnica (Universidade de Évora e Universidade de Aveiro, 22 a 27 de Outubro de 2000).* Évora Universidade de Évora.
Massachusetts Institute of Technology. (1868). *Fourth Annual Catalogue of the Officers and Students and Programme of the Course of Instruction of the Massachusetts Institute of Technology, 1868–69*. Boston: A.A. Kingman.
Massachusetts Institute of Technology. (1870). *Fifth Annual Catalogue of the Officers and Students and Programme of the Course of Instruction of the Massachusetts Institute of Technology, 1869–70*. Boston: A.A. Kingman.
Mayer, A. (1874). Researches in Acoustics: Six Experimental Methods of Sonorous Analysis Described and Discussed. *Philosophical Magazine*, 513–525.<
Mayer, A. (1878). *Annalen der Physik, 3*, 328.
Mayer, A. (1894). Researches in Acoustics. *American Journal of Physics, XLVII(277)*, 1–29.
Mayer, A. (1896). Researches in Acoustics. *American Journal of Science, 1*, 81–105.
McConnell, A. (1994). From Craft Workshop to Big Business: The London Scientific Instrument Trade's Response to Increasing Demand 1750–1820. *London Journal, 19*, 26–53.
McConnell, A. (2007). *Jesse Ramsden (1735–1800): London's leading scientific instrument maker*. Aldershot, England, Burlington, VT: Ashgate.
McDonald, P. J. (2003). Demonstration by Simulation: The Philosophical Significance of Experiment in Helmholtz's Theory of Perception'. *Perspectives on Science, 11*, 170–207.
McKendrick, J. G. (1899). *Hermann Ludwig Ferdinand von Helmholtz*. London: T. Fisher Unwin.
Melde, E. F. (1860a). Ueber die Erregung stehender Wellen eines fadenförmigen Körpers. *Annalen der Physik und Chemie, 109*, 193–215.
Melde, E. F. (1860b). Ueber die Erregung stehender Wellen eines fadenförmigen Körpers. *Annalen der Physik und Chemie, 111*, 513–537.

Melo, S. C. A. d. S. (2002). *A Óptica no Contexto do Ensino da Física na Faculdade de Philosophia da Universidade de Coimbra na Segunda Metade do Século XIX Dissertação de Mestrado na Especialidade do Ensino da Física*Coimbra: Universidade de Coimbra Faculdade de Ciências e Tecnologia.

Merritt, E. (1894). On a Method of Photographing the Manometric Flame with Applications to the Study of Vowel A. *Physical Review, 1*, 166–176.

Mertens, J. (1998). From the Lecture Room to the Workshop: John Frederic Daniell, the Constant Battery, and Electrometallurgy around 1840. *Annals of Science, 55*, 241–261.

Merz, J. T. (1976). *A History of European Thought in the Nineteenth Century* (vol. 1). Gloucester, MA: Peter Smith.

Middleton, W. E. K. (1941). *Visibility in Meteorology: The Theory and Practice of the Measurement of Visual Range* (2nd ed.). Toronto: University of Toronto Press.

Millant, R. (1972). *J. B. Vuillaume: His Life and Work* (A. Hill, Trans.). London: W.E. Hill & Sons.

Millburn, J. R. (2000). *Adams of Fleet Street: instrument makers to King George III*. Aldershot; Brookfield, VT: Ashgate.

Miller, D. C. (1903). *Laboratory Physics: A Student's Manual for College and Scientific Schools*. Boston: Ginn & Company.

Miller, D. C. (1916). *The Science of Musical Sounds*. New York: MacMillan.

Miller, D. C. (1935). *Anecdotal History of the Science of Sound*. New York: MacMillan Company.

Mills, M. (2008). *The Dead Room: Deafness and Communication Engineering*. Doctoral Thesis. Harvard, History of Science.

Mills, M. (2009). "Another Etymology for "Bionic": Hearing Aids and Disability History at Kent State," *Rittenhouse, 22*(1), 47–67.

Milner, J. (1988). *The Studios of Paris: The Capital of Art in the Late Nineteenth Century*. New Haven: Yale University Press.

Mody, C. C. M. (2004). *Crafting the Tools of Knowledge: The Invention, Spread, and Commercialization of Probe Microscopy, 1960–2000*. Cornell University.

Moigno. (1872). *L'art des projections*. Paris: Bureau du Journal les Mondes.

Moigno, F. N. M. (1855). La Retraite de Marloye. *Cosmos: Revue encyclopédique hebdomadaire des progrès des sciences et de leurs applications aux arts et à l'Industrie, 7*, 498.

Moigno, F. N. M. (1859a). Phonautographe et fixation graphique de la voix. *Cosmos: Revue encyclopédique hebdomadaire des progrès des sciences et de leurs applications aux arts et à l'Industrie, 14, 320*, 417.

Moigno, F. N. M. (1859b). Sur la phonautographe. *Cosmos: Revue encyclopédique hebdomadaire des progrès des sciences et de leurs applications aux arts et à l'Industrie, 15*, 677–679.

Moigno, F. N. M. (1860). Vibrations des Membranes. *Cosmos: Revue encyclopédique hebdomadaire des progrès des sciences et de leurs applications aux arts et à l'Industrie, 17, 379–380.*

Moigno, F. N. M. (1861a). Séance, Lundi, le 10 Juin 1861. Cosmos: Revue encyclopédique hebdomadaire des progrès des sciences et de leurs applications aux arts et à l'Industrie, 18, 667–669.

Moigno, F. N. M. (1861b). Séance, Lundi, le 24 Juin 1861. *Cosmos: Revue encyclopédique hebdomadaire des progrès des sciences et de leurs applications aux arts et à l'Industrie, 18,* 780.

Moigno, F. N. M. (1861c). Séance, Lundi, le 15 juillet 1861. *Cosmos: Revue encyclopédique hebdomadaire des progrès des sciences et de leurs applications aux arts et à l'Industrie, 19,* 72.

Moigno, F. N. M. (1862a). Diapason Dynamoscopique. *Cosmos: Revue encyclopédique hebdomadaire des progrès des sciences et de leurs applications aux arts et à l'Industrie, 20,* 569–570.

Moigno, F. N. M. (1862b). Phonoscope. *Cosmos: Revue encyclopédique hebdomadaire des progrès des sciences et de leurs applications aux arts et à l'Industrie, 20*, 700–701.

Moigno, F. N. M. (1862c). Séance, Lundi le 29 Septembre 1862. *Cosmos: Revue encyclopédique hebdomadaire des progrès des sciences et de leurs applications aux arts et à l'Industrie, 21,* 377.

Moigno, F. N. M. (1864). Nouveau Stéthoscope. *Cosmos: Revue encyclopédique hebdomadaire des progrès des sciences et de leurs applications aux arts et à l'Industrie, 25*, 319–320.

Moigno, F. N. M. (1865). Médailles décernées par la société d'encouragement dans un séance publique du 14 Juin 1865. *Les Mondes, revue hebdomadaire des sciences, 8*, 533–535.

Moigno, F. N. M. (1868). Hommage rendu au mérite. *Les Mondes, revue hebdomadaire des sciences, 17*, 280.

Moigno, F. N. M. (1871). Calomnie. *Les Mondes, revue hebdomadaire des sciences, 26*, 601–602.

Mollan, R. C. (1990). *Irish National Inventory of Historic Scientific Instruments: Interim Report, 1990*. Dublin: Royal Dublin Society: EOLAS, the Irish Science and Technology Agency.

Moniteur Scientifique. (1901). Nécrologie Mort de M. R. Koenig. *Moniteur Scientifique, 15*, 736.

Morrison-Low, A. D. (2007). *Making Scientific Instruments in the Industrial Revolution*. Aldershot, UK; Burlington, VT: Ashgate.

Morse, J. T. (Ed.). (1896). *Life and Letters of Oliver Wendell Holmes*. Boston.

Moyer, A. E. (1997). *Joseph Henry: the Rise of an American Scientist*. Washington D.C.: Smithsonian Institution Press.

Multhauf, R. P. (1961). *The Introduction of Self-Registering Meteorological Instruments*. Washington D.C.: Smithsonian Institution.

Murray, D. J., & Bahar, F. (1998). Gestalt Theory and Evolutionary Psychology. In R. W. Rieber & K. Salzinger (Eds.), *Psychology: Theoretical-Historical Perspectives* ,(2nd ed.) (pp. 255–287). Washington D.C.: American Psychological Association.

Murray D. J. (1995). Gestalt Psychology and Cognitive Revolution. New York: Harvester Wheatseaf.

Murray D. J. (1988). A History of Western Psychology (2nd ed.). Englewood Cliffs, N.J.: Prentice Hall.

Neumann-Redlin von Meding, E. (2001). Feinmechaniker und Wissenschaftler für akustische Präzisionsinstrumente und Grundlagen der Tonlehre.

Neumann, H. (1932a). Ein berühmter Königsberger Erfinder, Dr. h. c. Koenig – zu seinem 100. Gerburstag. Königsberger Hartung'sche Zeitung. *Blatt Abendausgabe Stadtspiegel*.

Neumann, H. (1932b, November 25). Rudolph Koenig: Zum 100. Geburstag eines Königsbergers. *Königsberger Allgemeine Zeitung Abendsgabe*.

Neumann, H. (1932c). *Rudolph Koenig zu Seinen Hundertsten Geburtstag – NFA*. Unpublished manuscript, Bückeburg.

Newman, E. B., Stevens, S. S., & Davis, H. (1937). Factors in the Production of Aural Harmonics and Combination Tones. *Journal of the Acoustical Society of America, 9*, 107–118.

Niaudet-Breguet. (1866). Application du diapason à l'horlogerie. Séance de lundi 10 décembre 1866. *Comptes Rendus de l'Académie de Sciences, 63*, 991–992.

Nichols, E. L., & Merritt, E. (1898). The Photography of Manometric Flames. *Physical Review, 7*, 93–101.

Nochlin, L., & Courbet, G. (2007). *Courbet*. New York, N.Y.: Thames & Hudson.

O'Connell, Ann (2008) Herman Finckh: German Emigrant: Artisan Jeweller and Horticulturist. *Ashfield History*, no. 7, 63–98.

Olesko, K. M. (1991). *Physics as a Calling: Discipline and Practice in the Königsberger Seminar for Physics*. Ithaca, NY, London: Cornell University Press.

Olesko, K. M. (1994). Civic Culture and Calling in the Königsberg Period. In V. L. Krüger (Ed.), *Universalgenie Helmholtz: Rückblick nach 100 Jahren*, Berlin: Akademie Verlag.

Olmstead, J. M. D., & Olmstead, E. H. (1952). *Claude Bernard & the Experimental Method in Medicine*. London: Abelard – Schuman.

Opelt, F. W. (1852). *Allgemeine Theorie der Musik*. Leipzig.

Opelt, F. W. (1855). *Forschritte der Physik im Jahre 1852*. Berlin.

Ø rsted, H. C. (1998). On Acoustical Figures. In K. Jelved, A. D. Jackson & O. Knudson (Eds.), *Selected Scientific Works of Hans Christian Ørsted*, Princeton: Princeton University Press.

Oudinet, J. (1882). *Principes de la construction des instruments de précision*. Paris.

Pantalony. (2008). What is it? Twentieth-Century Artifacts out of Context. *History of Science Society Newsletter, 37*(3), 24, 18–19.

Pantalony, D. (2001). Analysing Sound in the Nineteenth Century: The Koenig Sound analyser. *Bulletin of the Scientific Instrument Society,#68*, 16–21.

Pantalony, D. (2003). Fine Tuning in Rudolph Koenig's Workshop: The Making of the 1876 Grand Tonometer. *Bulletin of the Scientific Instrument Society, 79*, 23–26.

Pantalony, D. (2004a). Americans in Europe: The Purchasing Trip of Ira and Charles Young in 1853. *Bulletin of the Scientific Instrument Society, 76*, 23–27.

Pantalony, D. (2004b). Seeing a Voice: Rudolph Koenig's Instruments for Studying Vowels. *American Journal of Psychology, 117*, 425–442.

Pantalony, D. (2005a). Rudolph Koenig's Workshop of Sound: Instruments, Theories, and the Debate over Combination Tones. *Annals of Science, 62*, 57–82.

Pantalony, D., Kremer, R. L., & Manasek, F. J. (2005b). *Study, Measure, Experiment: Stories of Scientific Instruments at Dartmouth College*. Norwich, Vt.: Terra Nova Press.

Paul, H. W. (1985). *From Knowledge to Power: The Rise of the Science Empire in France, 1860–1939*. Cambridge, New York: Cambridge University Press.

Paulin, V. (1938). Home work in France: Its origins, evolutions, and future. *International Labour Review, 37*, 192–225.

Phillips, V. J. (1987). *Waveforms: A History of Early Oscillography*. Bristol: Adam Hilger.

Pietsch, L. (1901). *Knaus*. Bielefeld: Velhagan & Klasing.

Pilcher, V. E., & Union College (Schenectady N.Y.). (1994). *Early Science and the First Century of Physics at Union College, 1795–1895*. Schenectady, N.Y.

Pinch, T. J., & Trocco, F. (2002). *Analog Days: The Invention and Impact of the Moog synthesiser*. Cambridge, MA: Harvard University Press.

Pisko, F. J. (1865). *Die Neuren Apparate der Akustik*. Wien: Carl Gerold's Sohn.

Plomp, R. (1965). Detectability threshold for combination tones. *Journal of the Acoustical Society of America, 37*(6), 1110–1115.

Plomp, R. (1967). Beats and Mistuned Consonances. *Journal of the Acoustical Society of America, 42*, 462–474.

Plomp, R., & Steeneken, H. J. M. (1969). Effect of Phase on the Timbre of Complex Tones. *Journal of the Acoustical Society of America, 46*, 409–421.

Politzer, A. (1864). Untersuchung über Schallfortpflanzung und Schallleitung im Gehörorgane, im gesunden und kranken Zustande: I: Experimental-physiologisches Theil. *Archive für Ohrenheilkunde, 1*, 59–73.

Politzer, A. (1883). *A Text-Book of the Diseases of the Ear and Adjacent Organs*(J. P. Cassells, Trans.). Philadelphia: Lea.

Politzer, A. (1913). *Geschichte der Ohrenheilkunde* (vol. Band II). Stuttgart: Enke.

Prescott, S. C. (1954). *When MIT was "Boston Tech" 1861–1916*. Cambridge, MA.: The Technology Press.

Preyer, W. (1879). *Akustische Untersuchungen*. Jena.

Privat-Deschanel, A. (1877). *[Elementary Treatise on Natural Philosophy... Translated and edited, with extensive additions, by J. D. Everett.]* (4th ed.). London.

Queen & Co., J. W. (1884). *Priced and Illustrated Catalogue of Physical Instruments, Chemical Apparatus, Chemicals and School Apparatus Generally*. Philadelphia.

Radau, R. (1862a). Acoustique. *Cosmos: Revue encyclopédique hebdomadaire des progrès des sciences et de leurs applications aux arts et à l'Industrie, 20*, 621–624; 658–666; 700–701.

Radau, R. (1862b). Acoustique. *Cosmos: Revue encyclopédique hebdomadaire des progrès des sciences et de leurs applications aux arts et à l'Industrie, 21*, 100–112; 147–149.

Radau, R. (1870). *Acoustique*. Paris: Librairie de L. Hachette.

Rayleigh, B. (1877). Absolute Pitch. *Nature, 17*, 12–14.

Rayleigh, B. (1878). *The Theory of Sound* (vol. 2). London: MacMillan and Co.

Rayleigh, B. (1896). *The Theory of Sound* .(2nd ed.). New York: MacMillan and Co.

Rees, A. (1819). *The Cyclopedia; or, Universal Dictionary of the Arts, Sciences, and Literature*. London: Longman, Hurst, Rees, Orme, & Brown.

Regnault, V. (1868). Expériences sur la Vitesse de Propagation des Sons. *Mémoires de l'Académie des Sciences de l'institut Impérial de France, 37.*

Reid, J. S. (1987). A Forgotten Demonstration by Charles Wheatstone. *Bulletin of the Scientific Instrument Society, 12*, 11.
Rezneck, S. (1962). An American Physicist's Year in Europe: Henry A. Rowland, 1875–1876. *American Journal of Physics, 30*, 877–886.
Richardson, E. G. (1927). *Sound; a Physical Text-Book*. New York, London: Longmans E. Arnold & Co.
Richardson, E. G. (1947). *Sound: A Physical Text-Book*. New York: Longmans, Green & Co.
Richardson, E. G. (1953). *Sound: A Physical Text-Book*. London: Edward Arnold.
Rijke, P. L. (1859). Notice of a new method of causing a vibration of air contained in a tube open at both ends. *Philosophical Magazine, 17*, 419–422.
Ringer, F. K. (1969). *The Decline of the German Mandarins: The German Academic Community, 1890–1933*. Cambridge, MA: Harvard University Press.
Rocke, A. J. (2001). *Nationalizing Science: Adolphe Wurtz and the Battle for French Chemistry*. Cambridge, MA: The MIT Press.
Rodrigues (Direcção), M. A. (1992). *Memoria Professorum Universitatis Conimbrigensis, 1772–1937* (vol. 2). Coimbra: Arquivo da Universidade de Coimbra.
Roederer, J. G. (1979). *Introduction to the Physics and Psychophysics of Music* (2nd ed.). New York: Springer-Verlag.
Rossing, T. D. (1990). *The Science of Sound* (2nd ed.). Reading, MA: Addison-Wesley.
Rousselot, P. J. (1908). *Principes de Phonétique Expérimentale* (vol. 2). Paris, Leipzig: Welter.
Rowland, H. A. (1883). A Plea for Pure Science. *Science, 2*(29), 242–250.
Rowland, H. A. (1902). The Physical Papers of Henry Augustus Rowland [microform /]. Baltimore: Johns Hopkins Press.
Rücker, A. W., & Edser, E. (1895). On the Objective Reality of Combination Tones. *Philosophical Magazine, 39*, 341–357.
Saint-Edme, E. (1864). Acoustique. *Cosmos: Revue encyclopédique hebdomadaire des progrès des sciences et de leurs applications aux arts et à l'Industrie, 24*, 439–441; 457–462.
Sanders, J. (1997). The Clarendon Laboratory Archive in Oxford. *Bulletin of the Scientific Instrument Society,54*, 10–14.
Sandys, W., & Forster, S. A. (1864). *The History of the Violin, and Other Instruments Played on with the Bow from the Remotest Times to the Present. Also, an Account of the Principal makers, English and foreign, with numerous illustrations*. London: J. R. Smith [etc.].
Savart, F. (1820). *Annales de Chimie et de Physique, XIV*, 113.
Savart, F. (1824). *Annales de Chimie et de Physique, XXV, 138*, 225.
Savart, F. (1827). Recherches sur les vibrations normales. *Annales de Chimie et de Physique, 36*, 187–208.
Savart, F. (1830). *Annales de Chimie et de Physique, XLIV*, 337.
Savart, F. (1831). Note sur la limite de la perception des sons graves. *Annales de Chimie et de Physique, 47*, 69–74.
Savart, F. (1840). *l'Institut, 8*, 55–56, 69–71, 91–92, 122–123.
Schafer, R. M. (1977). *The Tuning of the World*. Toronto: McClelland and Stewart.
Schaffer, S. (1988). Astronomers Mark Time: Discipline and the Personal Equation. *Science in Context, 2*,115–145.
Schaffgotsch, F. (1858). Der Tonflammenapparat. *Annalen der Physik, 102*, 627–629.
Scheibler, H. (1834). *Physikalische und Musikalische Tonmesser*. Essen: Bädeker.
Schmidgen, H. (2005). Physics, ballistics, and psychology: A history of the chronoscope in/as context, 1845–1890. *History of Psychology, 8*, 46–78.
Schmidgen, H. (2007). The Donders Machine: Matter, Signs, and Time in a Physiological Experiment, ca. 1865. *Configurations, 13*(2), 211–256.
Scott de Martinville, É.-L. (1857). *Brevet d'Invention, no. 31470. Un procédéau moyen duquel on peut écrire et dessiner par le son (acoustique), multiplier graphiquement les résultats obtenus et en faire des applications industrielles*. Paris: Unpublished manuscript in INPI.
Scott de Martinville, É.-L. (1859a). Phonautographe et fixation graphique de la voix. *Cosmos: Revue encyclopédique hebdomadaire des progrès des sciences et de leurs applications aux arts et à l'Industrie, 14*, 314–320.

Scott de Martinville, É.-L. (1859b). *Certificat d'Addition à un Brevet d'Invention du mars 28 1857, no. 31470. Un procédéau moyen duquel on peut écrire et dessiner par le son (acoustique), multiplier graphiquement les résultats obtenus et en faire des applications industrielles*. Paris: Unpublished manuscript in INPI.

Scott de Martinville, É.-L. (1861). Inscription Automatique des Sons de l'air au Moyen d'une Oreille Artificielle. *Comptes Rendus Hebdomadaires des Séances de l'Académie de Sciences, 53*, 108–111.

Scott de Martinville, É.-L. (1878). *Le Problème de la parole s'écrivant elle-meme, la France – l'Amérique*. Paris: Chez l'Auteur, Marchand d'Estampes et Libraire.

Scripture, E. W. (1902). *The Elements of Experimental Phonetics*. New York: Scribner's.

Scripture, E. W. (1906). *Researches in Experimental Phonetics: The Study of Speech Curves*. Washington D.C.: Carnegie Institution.

Secchi, P. A. (1850). Sopra una nuova macchina per rappresentare i moti vibratorii delle ondulazioni luminose inventata dal Sig. Wheatstone. *Correspondenza Scientifica di Roma*.

Seebeck, A. (1840). Ueber die Erregung von Tönen mittelst der Wärme. *Annalen der Physik*, 1–44.

Seebeck, A. (1841). Beobachtungen über einige Bedingungen der Enstehung von Tönen. *Annalen der Physik, 2nd ser. liii*, 417–436.

Seebeck, A. (1843). Ueber die Sirene. *Annalen der Physik, 2nd ser. lx*449–481.

Shankland, R. S. (1970). Karl Rudolph Koenig. In *Dictionary of Scientific Biography*. New York: Scribner.

Shapin, S., Schaffer, S., & Hobbes, T. (1985). *Leviathan and the Air-Pump: Hobbes, Boyle, and the Experimental Life: Including a Translation of Thomas Hobbes, Dialogus physicus de natura aeris by Simon Schaffer*. Princeton, NJ: Princeton University Press.

Sherman, R. (1988). Charles Page, Daniel Davis, and their Electromagnetic Apparatus. *Rittenhouse, 2*(6), 34–47.

Sherman, R. (1991). "That pleasing and noble branch of philosophy:" The electrical lecture of Isaac Greenwood III. *Rittenhouse, 5*(2), 46–52.

Shulman, S. (2008). *The Telephone Gambit: Chasing Alexander Graham Bell's Secret* (1st ed.). New York: W. W. Norton & Co.

Sibum, H. O. (2000). Experimental history of science. In S. Lindqvist (Ed.), *Museums of modern science*, (pp. 77–86). Canton, MA: Science History Publications/USA.

Siemens, D. (1879). *Science Lectures at South Kensington*. London: MacMillan.

Silverman, R. J. (1992). *Instrumentation, representation, and perception in modern science: Imitating function in the nineteenth century*. University of Washington.

Silverman, W. A. (1957). *The Violin Hunter: The Life Story of Luigi Tarisio the Great Collector of Violins*. London: William Reeves.

Simon, J. (2004). *Adolphe Ganot (1804–1887) and his textbooks of physics*. Oxford, UK: University of Oxford.

Skrzeczka, R. F. L. d. D. (1849). *Zur öffentlichen Prüfung im Kneiphöfischen Stadt-Gymnasium*. Königsberg: E.J. Dalkowski.

Smith, C., & Wise, M. N. (1989). *Energy and Empire: A Bibliographical Study of Lord Kelvin*. Cambridge: Cambridge University Press.

Smith, C. S. (1981). *A Search for Structure: Selected Essays on Science, Art, and History*. Cambridge, MA: MIT Press.

Sondhauss, C. F. (1850). Ueber den Brummfreisel und das Schwingungsgesetz der kubischen Pfeifen. *Annalen der Physik (Poggendorff's), 81*, 235–257; 347–373.

South Kensington Museum Conferences held in connection with the Special Loan Collection of Scientific Apparatus, 1876 (vol. Physics and Mechanics). London: Chapman and Hall.

Spaight, J. T. (2004). Herschel's Commercial Telescopes. *Journal for the History of Astronomy, 35*, 45–69.

Spofford, H. P. (1869, Dec. 18). The Institute of Technology. *Frank Leslie's Illustrated Newspaper*, pp. 228, 234–235.

Spottiswoode, W. (1879). On Beats and Combination Tones. *Proceedings of the Musical Association, 5*, 118–130.

Staley, R. (Ed.). (1994). *The Physics of Empire. Public Lectures.* Cambridge: Whipple Museum Publications.
Stephenson, R. J., & Schoepfle, G. K. (1946). A Study of Manometric Flames. *American Journal of Physics, 14*, 294–299.
Stevens, S. S., & Davis, H. (1938). *Hearing: Its Psychology and Physiology*. New York: Wiley.
Sterne, J. (2003). *The Audible Past: Cultural Origins of Sound Reproduction*. Durham: Duke University Press.
Sweetnam, G. K. (2000). *The Command of Light: Rowland's School of Physics and the Spectrum.* Philadelphia: American Philosophical Society.
Taub, L. C., & Willmoth, F. (2006). *The Whipple Museum of the History of Science: instruments and interpretations, to celebrate the sixtieth anniversary of R.S. Whipple's gift to the University of Cambridge*. Cambridge: Cambridge University Press.
Terquem, A. (1859). *Etude des vibrations longitudinales des verges prismatique libres aux deux extrémités*. Paris: Mallet-Bachelier.
Terquem, A. (1870). Étude sur le Timbre des Sons Produits par des Chocs Discontinus et en Particulier par la Sirène. *Annales Scientifiques de l'École Normale Supérieure, (ser.1) 7*, 269–365.
Terquem, A. (1871). Mémoire sur les Sons Produits par les Ébranlements Discontinus. *Comptes Rendus de l'Académie de Sciences, 73*, 165–168.
Thompson, E. A. (2002). *The Soundscape of Modernity: Architectural Acoustics and the Culture of Listening in America, 1900–1933*. Cambridge, MA: MIT Press.
Thompson, J. S., & Thompson, H. G. (1920). *Silvanus Phillips Thompson: His life and Letters*. New York: E.P. Dutton and Company.
Thompson, S. P. (1878). Phenomena of Binaural Audition, Part II. *Philosophical Magazine, VI, ser. 5*, 383–391.
Thompson, S. P. (1881). Phenomena of Binaural Audition, Part III. *Philosophical Magazine, XIL, ser. 5, 352–355.*
Thompson, S. P. (1883). *Philipp Reis: Inventor of The Telephone A Biographical Sketch*. London: E. & F. N. Spon.
Thompson, S. P. (1886). Note on a Mode of Maintaining Tuning Forks by Electricity. *Philosophical Magazine, 22*, 216–217.
Thompson, S. P. (1891). The researches of Dr. R. Koenig on the physical basis of musical sounds. *Nature, 43*, 199–203; 224–227; 249–253.
Thompson, S. P. (1901). Rudolph Koenig. *Nature, 64*(1669), 630–632.
Thompson, W. (1882). The Beats of Mistuned Consonances. *Philosophical Magazine, 13*, 68–70.
Thomson, W. (1877–8). On Beats of Imperfect Harmonics. *Proceedings of the Royal Society of Edinburgh.*
Threlfall, R. (1898). *On laboratory arts*. London: MacMillan.
Timbs, J. (1863). *The International Exhibition. The Science, & Art of the Age: or, The International Exhibition of 1862 Popularly Described from its Origin to its Close; Including Details of the Principle Objects and Articles Exhibited.* London: Lockwood & Co.
Titchener, E. B. (1899). The Equipment of a Psychological Laboratory. *American Journal of Psychology, 11*, 250–265.
Titchener, E. B. (1915). *Experimental Psychology: A Manual of Laboratory Practice* (vol. 2). London: MacMillan.
Toepler, A. (1866). *Annalen der Physik, 127*, 556.
Toepler, A., & Boltzmann, L. (1870). *Annalen der Physik, 141*(321–352).
Trevelyan, A. (1832). *Edinburgh Journal of Science, VI*, 141–143.
Trevelyan, A. (1833). On the Vibration of Heated Metals. *Philosophical Magazine, 3*(17), 321–332.
Trevelyan, A. (1834). Notice regarding some Experiments on the Vibration of Heated Metals. *Transactions of the Royal Society of Edinburgh, XII*, 137–146.
Trevelyan, A. (1835). Further Notice of the Vibrations of Heated Metals; with description of a new and convenient Apparatus for experimenting with.. *Philosophical Magazine, VI*, 85–86.
Turner, G. L'E. (1995). The Florentine Workshop of Giovan Battista Giusti, 1556-c. 1575. *Nuncius: Annali di Storia della Scienza, 10*, 131–172.

Turner, G. L'E. (1996). *The Practice of Science in the Nineteenth Century: Teaching and Research Apparatus at the Teyler Museum*. Haarlem: Teyler Museum.
Turner, R. S. (1994). *In the Eye's Mind: Vision and the Helmholtz-Hering Controversy*. Princeton, NJ: Princeton University Press.
Turner, R. S. (1971). The Growth of Professional Research in Prussia, 1815 to 1848 – Causes and Context. *HSPS, 3*, 137–182.
Turner, R. S. (1977). The Ohm-Seebeck Dispute, Hermann von Helmholtz, and the Origins of Physiological Acoustics. *The British Journal for the History of Science, 10*(34), 1–24.
Turner, S. (1996). Demonstrating harmony: Some of the many devices used to produce Lissajous curves before the oscilloscope. *Rittenhouse, 11*(2), 33–51.
Turner, S. R. (1973). Hermann von Helmholtz. In Charles C. Gillispie (Ed.), *Dictionary of Scientific Biography*. New York: Scribner.
Turpin, J. (1883). Some practical bearings of the study of acoustics upon music as an art. *Proceedings of the Musical Association, 9*, 71–93.
Tyndall, J. (1869). *Le Son* (M. L. A. Moigno, Trans.). Paris: Gautier-Villars.
Tyndall, J. (1896). *Sound* (3rd ed.). New York: D. Appleton and Co.
Tyndall, J. (1898). *Sound* (7th ed.). London: Longmans, Green and Co.
Ullman, D. (1986). Helmhotz-Koenig-Waetzmann und die Natur der Kombinatiöne. *Centaurus, 29*, 40–52.
United States Centennial Commission. (1876). *International Exhibition, 1876. Official Catalogue* (2nd and revised ed.). Philadelphia: John R. Nagle.
United States. Centennial Commission, & Walker, F. A. (1880). *International Exhibition, 1876. Reports and awards*. Washington, DC: Govt. Printing Office.
United States. Commission to the Paris Exposition 1867. [from old catalog], & Blake, W. P. (1870). *Reports of the United States commissioners to the Paris universal exposition, 1867*. Washington: Govt. Printing Office.
Universidade de Coimbra. (1868). *Annuario da no Anno Lectivo de 1868–1869*. Coimbra: Imprensa da Universidade.
University of Coimbra Faculty of Science and Technology Physics Museum. (1997). *Ingenuity and Art: A Collection of Instruments of the Real Gabinete de Fisica*. Coimbra, Lisboa: Calouste Gulbenkian Foundation.
Viegas, A. d. S. (1867). Viagem Scientífica do Dr. Antonio dos Santos Viegas – Pimeiro relatório:Dezembro de 1866 a Maio de 1867. Diário de Lisboa. *Folha official do governo portuguez*(229), 2966–2974.
Viegas, A. d. S. (1889). *Faculdade de Philosophia Programma da 3.a Cadeira, Physica, 1.a Parte, para o Anno Lectivo de 1889 a 1890*. Coimbra: Imprensa da Universidade.
Violle, J. L. G. (1883). *Cours de physique*. Paris: G. Masson.
Vogel, S. (1993). Sensation of tone, perception of sound, and empiricism. In D. Cahan (Ed.), *Hermann von Helmholtz and the Foundations of Nineteenth Century Science* (pp. 259–287). Berkley: University of California Press.
Wackernagel, M. (1981). *The World of the Florentine Renaissance Artist: Projects and Patrons, Workshop and Art Market*. Princeton, NJ: Princeton University Press.
Waetzmann, E. (1920a). Verzerrung von Schwingungen infolge unsymmetrischer Verhälnisse. *Zeitschrift für Physik, 1*, 271–282.
Waetzmann, E. (1920b). Versuch einer Versöhnung der Helmholtzschen Theorie der Kombinationstöne und der R. Königschen der Stosstöne. *Zeitschrift für Physik, 1*, 416–425.
Walker, F. A. (Ed.). (1880). *United States Centennial Commission. International Exhibition 1876. Reports and Awards. Groups XXI–XXVII* (Vol. 7). Washington, DC: Government Printing House.
Warner, D. J. (1992). Physics as a moral discipline: Undergraduate laboratories in the late nineteenth century. *Rittenhouse, 6*(24), 116–128.
Warner, D. J. (1993). French instruments in the United States. *Rittenhouse, 8*(1), 1–32.
Warner, D. J., & Ariail, R. B. (1995). *Alvan Clark & Sons, Artists in Optics* (2nd English ed.).

Richmond, VA: Willmann-Bell, in association with National Museum of American History, Smithsonian Institution.
Warner, J. H. (1998). *Against the Spirit of System: The French Impulse in Nineteenth-Century American Medicine*. Princeton, NJ: Princeton University Press.
Wertheim, G. (1848). Mémoire sur les sons produits par le courant électrique. *Annales de Chimie et de Physique, 23*, 302–327.
Wertheim, G. (1851). Mémoire sur les vibrations sonnores de l'air. *Annales de Chimie et de Physique, 31*, 385–432.
Wheatstone, C. (1827). Description of the kaleidophone, or phonic kaleidoscope. *Quarterly Journal of Science, 1*, 344–351.
Wheatstone, C. (1833). On the figures obtained by Strewing Sand on vibrating surfaces, commonly called 'Acoustic Figures'. *Philosophical Transactions, 123*, 593–634.
White, R. (2000). *The Skule Story: The University of Toronto Faculty of Applied Science and Engineering, 1873–2000*. Toronto: University of Toronto Press.
Whiteing, R. (1867). The condition and habits of the French working classes. Special report. *Reports of the Artisans Selected by a Committee Appointed by the Council of the Society of Arts to Visit the Paris Universal Exhibition, 1867* (pp. 452–476). London: Bell and Daldy.
Williams, M. E. W. (1994). *The Precision Makers: A History of the Instruments Industry in Britain and France, 1870–1939*. London: Routledge.
Winkelmann, A. (Ed.). (1909). *Handbuch der Physik: Akustik* (Vol. 2). Leipzig: Johann Ambrosius Barth.
Wise, M. N. (Ed.). (1995). *The Values of Precision*. Princeton, NJ: Princeton University Press.
Wittje, R. (2006).*The rise of electroacoustics in Gemany during the interwar period*. Unpublished conference paper. Acoustics Conference, May 4–6, 2006. MIT, Cambridge, MA: Dibner Institute.
Wittje, R. (2003). *Acoustics, Atom Smashing and Amateur Radio: Physics and Instrumentation at the Norwegian Institute of Technology in the Interwar Period*. Trondheim: Norwegian University of Science and Technology.
Wolfe. (1867). Class X, musical instruments. In Authority of the Imperial Commission (Ed.), *Paris – Universal Exhibition, 1867. Complete Official Catalogue, Including the British and all other Sections. English Version, translated from the French Catalogue, Published by the Imperial Commission* (p. 167). London: J.M. Johnson & Sons.
Wood, A. B. (1964). *A Textbook of Sound*. London: G. Bell and Sons LTD.
Wylie, F. E. (1975). *M.I.T in Perspective: A Pictorial History of the Massachusetts Institute of Technology*. Boston, Toronto: Little, Brown and Company.
Zahm, J. A. (1892). *Sound and Music*. Chicago: A.C. McClurg & Co.
Zahm, J. A. (1900). *Sound and Music*. Chicago: A.C. McClurg & Co.
Zamminer, F. (1855). *Die Musik und die musikalischen Instrumente in ihrer Beziehung zu den Gesetzen der Akustik*. Giessen: Ricker.
Zamminer, F. (1856). Über die Schwingungsbewegung der Luft in Röhren. *Annalen der Physik* (Poggendorff's), *97*, 173–212.
Zhang, J., Harbottle, G., & Wang, C. (1999). Oldest playable musical instruments found at Jiahu early Neolithic site in China. *Nature, 401*, 366–368.
Zimmermann, E. (1928). *Psychologische, Physiologische Apparate*. Berlin: Liepzig.
Zoch, I. B. (1866). Ueber ein neues Verfahren zur Messung des Schnellgeschwindigkeit in Gasen. *Annalen der Physik un Chemie, CXXVIII*, 497–511.

Index

W

Y

Z

MIX
Papier aus verantwortungsvollen Quellen
Paper from responsible sources
FSC® C105338

If you have any concerns about our products,
you can contact us on
ProductSafety@springernature.com

In case Publisher is established outside the EU,
the EU authorized representative is:
Springer Nature Customer Service Center GmbH
Europaplatz 3, 69115 Heidelberg, Germany

Printed by Libri Plureos GmbH
in Hamburg, Germany